AP 微积分 AB & BC

简学琴 / 编著

- ☑ 丰富的微积分预备知识
- ☑ 全面的微积分 AB＆BC 知识点
- ☑ 规范的答题模板
- ☑ 详细的图形计算器操作

项目策划：王　锋
责任编辑：王　锋
责任校对：胡晓燕
封面设计：墨创文化
责任印制：王　炜

图书在版编目（CIP）数据

AP 微积分 AB & BC = AP Calculus AB & BC：英文 / 简学琴编著. — 成都：四川大学出版社，2021.11
ISBN 978-7-5690-3425-7

Ⅰ. ①A… Ⅱ. ①简… Ⅲ. ①微积分—高等学校—入学考试—美国—自学参考资料—英文 Ⅳ. ①O172

中国版本图书馆 CIP 数据核字（2020）第 158771 号

书　名	AP 微积分 AB & BC AP WEIJIFEN AB & BC
编　著	简学琴
出　版	四川大学出版社
地　址	成都市一环路南一段 24 号（610065）
发　行	四川大学出版社
书　号	ISBN 978-7-5690-3425-7
印前制作	成都完美科技有限责任公司
印　刷	郫县犀浦印刷厂
成品尺寸	185mm×260mm
印　张	19.75
字　数	481 千字
版　次	2021 年 11 月第 1 版
印　次	2021 年 11 月第 1 次印刷
定　价	78.00 元

◆ 版权所有 ◆ 侵权必究

◆ 读者邮购本书，请与本社发行科联系。
　电话：(028)85408408/(028)85401670/
　(028)86408023　邮政编码：610065
◆ 本社图书如有印装质量问题，请寄回出版社调换。
◆ 网址：http://press.scu.edu.cn

四川大学出版社
微信公众号

前　言

AP课程(Advanced Placement Course)全称为美国大学先修课程,是美国大学理事会(College Board)推出的在高中阶段开设的、难度相当于大学初级阶段学术标准与学业水平的一系列课程。良好的AP课程除了可以让学生提前获得大学学分和增加被录取概率,美国大学理事会的多项调查表明,AP课程还可以让学生更容易适应本科学习。AP课程中的数学类课程包括AP微积分AB、AP微积分BC和AP统计学,这些一直以来都是中国学生比较擅长的科目。

17世纪后半叶,牛顿和莱布尼兹在认识到求积问题与作切线问题之间的互逆关系后,建立了微积分基本定理。19世纪,柯西重新定义了极限和无穷小,建立了微积分基础。随着戴德金的实数理论和康托的集合理论的建立,微积分的严格基础体系宣告完成。微积分是近代数学中最伟大的成就。它极大地促进了数学学科的发展,产生了微分方程、复变函数、拓扑学等以微积分为基础的数学分支。微积分在物理、经济学、天文学等领域有着广泛的应用,它的发现被恩格斯看作人类精神的最高胜利。

作者从事AP微积分、AP统计学和A Level进阶数学的教学已有7年,接触国际数学教育之前,教过初中、高中数学和近十年的大学高等数学课程,通过认真比较研究国际微积分教学的差异,经过多轮的教学实践修订,最终编写此书。本书在编写过程中既汲取了国外教材图像直观的特点,又兼容了国内教材逻辑严密的要求,符合高中生的认知特点和AP微积分考试英文表达的规范,融入了作者多年教学的研究成果。

本书分为三个部分:第一个部分是微积分预备知识,第二个部分是AP微积分AB & BC考试的所有学习内容,第三个部分是图形计算器的使用。本书可作为AP Calculus AB & BC的教材或教学参考书。在使用本书时建议和美国大学理事会官网中AP Classroom提供的官方练习题配套使用。由于作者多年对图形计算器的教学有比较深入的研究,所以本书的第二部分配套了Casio fx-CG50图形计算器在考试中的使用技巧。

在教材编写过程中,参考了很多国内、国外的多个版本的微积分教材和其他书籍以及College Board在官方网站公开的真题,重要的书籍都列在了"参考文献"中,在此向各位原作者和College Board表示感谢!

在本课程的开发和校本教材的编写过程中,得到了四川外国语大学附属外国语学校领导和同事们的鼎力支持,不仅将校本教材用于多轮的AP微积分对比教学实验中,还将本课程作为学校数学创新课程进行重点打造,为作者录制微课提供了全面的帮助,最终本教材和配套的微课被评选为2019年重庆市普通高中精品选修课,同时作为重庆市普通高中数学课程创新基地成果得以公开出版。

由于作者水平有限,书中如有不当之处,恳求读者批评指正。

简学琴

2021 年 7 月

目　　录

Chapter 1　Prerequisites for Calculus(微积分预备知识) ········· 1
　1.1　Sets and Intervals(集合与区间) ········· 1
　1.2　Definitions of Functions(函数的定义) ········· 3
　1.3　Properties of Functions(函数的性质) ········· 5
　1.4　Function Operations(函数的运算) ········· 8
　1.5　Basic Elementary Functions(基本初等函数) ········· 10
　1.6　Parametric Equations(参数方程) ········· 20
　1.7　Polar Functions(极坐标方程) ········· 21
　1.8　Transformations of Functions(函数图像的变换) ········· 23
　　Practice Exercises(习题) ········· 24
　　习题参考答案 ········· 26

Chapter 2　Limits and Continuity(极限和连续) ········· 27
　2.1　Definitions of Limits(极限的定义) ········· 27
　2.2　The Precise Definition of a Limit(极限的严格定义) ········· 33
　2.3　Theorems on Limits(极限的定理) ········· 35
　2.4　Computing Limits(极限的计算) ········· 37
　2.5　Asymptotes(渐近线) ········· 46
　2.6　Continuity(连续性) ········· 51
　2.7　连续函数定理 ········· 55
　　Practice Exercises(习题) ········· 58
　　习题参考答案 ········· 60

Chapter 3　Definition of Derivative(导数的定义) ········· 61
　3.1　Definition of Derivative(导数的定义) ········· 61
　3.2　高阶导数 ········· 67

3.3 The Relationship between Differentiability and Continuity(可导与连续的关系) …… 68
3.4 不可导点的类型 …………………………………………………………… 70
Practice Exercises(习题) …………………………………………………… 72
习题参考答案 ………………………………………………………………… 73

Chapter 4 Computation of Derivative(导数的计算) …………………… 74
4.1 Arithmetic Operations on Derivative(导数的代数运算) ………………… 74
4.2 Derivative of Inverse Function(反函数的导数) ………………………… 77
4.3 Essential Formulas(基本公式) …………………………………………… 79
4.4 Chain Rule(链式法则) …………………………………………………… 79
4.5 Implicit Function Derivative(隐函数的导数) …………………………… 81
4.6 Logarithmic Differentiation(对数求导法) ……………………………… 83
4.7 Parametric Function Derivative(参数方程的导数) ……………………… 84
4.8 Polar Function Derivative(极坐标方程的导数) ………………………… 85
Practice Exercises(习题) …………………………………………………… 85
习题参考答案 ………………………………………………………………… 89

Chapter 5 Applications of Derivative(导数的应用) ……………………… 91
5.1 Average and Instantaneous Rates of Change(平均变化率与瞬时变化率) …… 91
5.2 Tangents and Normals(切线和法线) …………………………………… 92
5.3 The Mean Value Theorem for Derivatives(微分中值定理) …………… 93
5.4 Related Rates(相关变化率) ……………………………………………… 97
5.5 L'Hôpital's Rule(洛必达法则) …………………………………………… 99
5.6 Monotony of Functions(函数的单调性) ………………………………… 102
5.7 Concavity and the Point of Inflection(凹凸性与拐点) ………………… 106
5.8 Curve Sketching(函数图形的描绘) ……………………………………… 108
5.9 Absolute Minimum Value and Absolute Maximum Value(最大值与最小值) …… 111
5.10 Motion Problems(运动问题) …………………………………………… 114
Practice Exercises(习题) …………………………………………………… 117
习题参考答案 ………………………………………………………………… 121

Chapter 6 Differential and Approximation(微分与近似计算) ………… 122
6.1 Differentials(微分) ………………………………………………………… 122
6.2 Approximating a Derivative Value(导数的近似计算) ………………… 124
6.3 Local Linear Approximation(局部线性近似) …………………………… 126
6.4 Newton's Method(牛顿法) ……………………………………………… 127

Practice Exercises(习题) ………………………………………………………… 129
　　习题参考答案 ……………………………………………………………………… 129

Chapter 7　Antidifferentiation(不定积分) ………………………………………… 130
　7.1　Definition of Antidifferentiation(不定积分的定义) ……………………… 130
　7.2　Integral by Substitution(换元积分法) …………………………………… 134
　7.3　Integral by Parts(分部积分法) …………………………………………… 141
　7.4　Indefinite Integral of Rational Functions(有理函数的不定积分) ……… 144
　　Practice Exercises(习题) ………………………………………………………… 147
　　习题参考答案 ……………………………………………………………………… 148

Chapter 8　Definite Integrals(定积分) ……………………………………………… 152
　8.1　Riemann Sums and Definite Integrals(黎曼和与定积分) ……………… 152
　8.2　Approximation of Definite Integral(定积分的近似计算) ……………… 157
　8.3　Properties of Definite Integrals(定积分的性质) ………………………… 160
　8.4　Fundamental Theorem of Calculus(微积分基本定理) ………………… 162
　8.5　Operations on Definite Integrals(定积分的计算) ……………………… 166
　8.6　Improper Integral(反常积分) ……………………………………………… 170
　　Practice Exercises(习题) ………………………………………………………… 174
　　习题参考答案 ……………………………………………………………………… 177

Chapter 9　Applications of the Integral to Geometry(定积分的几何应用) ……… 180
　9.1　The Element Method of Definite Integrals(定积分的元素法) ………… 180
　9.2　Area between Two Curves(由两条曲线所围成的图形的面积) ………… 180
　9.3　Volumes by Slicing(切片法求体积) ……………………………………… 185
　9.4　Length of a Plan Curve(平面曲线的弧长) ……………………………… 192
　　Practice Exercises(习题) ………………………………………………………… 194
　　习题参考答案 ……………………………………………………………………… 197

Chapter 10　Differential Equations(微分方程) …………………………………… 199
　10.1　Definitions of Differential Equations(微分方程的相关概念) ………… 199
　10.2　Separable Differential Equations(可分离变量的微分方程) ………… 200
　10.3　Numerical and Graphical Methods(微分方程的数值和图像解法) … 201
　10.4　Applications of First-Order Differential Equations(一阶微分方程的应用) ……… 208
　　Practice Exercises(习题) ………………………………………………………… 215
　　习题参考答案 ……………………………………………………………………… 216

Chapter 11　Sequences and Series（序列和级数） ······ 217
　11.1　Sequences（序列） ······ 217
　11.2　Series（级数） ······ 218
　11.3　Power Series（幂级数） ······ 226
　Practice Exercises（习题） ······ 233
　习题参考答案 ······ 235

Chapter 12　图形计算器的使用 ······ 238

Chapter 13　AP 微积分模拟试题及解析 ······ 255
　本套模拟试题评分标准 ······ 285

Chapter 14　主要公式及定理 ······ 287
　14.1　Functions（函数） ······ 287
　14.2　Limits and Continuity（极限与连续） ······ 290
　14.3　Definition of Derivative（导数的定义） ······ 291
　14.4　Computation of Derivative（导数的计算） ······ 291
　14.5　Applications of Derivative（导数的应用） ······ 292
　14.6　Differential and Approximation（微分与近似计算） ······ 294
　14.7　Antidifferentiation（不定积分） ······ 295
　14.8　Definite Integrals（定积分） ······ 296
　14.9　Applications of the Integral to Geometry（定积分的几何应用） ······ 299
　14.10　Differential Equations（微分方程） ······ 301
　14.11　Sequences and Series（序列和级数） ······ 302

参考文献 ······ 306

Chapter 1　Prerequisites for Calculus(微积分预备知识)

微积分的研究对象是函数,本章将主要研究函数的定义、函数的运算和基本初等函数的性质等 AP 微积分所需要的预备知识.

1.1　Sets and Intervals (集合与区间)

1.1.1　Set(集合)

A **set**(集合) is a collection of objects, and these objects are the **elements**(元素) of the set.

集合是指具有某种特定性质的事物的总体,用大写字母 $A,B,C,\cdots$ 表示.

组成集合的事物称为集合的元素,用小写字母 $a,b,c,\cdots$ 表示. a 是集合 M 的元素,记作 $a\in M$;a 不是集合 M 的元素,表示为 $a\notin M$(或 $a\overline{\in}M$).

按元素的个数分类,集合可以分为有限集和无限集.若集合中含有有限个元素,则称为有限集,反之称为无限集.

集合的表示方法通常有列举法和描述法两种.列举法,就是把集合的全体元素一一列举出来.例如,掷一枚骰子,掷得的点数构成的集合可以表示为 $A=\{1,2,3,4,5,6\}$,列举法通常用于有限集.若集合 M 是由具有某种性质 P 的元素 x 的全体所组成,则集合 M 可用描述法表示为:

$$M=\{x\mid x \text{ 具有性质 } P\}.$$

例如:$M=\{(x,y)\mid x,y \text{ 为实数}, x^2+y^2=1\}$.

[注]几个常见的数集:

N——表示所有自然数构成的集合,称为自然数集.

R——表示所有实数构成的集合,称为实数集.

Z——表示所有整数构成的集合,称为整数集.

Q——表示所有有理数构成的集合,称为有理数集.

1.1.2 Intervals(区间)

1.1.2.1 Finite Intervals(有限区间)

设 $a<b$，称数集 $\{x\,|\,a<x<b\}$ 为 **open interval**(开区间)，记为 (a,b)，即
$$(a,b)=\{x\,|\,a<x<b\}.$$

类似的还有：

$[a,b]=\{x\,|\,a\leqslant x\leqslant b\}$ 称为 **closed interval**(闭区间)；

$[a,b)=\{x\,|\,a\leqslant x<b\}$，$(a,b]=\{x\,|\,a<x\leqslant b\}$ 称为 **half-open intervals**(半开区间).

其中 a 和 b 称为区间 (a,b)，$[a,b]$，$[a,b)$，$(a,b]$ 的 **endpoints**(端点)，$b-a$ 称为区间长度，若区间长度为有限值，则称为有限区间，反之称为无限区间.

区间在数轴上的表示，见图 1.1 和图 1.2.

图 1.1　开区间和闭区间

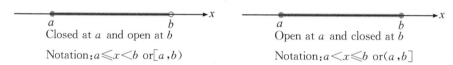

图 1.2　半开半闭区间

1.1.2.2 Infinite Intervals(无限区间)

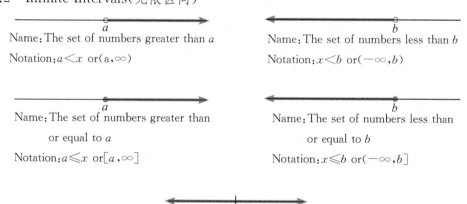

图 1.3　无限区间

[**注**] 在美国教材中正无穷通常省略正号，记作 ∞，只有在同时表示正、负无穷时用 $\pm\infty$.

1.2 Definitions of Functions(函数的定义)

1.2.1 Definitions of Functions(函数的定义)

A **function(函数)** from a set D to a set R is a rule that assigns a unique element in R to each element in D.

To indicate that y comes from the function acting on x, we use Euler's elegant function notation

$$y=f(x) \quad (\text{read}: f \text{ of } x)$$

Here x is the **independent variable(自变量)** and y is the **dependent variable(因变量)**.

1.2.2 The Vertical Line Test(垂线测试法)

A curve in the xy-plane is the graph of some function $f(x)$ if and only if no vertical line intersects the curve more than once.

$y=f(x)$ 为一个函数的充要条件是作垂直于 x 轴的任意直线与函数图像至多有一个交点. 图 1.4 中所示的图像均不是函数, 对于圆的关系式 $x^2+y^2=25$ 可以将其分割为上半圆 $y=\sqrt{25-x^2}$ 和下半圆 $y=-\sqrt{25-x^2}$, 则可以看作两个函数.

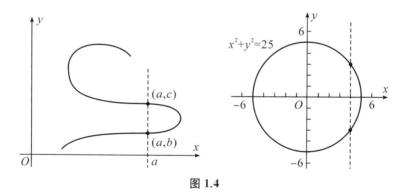

图 1.4

[注]

(1)AP 微积分的研究对象是单值实函数, 即每一个 x 只有唯一的 y 与之对应, 且 x,y 均为实数.

(2)函数与自变量的字母无关, 如 $y=\pi x^2$ 与 $y=\pi r^2$ 表示同一个函数.

1.2.3 Domain and Range(定义域和值域)

A function from a set D to a set R, the set D of all input values is the **domain(定义域)** of the function, and the set R of all output values is the **range(值域)** of the function.

常见的初等函数的定义域(详见 1.5 节初等函数)如下:

(1) $y=\sqrt[n]{x}$ (n 为偶数)   Domain: $\{x \mid x \geqslant 0\}$

(2) $y=\dfrac{1}{x}$ Domain: $\{x \mid x \neq 0\}$

(3) $y=\ln x$ Domain: $\{x \mid x > 0\}$

$y = \log_a x \, (a > 0, a \neq 1)$ Domain: $\{x \mid x > 0\}$

(4) $y = \tan x$ Domain: $\{x \mid x \neq k\pi + \frac{\pi}{2}, k \in \mathbf{Z}\}$

$y = \cot x$ Domain: $\{x \mid x \neq k\pi, k \in \mathbf{Z}\}$

$y = \arcsin x = \sin^{-1} x$ Domain: $[-1, 1]$

$y = \arccos x = \cos^{-1} x$ Domain: $[-1, 1]$

Example 1.1 求定义域.

Find the domain of each of these functions:

a. $f(x) = x^3$

b. $f(x) = \dfrac{1}{(x-1)(x-3)}$

c. $f(x) = \tan x$

d. $f(x) = \sqrt{x^2 - 5x + 6}$

Solution

a. The domain of $f(x)$ is the interval $(-\infty, \infty)$.

b. The domain of $f(x)$ is $\{x \mid x \neq 1 \text{ and } x \neq 3\} = (-\infty, 1) \cup (1, 3) \cup (3, \infty)$.

c. $f(x) = \tan x = \dfrac{\sin x}{\cos x}$, $\cos x \neq 0$,

The domain of $f(x)$ is $\{x \mid x \neq k\pi + \frac{\pi}{2}, k \in \mathbf{Z}\}$.

d. The domain of $f(x)$ is the interval $(-\infty, 2] \cup [3, \infty)$.

1.2.4 Piecewise Functions(分段函数)

A piecewise function defined by applying different formulas to different parts of their domains.

分段函数是对于不同自变量的取值范围对应着不同的函数表达式的函数.最典型的例子是 **The absolute value function**(绝对值函数)函数 $y = |x|$ 和 **The greatest-integer function** (最大整数函数,常称为取整函数) $y = [x]$,在 AP 微积分的考试中需要牢固掌握这两个函数.

The absolute value function(绝对值函数) $y = |x|$ 的图像如图 1.5 所示.

$$y = |x| = \begin{cases} -x, & x < 0 \\ x, & x \geq 0 \end{cases}$$

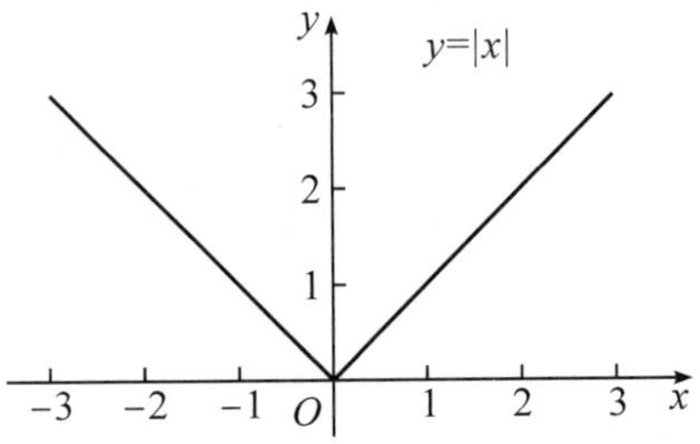

图 1.5

The absolute value function has domain $(-\infty,\infty)$ and range $[0,\infty)$.

对于 **The greatest-integer function** $y=[x]$，其中$[x]$表示不超过 x 的最大整数，例如 $\left[\dfrac{2}{3}\right]=0,[e]=2,[-1]=-1,[-2.5]=-3$.取整函数 $y=[x]$ 的图像如图 1.6 所示，在 x 取整数值的位置图像发生了跳跃.

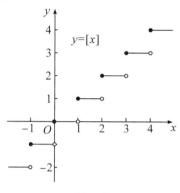

图 1.6

Example 1.2 求分段函数的定义域.

$$\text{Graph } y=f(x)=\begin{cases} -x, & x<0 \\ x^2, & 0\leqslant x\leqslant 1 \\ 1, & x>1 \end{cases}$$

Solution

The values of f are given by three separate formulas: $y=-x$ when $x<0$, $y=x^2$ when $0\leqslant x\leqslant 1$ and $y=1$ when $x>1$. However, the function is just one function, whose domain is the entire set of real numbers(见图 1.7).

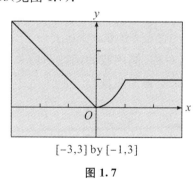

[-3,3] by [-1,3]

图 1.7

1.3　Properties of Functions(函数的性质)

1.3.1　Monotony(单调性)

Let $f(x)$ be a function defined on an interval I and let x_1 and x_2 be any two points in I.

(1) If $f(x_1)<f(x_2)$ whenever $x_1<x_2$, then f is said to be **increasing**(单调递增) on I.

(2) If $f(x_2) < f(x_1)$ whenever $x_1 < x_2$, then f is said to be **decreasing(单调递减)** on I.

A function that is increasing or decreasing on I is called **monotonic** on I.

在讨论函数的单调性时往往要说明所对应的定义区间,比如绝对值函数在其定义域 $(-\infty,\infty)$ 是不单调的,但是在区间 $(-\infty,0]$ 是单调递减的函数,在区间 $[0,\infty)$ 是单调递增的函数.

1.3.2 Boundedness(有界性)

A function $f(x)$ is **bounded below(有下界)** if there is some number b that is less than or equal to every number in the range of $f(x)$. Any such number b is called a **lower bound(下界)** of $f(x)$.

若存在某个实数 b,使得 $f(x) \geqslant b$,则称 $f(x)$ 有下界,实数 b 就是 $f(x)$ 的一个下界.

A function $f(x)$ is **bounded above(有上界)** if there is some number B that is greater than or equal to every number in the range of $f(x)$. Any such number B is called an **upper bound(上界)** of $f(x)$.

若存在某个实数 B,使得 $f(x) \leqslant B$,则称 $f(x)$ 有上界,实数 B 就是 $f(x)$ 的一个上界.

A function $f(x)$ is **bounded(有界)** if it is bounded both above and below.

既有下界又有上界的函数称为有界函数,正弦函数、余弦函数和反三角函数在其定义域都为有界函数.讨论函数的有界性时必须指明对应的定义区间,比如 $f(x) = \dfrac{1}{x}$ 在其定义域既无上界也无下界,但是在区间 $(-\infty,0)$ 有上界,在区间 $(0,\infty)$ 有下界.

1.3.3 Symmetry(对称性)

1.3.3.1 Definition(定义)

A function $f(x)$ is $\begin{cases} \text{odd} \\ \text{even} \end{cases}$ if, for all x in the domain of f, $\begin{cases} f(-x) = -f(x) \\ f(-x) = f(x) \end{cases}$.

The graph of an **odd function(奇函数)** is symmetric about the origin; the graph of an **even function(偶函数)** is symmetric about the y-axis.

奇函数的函数图像关于原点对称,偶函数的函数图像关于 y 轴对称.比如,$y = x^3$,$y = \sin x$ 为奇函数,它们的函数图像关于原点对称;$y = |x|$,$y = x^2$,$y = \cos x$ 为偶函数,它们的函数图像关于 y 轴对称.

1.3.3.2 函数奇偶性的运算法则

奇±奇=奇;偶±偶=偶;奇×奇=偶;奇×偶=奇;偶×偶=偶.

Example 1.3 函数奇偶性的判断.

Tell whether each of the following functions is odd, even, or neither.

a. $f(x) = x^2 - 3$;

b. $g(x) = x^2 - 2x - 2$;

c. $h(x) = \dfrac{x^3}{4 - x^2}$.

Solution

a. Since $f(-x) = (-x)^2 - 3 = x^2 - 3 = f(x)$ for all x in the domain of $f(x)$,

So $f(x)$ is an even function.

b. $g(-x)=(-x)^2-2(-x)-2=x^2+2x-2$

$g(x)=x^2-2x-2 \qquad -g(x)=-x^2+2x+2$

So $g(-x)\neq g(x)$ and $g(-x)\neq -g(x)$.

We conclude that $g(x)$ is neither odd nor even.

c. Since $h(-x)=\dfrac{(-x)^3}{4-(-x)^2}=\dfrac{-x^3}{4-x^2}=-h(x)$ for all x in the domain of $h(x)$,

So the function $h(x)$ is odd.

1.3.4 Periodicity(周期性)

1.3.4.1 Definition(定义)

A function is periodic if there is a positive number p such that $f(x+p)=f(x)$ for every value of x. The smallest such value of p is the **period(周期)** of $f(x)$.

通常我们所说的周期函数的周期指的是最小正周期,$y=\sin x$ 和 $y=\cos x$ 都是周期为 2π 的周期函数.并不是所有的周期函数都有最小正周期,如:

Dirichlet function 狄利克雷函数

$$D(x)=\begin{cases}0, x\in \mathbf{R}/\mathbf{Q}\\ 1, x\in \mathbf{Q}\end{cases}$$

$D(x)$为周期函数,任意有理数都是它的周期,所以 $D(x)$ 没有最小正周期.

1.3.4.2 Period of a Sinusoid(正弦型函数的周期)

The period of the sinusoid $f(x)=a\sin(bx+c)+d$ is $\dfrac{2\pi}{|b|}$.

同理,余弦型函数 $f(x)=a\cos(bx+c)+d$ 的最小正周期为 $\dfrac{2\pi}{|b|}$.

Example 1.4 求函数的最小正周期.

Find the period of each function.

a. $y=\sin(1-2x)$;

b. $y=\sin^2 x$;

c. $y=3-|\sin 4x|$;

d. $y=2\cos\dfrac{x}{2}-3\sin\dfrac{x}{3}$.

Solution

a. $y=\sin(1-2x)$的最小正周期为 $\dfrac{2\pi}{|-2|}=\pi$.

b. 因为 $y=\sin^2 x=\dfrac{1-\cos 2x}{2}=\dfrac{1}{2}-\dfrac{1}{2}\cos 2x$,$y=\cos x$ 的最小正周期为 2π,

所以 $y=\sin^2 x$ 的最小正周期为 $\dfrac{2\pi}{2}=\pi$.

c. 因为 $y=|\sin x|$ 的最小正周期为 π,所以 $y=3-|\sin 4x|$ 的最小正周期为 $\dfrac{\pi}{4}$.

d.因为 $y=2\cos\frac{x}{2}$ 的最小正周期为 4π,$y=3\sin\frac{x}{3}$ 的最小正周期为 6π,

所以 $y=2\cos\frac{x}{2}-3\sin\frac{x}{3}$ 的最小正周期为 4π 和 6π 的最小公倍数 12π.

1.4 Function Operations(函数的运算)

1.4.1 Arithmetic Operation on Functions(函数的四则运算)

Given function $f(x)$ and $g(x)$, we define
$$(f+g)(x)=f(x)+g(x)$$
$$(f-g)(x)=f(x)-g(x)$$
$$(fg)(x)=f(x)g(x)$$
$$(f/g)(x)=f(x)/g(x) \quad (g(x)\neq 0)$$

For the function $f+g$, $f-g$ and f/g, we define the domain to be intersection of the domain of f and g, and for the function f/g, we define the domain to be intersection of the domain of f and g but with the points where $g(x)=0$ excluded.

1.4.2 Composite Functions(复合函数)

We say that the function $f(g(x))$ (read "f of g of x") is the composite of g and f. It is made by **composing** g and f in the order of first g, then f.

$y=f(g(x))$ 称为先 g 后 f 的复合函数,它是由函数 $u=g(x)$ 与函数 $y=f(u)$ 复合而成,其中 $u=g(x)$ 称为内层函数,$y=f(u)$ 称为外层函数.复合函数 $y=f(g(x))$ 的定义域是函数 $u=g(x)$ 的定义域的子集,复合函数自变量对应的 $u=g(x)$ 的值 $(f\circ g)(x)$ 要在外层函数 $y=f(u)$ 的定义域内.复合函数 $y=f(g(x))$ 也可记作: $y=(f\circ g)(x)$,即 $f(g(x))=(f\circ g)x$.

Example 1.5 函数的复合.

$f(x)=x+5, g(x)=x^2-3$ find

a. $f(g(x))$; b. $f(g(0))$; c. $g(f(x))$;
d. $g(f(0))$; e. $f(f(x))$; f. $g(g(-2))$.

Solution

a. $f(g(x))=(x^2-3)+5=x^2+2$.

b. $f(g(0))=0^2+2=2$.

c. $g(f(x))=(x+5)^2-3=x^2+10x+22$.

d. $g(f(0))=0^2+10\times 0+22=22$.

e. $f(f(x))=(x+5)+5=x+10$.

f. 因为 $g(g(x))=(x^2-3)^2-3$,

所以 $g(g(-2))=[(-2)^2-3]^2-3=1^2-3=-2$.

[注]一般情况下,$f\circ g=f(g(x))\neq g(f(x))=g\circ f$,所以求复合函数时,一定要注意内层函数和外层函数的顺序.

1.4.3　Inverse Functions(反函数)

1.4.3.1　Definition(定义)

A function $f(x)$ is one-to-one on a domain D if $f(a) \neq f(b)$ whenever $a \neq b$.

The function defined by reversing a one-to-one function f is the inverse of f. The symbol for the inverse of f is f^{-1}, read "f inverse." The -1 in f^{-1} is not an exponent; $f^{-1}(x)$ does not mean $\dfrac{1}{f(x)}$.

$$\text{domain of } f^{-1} = \text{range of } f; \qquad \text{range of } f^{-1} = \text{domain of } f$$

若 $y=f(x)$ 在定义域内每一个自变量对应唯一的函数值,同时反过来每一个函数值只有唯一的自变量与之对应,这样的函数称为 **one-to-one function(一对一函数)**.若 $y=f(x)$ 是一对一函数,则由它解出的 $x=f^{-1}(y)$ 称为 $y=f(x)$ 的反函数,由于函数与字母无关,并且一般用 x 表示自变量,用 y 表示因变量,所以 $y=f(x)$ 的反函数用 $y=f^{-1}(x)$ 表示. $y=f(x)$ 称为直接函数,$y=f^{-1}(x)$ 称为反函数.直接函数的定义域就是反函数的值域,直接函数的值域就是反函数的定义域.

Since $f(a)=b$ if and only if $f^{-1}(b)=a$, the point (a,b) is on the graph of f if and only if the point (b,a) is on the graph of f^{-1}. But we get the point (b,a) from (a,b) by reflecting about the line $y=x$(见图 1.8).

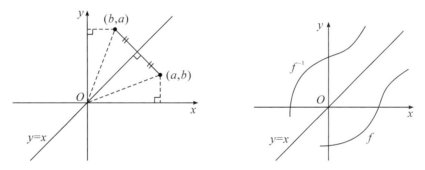

图 1.8　互为反函数的两个函数的图像关于 $y=x$ 对称

1.4.3.2　Inverse Properties(反函数的性质)

(1) $(f \circ f^{-1})(x) = (f^{-1} \circ f)(x) = x$.

(2) $f(x)=\mathrm{e}^x$ and $f^{-1}(x)=\ln x$ are inverses of each other.

$\ln \mathrm{e}^x = \mathrm{e}^{\ln x} = x\ (x>0)$(此等式常用于幂指函数和指数函数之间的相互转化).

Example 1.6　*反函数的性质.*

$f(x)=x^2, x \leqslant 0$, find $f^{-1}(x)$ and verify that
$$(f \circ f^{-1})(x) = (f^{-1} \circ f)(x) = x.$$

Solution

$y=x^2, x \leqslant 0$

$y=-\sqrt{x}$

Interchange x and y.

$y = f^{-1}(x) = -\sqrt{x}$ or $-x^{1/2}$

For $x \geq 0$ (the domain of f^{-1}), $(f \circ f^{-1})(x) = f(-\sqrt{x}) = (-\sqrt{x})^2 = x$

For $x \leq 0$ (the domain of f), $(f^{-1} \circ f)(x) = f^{-1}(x^2) = -\sqrt{x^2} = x$.

1.5 Basic Elementary Functions(基本初等函数)

1.5.1 Power Functions(幂函数)

1.5.1.1 Definition(定义)

Any function that can be written in the form

$$f(x) = k \cdot x^a, \text{where } k \text{ and } a \text{ are nonzero constants,}$$

is a power function. The constant a is the power, and k is the constant of variation, or constant of proportion. We say $f(x)$ varies as the power of x, or $f(x)$ is proportional to the power of a th power of x.

1.5.1.2 Graphs of Power Functions(幂函数的图像)

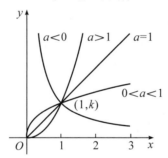

图 1.9 The graphs of $f(x) = k \cdot x^a$ for $k > 0$ and $x \geq 0$

作幂函数的图像时往往先确定第一象限的图像,然后根据定义域和函数的奇偶性确定另一个分支是否数形,以及另一个分支和第一象限的图像的对称关系.熟练掌握幂函数的图像有利于以图形结合的方法解决问题.

常见的幂函数有：

The Identity Function: $y = x$ 　　　　The Squaring Function: $y = x^2$

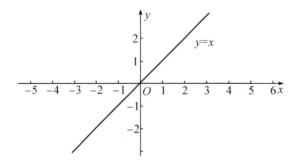
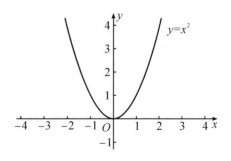

The Cubic Function: $y=x^3$ The Square Root Function: $y=\sqrt{x}=x^{1/2}\ (x\geqslant 0)$

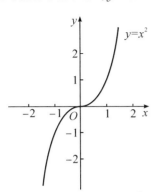
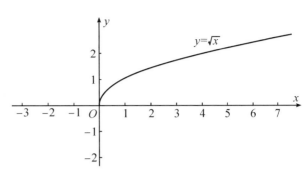

The Reciprocal Function: $y=\dfrac{1}{x}\ (x\neq 0)$

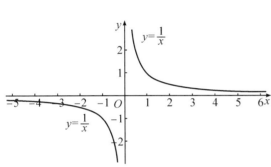
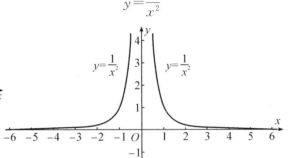

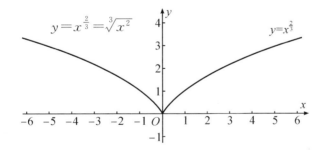

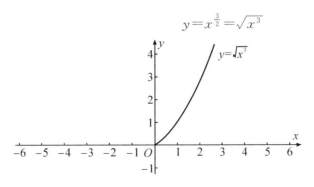

1.5.2 Polynomial Functions(多项式函数)

1.5.2.1 Definition(定义)

Let n be a nonnegative integer and let $a_0, a_1, a_2, \cdots, a_{n-1}, a_n$ be real numbers with $a_n \neq 0$. The function given by

$$f(x) = a_n x^n + a_{n-1} x^{n-1} + \cdots + a_2 x^2 + a_1 x + a_0$$

is a **polynomial function of degree** n(n 次多项式函数).**The leading coefficient**(首项系数) is a_n.

1.5.2.2 Polynomial Functions of No and Low Degree(无次数或低次的多项式函数)

Zero function: $f(x) = 0$(没有次数)

Constant function: $f(x) = a\,(a \neq 0)$(次数为 0)

Linear function: $f(x) = ax + b\,(a \neq 0)$(次数为 1)

Quadratic function: $f(x) = ax^2 + bx + c\,(a \neq 0)$(次数为 2)

1.5.2.3 Linear Functions(线性函数)

线性函数的函数图像为一条直线,其函数表达式常见的有以下几种形式:

Point-Slope Form(点斜式):

Through $P_1(x_1, y_1)$ and with slope m: $y = m(x - x_1) + y_1$

Slope-Intercept Form(斜截式):

With slope m and y-intercept b: $y = mx + b$

Two-Point Form(两点式):

Through $P_1(x_1, y_1)$ and $P_2(x_2, y_2)$: $y = \dfrac{y_2 - y_1}{x_2 - x_1}(x - x_1) + y_1$

Intercept Form(截距式):

With x and y-intercept a and b, respectively: $\dfrac{x}{a} + \dfrac{y}{b} = 1$

x-**intercept**(x-截距)即函数图像与 x 轴交点的横坐标,若 x-intercept$= a$,即 $x = a$ 时 $y = 0$.

y-**intercept**(y-截距)即函数图像与 y 轴交点的纵坐标,若 y-intercept$= b$,即 $y = b$ 时 $x = 0$.

General Form(一般式):

$Ax + By + C = 0$, where A and B are not both zero.

If $B \neq 0$, the slope is $-\dfrac{A}{B}$; the y-intercept is $-\dfrac{C}{B}$; the x-intercept is $-\dfrac{C}{A}$.

1.5.2.4 Quadratic Functions(二次函数)

二次函数的图像是一条抛物线,常见的二次函数有以下几种形式:

Standard Form(标准式): $y = ax^2 + bx + c\,(a \neq 0)$

Vertex Form(顶点式): $y = a(x - h)^2 + k$

The graph of f is a parabola with vertex (h, k) and axis $x = h$, where $h = -\dfrac{b}{2a}$.

If $a>0$, the parabola opens upward, and if $a<0$, it opens downward.

Initial value $=y$-intercept$=f(0)=c$.

x-intercepts$=\dfrac{b\pm\sqrt{b^2-4ac}}{2a}$.

1.5.3 Rational Functions(有理函数)

1.5.3.1 Definition(定义)

Let f and g be polynomial functions with $g(x)\neq 0$. Then the function given by
$$y=\frac{f(x)}{g(x)}=\frac{a_n x^n+a_{n-1}x^{n-1}+\cdots+a_2 x^2+a_1 x+a_0}{b_m x^m+b_{m-1}x^{m-1}+\cdots+b_2 x^2+b_1 x+b_0}$$
is a **rational function(有理函数)**.

The domain of a rational function is the set of all real numbers except the zeros of its denominator.

1.5.3.2 Division Algorithm for Polynomials(多项式除法)

Let $f(x)$ and $d(x)$ be polynomials with the degree of f greater than or equal to the degree of d, and $d\neq 0$. Then there are unique polynomials $q(x)$ and $r(x)$, called the quotient and remainder, such that
$$f(x)=d(x)\cdot q(x)+r(x) \qquad ①$$
where either $r(x)=0$ or the degree of r is less than the degree of d.

The function $f(x)$ in the division algorithm is the dividend, and $d(x)$ is the divisor. If $r(x)=0$, we say $d(x)$ divides evenly into $f(x)$.

The summary statement ① is sometimes written infraction form as follows:
$$\frac{f(x)}{d(x)}=q(x)+\frac{r(x)}{d(x)} \qquad ②$$

Example 1.7 长除法.

Use long division to find the quotient and remainder when $2x^4-x^3-2$ is divided by $2x^2+x+1$. Write a summary statement in both polynomial and fraction form.

Solution

$$\begin{array}{r}
x^2 - x \\
2x^2+x+1{\overline{\smash{\big)}\,2x^4 - x^3 + 0x^2 + 0x - 2}}\\
\underline{2x^4 + x^3 + x^2}\\
-2x^3 - x^2 + 0x - 2\\
\underline{-2x^3 - x^2 - x}\\
x - 2
\end{array}$$

The division algorithm yields the polynomial form
$$2x^4-x^3-2=(2x^2+x+1)(x^2-x)+(x-2)$$

Using the equation ②, we obtain the fraction form
$$\frac{2x^4-x^3-2}{2x^2+x+1}=x^2-x+\frac{x-2}{2x^2+x+1}.$$

1.5.4 Exponential Functions(指数函数)

1.5.4.1 Definition(定义)

In general, an **exponential function** is a function of the form
$$y=a^x \ (a>0, \text{ and } a\neq 1)$$
If $x=n$, a positive integer, then
$$a^n=\underbrace{a\cdot a\cdot\cdots\cdot a}_{n \text{ factors}}$$
If $x=0$, then $a^0=1$, and if $x=-n$, where n is a positive integer, then
$$a^{-n}=\frac{1}{a^n}$$
If x is a rational number, $x=\dfrac{p}{q}$, where p and q are integers and $q>0$, then
$$a^x=a^{\frac{p}{q}}=\sqrt[q]{a^p}=(\sqrt[q]{a})^p$$

1.5.4.2 Graph of a Exponential Functions(指数函数的图像)

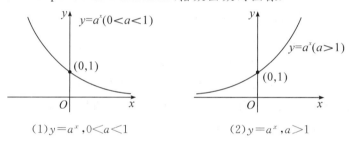

(1) $y=a^x$, $0<a<1$ (2) $y=a^x$, $a>1$

图 1.10

1.5.4.3 Laws of Exponents(指数的性质)

If a and b are positive numbers and m and n any real numbers, then
$$a^m\cdot a^n=a^{m+n} \qquad a^m\div a^n=a^{m-n} \qquad (a^m)^n=a^{m\cdot n} \qquad (ab)^n=a^n b^n$$

1.5.5 Logarithmic Functions (对数函数)

1.5.5.1 Definition(定义)

Since $y=a^x$ is one-to-one, it has an inverse, $f^{-1}(x)=\log_a x$, called the **logarithmic function with base** a. We note that
$$y=\log_a x \qquad \text{if and only if} \qquad a^y=x$$
The domain of $y=\log_a x$ is the set of positive reals, its range is the set of all reals.

Logarithms with base e and base 10 are so important in applications that calculators have special keys for them. They also have their own special notation and names:
$$\log_e x=\ln x$$
$$\log_{10} x=\lg x$$
The function $y=\ln x$ is called **the natural logarithm function**(自然对数函数) and $y=\lg x$ is often called **the common logarithm function**(常用对数函数).

1.5.5.2 Graph of a Logarithmic Functions(对数函数的图像)

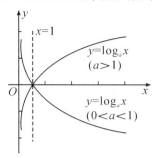

图 1.11

1.5.5.3 Properties of Logarithms(对数的性质)

The logarithmic function $y=\log_a x (a>0, a\neq 1)$ has the following properties:

$$\log_a 1=0 \qquad \log_a a=1 \qquad \log_a(mn)=\log_a m+\log_a n$$

$$\log_a \frac{m}{n}=\log_a m-\log_a n \qquad \log_a x^m=m\log_a x$$

1.5.6 Trigonometric Functions(三角函数)

1.5.6.1 Definition(定义)

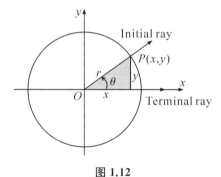

图 1.12

Let θ be any angle in standard position and let $P(x,y)$ be any point on the terminal side of the angle (except the origin). Let r denote the distance $P(x,y)$ from to the origin, i.e., let $r=\sqrt{x^2+y^2}$ (见图 1.12). Then

$$\text{sine}: \sin\theta=\frac{y}{r} \qquad\qquad \text{cosecant}: \csc\theta=\frac{r}{y}$$

$$\text{cosine}: \cos\theta=\frac{x}{r} \qquad\qquad \text{secant}: \sec\theta=\frac{r}{x}$$

$$\text{tangent}: \tan\theta=\frac{y}{x} \qquad\qquad \text{cotangent}: \cot\theta=\frac{x}{y}$$

1.5.6.2 Graphs of Trigonometric Functions(三角函数的图像)

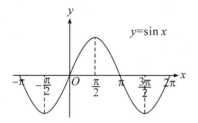

Domain: $-\infty < x < \infty$
Range: $-1 \leqslant y \leqslant 1$
Period: 2π

(a)

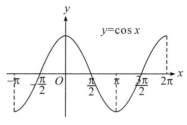

Domain: $-\infty < x < \infty$
Range: $-1 \leqslant y \leqslant 1$
Period: 2π

(b)

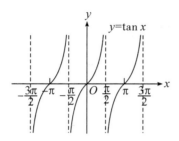

Domain: $x \neq \pm\frac{\pi}{2}, \pm\frac{3\pi}{2}, \cdots$
Range: $-\infty < y < \infty$
Period: π

(c)

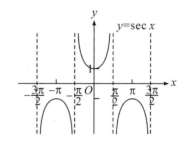

Domain: $x \neq \pm\frac{\pi}{2}, \pm\frac{3\pi}{2}, \cdots$
Range: $y \leqslant -1$ and $y \geqslant 1$
Period: 2π

(d)

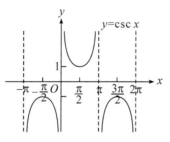

Domain: $x \neq 0, \pm\pi, \pm 2\pi, \cdots$
Range: $y \leqslant -1$ and $y \geqslant 1$
Period: 2π

(e)

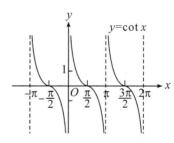

Domain: $x \neq 0, \pm\pi, \pm 2\pi, \cdots$
Range: $-\infty < y < \infty$
Period: π

(f)

图 1.13　Graphs of the (a) sine, (b) cosine, (c) tangent, (d) secant, (e) cosecant, and (f) cotangent functions using radian measure

1.5.6.3 Formulas from Trigonometry(三角公式)

Reciprocal Identities:

$$\sec x = \frac{1}{\cos x} \qquad \csc x = \frac{1}{\sin x} \qquad \cot x = \frac{1}{\tan x}$$

Pythagorean Identities:

$$\sin^2 x + \cos^2 x = 1 \qquad 1 + \tan^2 x = \sec^2 x \qquad 1 + \cot^2 x = \csc^2 x$$

Odd-Even Identities:

$$\sin(-x) = -\sin x \qquad \cos(-x) = \cos x \qquad \tan(-x) = -\tan x$$
$$\csc(-x) = -\csc x \qquad \sec(-x) = \sec x \qquad \cot(-x) = -\cot x$$

Sum and Difference Identities:

$$\sin(x+y) = \sin x \cos y + \cos x \sin y \qquad \sin(x-y) = \sin x \cos y - \cos x \sin y$$
$$\cos(x+y) = \cos x \cos y - \sin x \sin y \qquad \cos(x-y) = \cos x \cos y + \sin x \sin y$$
$$\tan(x+y) = \frac{\tan x + \tan y}{1 - \tan x \tan y} \qquad \tan(x-y) = \frac{\tan x - \tan y}{1 + \tan x \tan y}$$

和差化积：

$$\sin x + \sin y = 2 \sin \frac{x+y}{2} \cos \frac{x-y}{2}$$

$$\sin x - \sin y = 2 \cos \frac{x+y}{2} \sin \frac{x-y}{2}$$

$$\cos x + \cos y = 2 \cos \frac{x+y}{2} \cos \frac{x-y}{2}$$

$$\cos x - \cos y = -2 \sin \frac{x+y}{2} \sin \frac{x-y}{2}$$

积化和差：

$$\sin x \cos y = \frac{1}{2}[\sin(x+y) + \sin(x-y)]$$

$$\cos x \sin y = \frac{1}{2}[\sin(x+y) - \sin(x-y)]$$

$$\cos x \cos y = \frac{1}{2}[\cos(x+y) + \cos(x-y)]$$

$$\sin x \sin y = -\frac{1}{2}[\cos(x+y) - \cos(x-y)]$$

Cofunction Identities:

$$\cos\left(\frac{\pi}{2} - x\right) = \sin x \qquad \sin\left(\frac{\pi}{2} - x\right) = \cos x \qquad \tan\left(\frac{\pi}{2} - x\right) = \cot x$$
$$\sec\left(\frac{\pi}{2} - x\right) = \csc x \qquad \csc\left(\frac{\pi}{2} - x\right) = \sec x \qquad \cot\left(\frac{\pi}{2} - x\right) = \tan x$$

Double-Angle Identities:

$$\sin 2x = 2 \sin x \cos x$$
$$\cos 2x = \cos^2 x - \sin^2 x = 2 \cos^2 x - 1 = 1 - 2 \sin^2 x$$
$$\tan 2x = \frac{2 \tan x}{1 - \tan^2 x}$$

Power-Reducing Identities:

$$\sin^2 x = \frac{1 - \cos 2x}{2} \qquad \cos^2 x = \frac{1 + \cos 2x}{2} \qquad \tan^2 x = \frac{1 - \cos 2x}{1 + \cos 2x}$$

1.5.6.4 Trigonometric Functions of Important Angels(特殊角的三角函数)

θ	radians	$\sin\theta$	$\cos\theta$	$\tan\theta$
0°	0	0	1	0
30°	$\pi/6$	1/2	$\sqrt{3}/2$	$\sqrt{3}/3$
45°	$\pi/4$	$\sqrt{2}/2$	$\sqrt{2}/2$	1
60°	$\pi/3$	$\sqrt{3}/2$	1/2	$\sqrt{3}$
90°	$\pi/2$	1	0	—

1.5.7 Inverse Trigonometric Functions(反三角函数)

1.5.7.1 Definition(定义)

Inverse Sine Function(反正弦函数)

The unique angle y in the interval $\left[-\dfrac{\pi}{2},\dfrac{\pi}{2}\right]$ such that $\sin y = x$ is the inverse sine (or arcsine) of x, denoted $\sin^{-1}x$ or $\arcsin x$.

The domain of $y=\sin^{-1}x$ is $[-1,1]$ and the range is $\left[-\dfrac{\pi}{2},\dfrac{\pi}{2}\right]$.

Inverse Cosine Function(反余弦函数)

The unique angle y in the interval $[1,\pi]$ such that $\cos y = x$ is the inverse cosine (or arccosine) of x, denoted $\cos^{-1}x$ or $\arccos x$.

The domain of $y=\cos^{-1}x$ is $[-1,1]$ and the range is $[0,\pi]$.

Inverse Tangent Function(反正切函数)

The unique angle y in the interval $\left(-\dfrac{\pi}{2},\dfrac{\pi}{2}\right)$ such that $\tan y = x$ is the inverse tangent (or arctangent) of x, denoted $\tan^{-1}x$ or $\arctan x$.

The domain of $y=\tan^{-1}x$ is $(-\infty,\infty)$ and the range is $\left(-\dfrac{\pi}{2},\dfrac{\pi}{2}\right)$.

Note also that

$$\sec^{-1}x=\cos^{-1}\left(\dfrac{1}{x}\right) \qquad \csc^{-1}x=\sin^{-1}\left(\dfrac{1}{x}\right) \qquad \cot^{-1}x=\dfrac{\pi}{2}-\tan^{-1}x$$

1.5.7.2 Graphs of Inverse Trigonometric Functions(反三角函数的图像)

$y = \cos^{-1} x$

Domain: $-1 \leqslant x \leqslant 1$

Range: $0 \leqslant y \leqslant \pi$

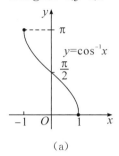

(a)

$y = \sin^{-1} x$

Domain: $-1 \leqslant x \leqslant 1$

Range: $-\dfrac{\pi}{2} \leqslant y \leqslant \dfrac{\pi}{2}$

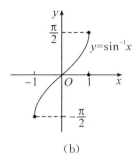

(b)

$y = \tan^{-1} x$

Domain: $-\infty < x < \infty$

Range: $-\dfrac{\pi}{2} < y < \dfrac{\pi}{2}$

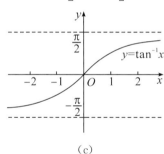

(c)

$y = \sec^{-1} x$

Domain: $x \leqslant -1$ or $x \geqslant 1$

Range: $0 \leqslant y \leqslant \pi$, $y \neq \dfrac{\pi}{2}$

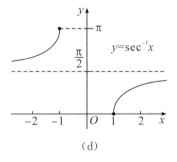

(d)

$y = \csc^{-1} x$

Domain: $x \leqslant -1$ or $x \geqslant 1$

Range: $-\dfrac{\pi}{2} \leqslant y \leqslant \dfrac{\pi}{2}$, $y \neq 0$

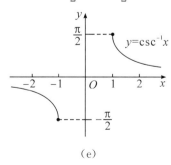

(e)

$y = \cot^{-1} x$

Domain: $-\infty < x < \infty$

Range: $0 < y < \pi$

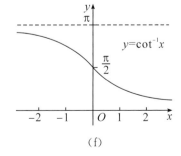

(f)

图 1.14

1.6 Parametric Equations(参数方程)

If x and y are given as functions
$$x=f(t), y=g(t)$$
over an interval of t-values, then the set of points $(x,y)=(f(t),g(t))$ defined by these equations is a parametric curve. The equations are parametric equations for the curve. The variable t is a **parameter(参数)** for the curve and its domain I is the parameter interval.

常见的参数方程：

(1)直线的参数方程：

过点(a,b)和点(c,d)的直线的参数方程为：
$$\begin{cases} x=a+t(c-a) \\ y=b+t(d-b) \end{cases}, -\infty<t<\infty$$

过点(a,b)且斜率为m的直线的参数方程为：
$$\begin{cases} x=a+t \\ y=b+mt \end{cases}, -\infty<t<\infty$$

(2)圆的参数方程：

以点(a,b)为圆心，r为半径的圆$(x-a)^2+(y-b)^2=r^2$的参数方程为：
$$\begin{cases} x=a+r\cos\theta \\ y=b+r\sin\theta \end{cases}, 0<\theta<2\pi$$

(3)椭圆的参数方程：

$\left(\dfrac{x}{a}\right)^2+\left(\dfrac{y}{b}\right)^2=1$ 对应的参数方程为：
$$\begin{cases} x=a\cos t \\ y=b\sin t \end{cases}, -\pi\leqslant t\leqslant\pi$$

Example 1.8

Find a Cartesian equation for a curve that contains the parameterized curve. What portion of the graph of the Cartesian equation is traced by the parametrised curve?

a. $x=3-3t, y=2t, 0\leqslant t\leqslant 1$

b. $x=3t, y=9t^2, -\infty<t<\infty$

c. $x=\sin(2\pi t), y=\cos(2\pi t), 0\leqslant t\leqslant 1$

d. $x=\cos(\pi-t), y=\sin(\pi-t), 0\leqslant t\leqslant\pi$

e. $x=4\cos t, y=2\sin t, 0\leqslant t\leqslant 2\pi$

Solution(计算出结果后，可用图形计算器作图验证)

$$x=3-3t \Rightarrow t=\dfrac{3-x}{3}$$

a. $y=2t=\dfrac{2}{3}(3-x)=-\dfrac{2}{3}x+2$

Initial point: (3,0)

Terminal point: (0,2)

The Cartesian equation is $y = -\dfrac{2}{3}x + 2$.

The portion traced by the curve is the segment from (3,0) to (0,2).

b. $y = 9t^2 = (3t)^2 = x^2$

No initial or terminal point.

The parameterized curve traces all of the parabola defined by $y = x^2$.

c. $x^2 + y^2 = \sin^2(2\pi t) + \cos^2(2\pi t) = 1$

Initial and terminal point: (0,1)

The parameterized curve traces all of the circle defined by $x^2 + y^2 = 1$.

d. $x^2 + y^2 = \cos^2(\pi - t) + \sin^2(\pi - t) = 1$

Initial point: (−1,0)

Terminal point: (1,0)

The parameterized curve traces the upper half of the circle defined by $x^2 + y^2 = 1$ (or all of the semicircle defined by $y = \sqrt{1 - x^2}$.

e. $\left(\dfrac{x}{4}\right)^2 + \left(\dfrac{y}{2}\right)^2 = \cos^2 t + \sin^2 t = 1$

Initial and terminal point: (4,0).

The parametrised curve traces all of the ellipse defined by $\left(\dfrac{x}{4}\right)^2 + \left(\dfrac{y}{2}\right)^2 = 1$.

1.7　Polar Functions(极坐标方程)

1.7.1　Polar Coordinate System(极坐标系)

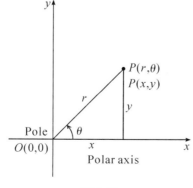

图 1.15

A **polar coordinate system**(极坐标系) is a plane with a point O, the **pole**(极点), and a ray from O, the **polar axis**(极轴), as shown in Figure 1.15. Each point $P(r,\theta)$ in the plane is assigned as **polar coordinates**(极坐标) follows: r is the **directed distance**(极径) from O to

P, and θ is the **directed angle(极角)** whose initial side is on the polar axis and whose terminal side is on the line OP.

1.7.2 Coordinate Conversion Equations(坐标系的转化)

Let the point P have polar coordinates (r,θ) and Cartesian coordinates (x,y). Then
$$x = r\cos\theta, \qquad y = r\sin\theta,$$
$$r^2 = x^2 + y^2, \qquad \tan\theta = \frac{y}{x}.$$

1.7.3 Polar Curves(极曲线)

Here are a few of the more common polar graphs and the θ-intervals that can be used to produce them.

Circles (圆):

$r=1, 0\leqslant\theta\leqslant 2\pi$ \qquad $r=2\sin\theta, 0\leqslant\theta\leqslant\pi$ \qquad $r=2\cos\theta, 0\leqslant\theta\leqslant\pi$

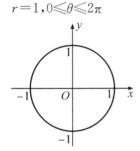
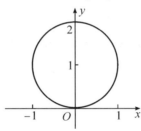
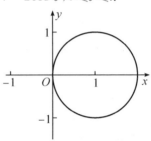

Rose Curves(玫瑰线):

$r=2\sin 3\theta, 0\leqslant\theta\leqslant\pi$ \qquad $r=2\cos 2\theta, 0\leqslant\theta\leqslant 2\pi$

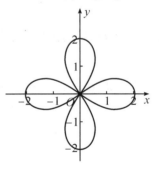

Limacon Curves(蚶线):

$r=1+3\sin\theta, 0\leqslant\theta\leqslant 2\pi$ \qquad $r=1+\cos\theta, 0\leqslant\theta\leqslant 2\pi$

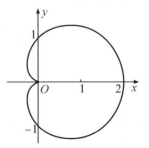

Lemniscate Curves(双纽线):

$r^2 = \sin 2\theta, 0 \leq \theta \leq \pi$ $r^2 = \cos 2\theta, 0 \leq \theta \leq \pi$

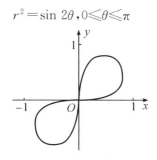 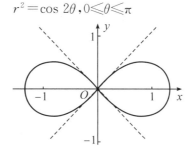

Spiral of Archimedes(阿基米德螺线):

$r = \theta, 0 \leq \theta \leq 2\pi$

1.8 Transformations of Functions(函数图像的变换)

1.8.1 Vertical and Horizontal Shifts(平移变换)

Suppose $c > 0$(如图 1.16). To obtain the graph of

$y = f(x) + c$, **shift the graph of** $y = f(x)$ **a distance** c **units upward.**

$y = f(x) - c$, **shift the graph of** $y = f(x)$ **a distance** c **units downward.**

$y = f(x - c)$, **shift the graph of** $y = f(x)$ **a distance** c **units to the right.**

$y = f(x + c)$, **shift the graph of** $y = f(x)$ **a distance** c **units to the left.**

1.8.2 Vertical and Horizontal Stretching and Reflecting(伸缩变换与对称变换)

Suppose $c > 0$(如图 1.17). To obtain the graph of

$y = cf(x)$, stretch the graph of $y = f(x)$ vertically by a factor of c.

$y = \dfrac{1}{c}f(x)$, shrink the graph of $y = f(x)$ vertically by a factor of c.

$y = f(cx)$, shrink the graph of $y = f(x)$ horizontally by a factor of c.

$y = f\left(\dfrac{x}{c}\right)$, stretch the graph of $y = f(x)$ horizontally by a factor of c.

$y = -f(x)$, reflect the graph of $y = f(x)$ about the x-axis.

$y = f(-x)$, reflect the graph of $y = f(x)$ about the y-axis.

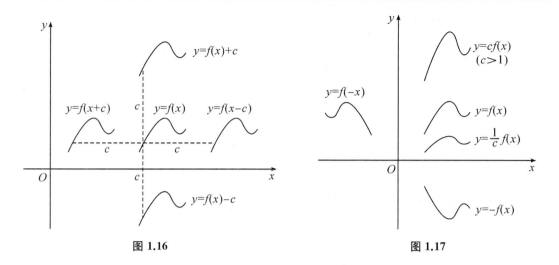

图 1.16　　　　　　　　　　　　图 1.17

Practice Exercises(习题)

1. The domain of the function defined by $f(x)=\ln(x^2-4)$ is the set of all real numbers x such that

   (A) $|x|<2$　　　(B) $|x|\leqslant 2$　　　(C) $|x|>2$　　　(D) $|x|\geqslant 2$

2. If the domain of the function f given by $f(x)=\dfrac{1}{1-x^2}$ is $\{x\,|\,|x|>1\}$, what is the range of f?

   (A) $\{x\,|-\infty<x<-1\}$　　　　　(B) $\{x\,|-\infty<x<0\}$

   (C) $\{x\,|-\infty<x<1\}$　　　　　(D) $\{x\,|-1<x<\infty\}$

3. The graph of $y^2=x^2+9$ is symmetric to which of the following?

   I. The x-axis

   II. The y-axis

   III. The origin

   (A) I only　　　(B) II only　　　(C) I and II　　　(D) I, II and III

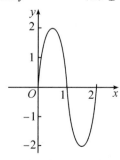

4. The figure above shows the graph of a sine function for one complete period. Which of the following is an equation for the graph?

   (A) $y=2\sin\left(\dfrac{\pi}{2}x\right)$　　(B) $y=\sin(\pi x)$　　(C) $y=2\sin(2x)$　　(D) $y=2\sin(\pi x)$

5. $4\cos\left(x+\frac{\pi}{3}\right)=$

(A) $2\cos x - 2\sqrt{3}\sin x$ (B) $2\cos x + 2\sqrt{3}\sin x$

(C) $2\sqrt{3}\cos x + 2\sin x$ (D) $4\cos x + 2$

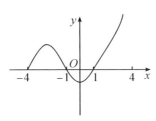

6. The graph of $y=f(x)$ is shown in the figure above. Which of the following could be the graph of $y=f(|x|)$?

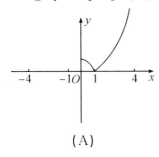

(A)

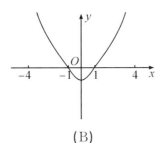

(B)

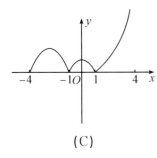

(C)

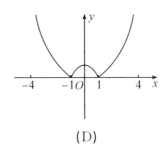

(D)

7. Let f and g be odd functions. If p, r, and s are nonzero functions defined as follows, which must be odd?

Ⅰ. $p(x)=f(g(x))$

Ⅱ. $r(x)=f(x)+g(x)$

Ⅲ. $s(x)=f(x)g(x)$

(A) Ⅰ only (B) Ⅱ only (C) Ⅰ and Ⅱ (D) Ⅰ, Ⅱ and Ⅲ

8. $\ln(x-2)<0$ if and only if

(A) $x<3$ (B) $0<x<3$ (C) $2<x<3$ (D) $x>2$

9. The set of all points (e^t, t), where t is a real number, is the graph of $y=$

(A) $e^{\frac{1}{x}}$ (B) $x e^{\frac{1}{x}}$ (C) $\frac{1}{\ln x}$ (D) $\ln x$

10. If $f(x)=\frac{4}{x-1}$ and $g(x)=2x$, then the solution of $f(g(x))=g(f(x))$ is

(A) $\left\{\dfrac{1}{3}\right\}$ (B) $\{2\}$ (C) $\{3\}$ (D) $\{-1,2\}$

11. If $f(x)=\dfrac{x}{x+1}$, then the inverse function, f^{-1}, is given by $f^{-1}(x)=$

(A) $\dfrac{x-1}{x}$ (B) $\dfrac{x+1}{x}$ (C) $\dfrac{x}{1-x}$ (D) $\dfrac{x}{x+1}$

习题参考答案

1.C 2.B 3.D 4.D 5.A 6.B 7.C 8.C 9.D 10.A 11.C

Chapter 2 Limits and Continuity(极限和连续)

2.1 Definitions of Limits(极限的定义)

极限的定义主要描述了两个变化过程,一个是自变量 x 的变化过程,另一个是因变量 y 的变化过程.极限的定义按照自变量的变化过程可以分为 x 趋近于常数 c 和 x 趋近于无穷大两种类型.

2.1.1 x Approaches c(x 趋近于常数 c)

2.1.1.1 Definitions(定义)

If the values of $f(x)$ can be made as close as we like to L by taking values of x sufficiently close to c (but not equal to c), then we write
$$\lim_{x \to c} f(x) = L$$
which is read "the limit of $f(x)$ as x approaches c is L" or "$f(x)$ approaches L as x approaches c". The expression can also be written as
$$f(x) \to L \text{ as } x \to c$$

如果 x 无限接近于常数 c(但不等于 c),函数 $f(x)$ 的值无限接近于常数 L,则称当 x 趋于 c 时,$f(x)$ 以 L 为极限.记作:
$$\lim_{x \to c} f(x) = L \quad 或 \quad f(x) \to L (当 x \to c \text{ 时}).$$

函数在 $x=c$ 处的极限存在与否与函数在 $x=c$ 处的函数值无关,也与函数在 $x=c$ 处是否有定义无关,见图 2.1.

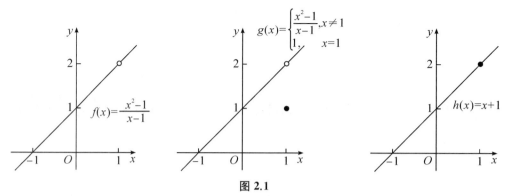

图 2.1

以上三个函数当 $x\to 1$ 时的极限均为 2，但是函数 $f(x)$ 在 $x=1$ 处没有定义；$g(x)$ 在 $x=1$ 处的函数值为 1，极限值不等于函数值；$h(x)$ 在 $x=1$ 处的函数值与极限值相等。

Example 2.1 根据函数图像求极限。

The graph of the function f is shown in the figure below find the value of $\lim\limits_{x\to 0}\sin(f(x))$.

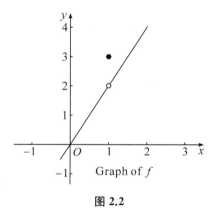

Graph of f

图 2.2

Solution

$\lim\limits_{x\to 0}\sin(f(x))=\sin(\lim\limits_{x\to 1}f(x))=\sin 2.$

[注] 极限运算和函数的复合运算可以交换顺序。

2.1.1.2　One-sided Limits（单侧极限）

(1) Right-hand limit or limit from above（右极限）。

If the values of $f(x)$ can be made as close as we like to L by taking values of x sufficiently close to c (but greater than c), then we write $\lim\limits_{x\to c^+}f(x)=L$ or $f(c^+)=L$.

若当 x 大于 c 且无限趋近于 c 时，$f(x)$ 无限接近于某一常数 L，则常数 L 叫作函数 $f(x)$ 当 $x\to c^+$ 时的右极限，记为 $\lim\limits_{x\to c^+}f(x)=L$ 或 $f(c^+)=L$。

类似地，可以定义左极限。

(2) Left-hand limit or limit from below（左极限）。

If the values of $f(x)$ can be made as close as we like to L by taking values of x sufficiently close to c (but smaller than c), then we write $\lim\limits_{x\to c^-}f(x)=L$ or $f(c^-)=L$.

若当 x 小于 c 且无限趋近于 c 时，$f(x)$ 无限接近于某一常数 L，则常数 L 叫作函数

$f(x)$当 $x \to c^-$ 时的左极限,记为 $\lim\limits_{x \to c^-} f(x) = L$ 或 $f(c^-) = L$.

Theorem 2.1

$\lim\limits_{x \to c} f(x) = L \Leftrightarrow \lim\limits_{x \to c^-} f(x) = L$ and $\lim\limits_{x \to c^+} f(x) = L$.

定理 2.1 极限与单侧极限的关系

$\lim\limits_{x \to c} f(x) = L$ 的充要条件是:当 $x \to c$ 时,$f(x)$ 的左极限和右极限相等,都等于 L.

定理 2.1 主要用于考察分段函数在分段点处的极限情况.

Example 2.2 极限与单侧极限.

For the function $f(x) = \begin{cases} x-1, & x<0 \\ 0, & x=0 \\ x+1, & x>0 \end{cases}$, find $\lim\limits_{x \to 0} f(x)$ or explain why it does not exist.

Solution

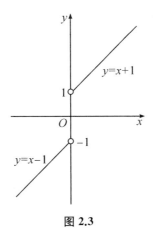

图 2.3

Since $\quad \lim\limits_{x \to 0^-} f(x) = \lim\limits_{x \to 0^-}(x-1) = -1$,

and $\quad \lim\limits_{x \to 0^+} f(x) = \lim\limits_{x \to 0^+}(x+1) = 1$,

thus $\quad \lim\limits_{x \to 0^-} f(x) \neq \lim\limits_{x \to 0^+} f(x)$.

The limit does not exist.

Example 2.3 The Greatest Integer Function(最大整数函数又称取整函数).

The greatest-integer function $y = [x]$, shown below, has different left-hand and right-hand limits at every integer. So, $[x]$ does have a limit at every nonintegral real number.

取整函数 $y = [x]$ 表示不大于 x 的最大整数,该函数在每个整数点处的左右极限都不同 (图 2.4),即取整函数在整数点处的极限不存在,仅在非整数点处存在极限.

For example, $\lim\limits_{x \to -1^-}[x] = 0$ but $\lim\limits_{x \to 1^+}[x] = 1$.

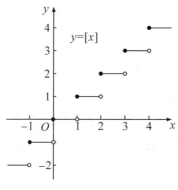

图 2.4

Example 2.4 图像法求极限.

For the function $f(t)$ graphed here, find the following limits or explain why they do not exist.

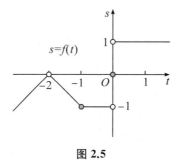

图 2.5

a. $\lim\limits_{t \to -2^-} f(t)$

b. $\lim\limits_{t \to -1} f(t)$

c. $\lim\limits_{t \to 0} f(t)$

Solution

a. $\lim\limits_{t \to -2^-} f(t) = 0$

b. $\lim\limits_{t \to -1} f(t) = -1$

c. Does not exist. As t approaches 0 from the left, $f(t)$ approaches -1. As t approaches 0 from the right, $f(t)$ approaches 1. There is no single number L that $f(t)$ gets arbitrarily close to as $t \to 0$.

Example 2.5 Prove that $\lim\limits_{x \to 0} |x| = 0$.

Solution

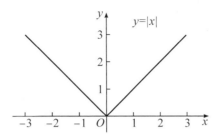

We examine both left- and right-hand limits of the absolute-value function as $x \to 0$.

Since
$$|x| = \begin{cases} -x, \text{if } x < 0 \\ x, \text{if } x > 0 \end{cases}$$

$$\lim_{x \to 0^-} |x| = \lim_{x \to 0^-} (-x) = 0 \text{ and } \lim_{x \to 0^+} |x| = \lim_{x \to 0^+} x = 0$$

thus
$$\lim_{x \to 0} |x| = 0$$

2.1.1.3 Infinite Limits(极限为无穷大)

Sometimes one-sided or two-sided limits fail to exist because the values of the function increase or decrease without bound.

有时函数值无限地增大或减小,此时函数单侧极限或极限不存在.

Example 2.6 Find $\lim\limits_{x \to 0} \dfrac{1}{x}$.

Solution

It is evident from the graph below that as x-values are taken closer and closer to 0 from the right, the values of $f(x) = \dfrac{1}{x}$ are positive and increase without bound; and as x-values are taken closer and closer to 0 from the left, the values of $f(x) = \dfrac{1}{x}$ are negative and decrease without bound. We describe these limiting behaviors by writing $\lim\limits_{x \to 0^-} \dfrac{1}{x} = -\infty$ and $\lim\limits_{x \to 0^+} \dfrac{1}{x} = \infty (\text{or } +\infty)$.

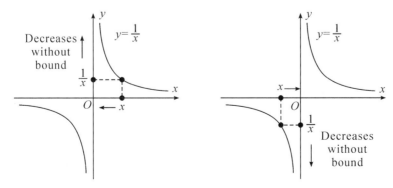

图 2.6

[注] (1) $\lim\limits_{x \to c} f(x) = \infty$ (or $-\infty$) 可以表述为：The function $f(x)$ is said to become infinite positively (or negatively) as x approaches c if $f(x)$ can be made **arbitrarily** large positively (or negatively) by taking x sufficiently close to c.

(2) If $\lim\limits_{x \to c} f(x) = \infty$ (or $-\infty$), the limit, technically does not exist. $\lim\limits_{x \to c} f(x) = \infty$ (or $-\infty$) 只是一种记号，表示函数在 $x \to c$ 时是一个无限增大或无限减少的变量，此时函数的极限是不存在的.

(3) 在大部分教材中 ∞ 表示 $\pm\infty$，但在 AP 考试中，除了容易引起混淆的情形外，∞ 表示 $+\infty$.

(4) 若 $\lim\limits_{x \to c} f(x) = 0$，则称 $f(x)$ 为 $x \to c$ 时的无穷小量. 从 $f(x) = \dfrac{1}{x}$ 的图像可以观察出，当分母 $x \to 0^-$ 时，$f(x) \to -\infty$，当 $x \to 0^+$ 时，$f(x) \to \infty$.

2.1.2　x Approaches Infinity (x 趋近于无穷大)

2.1.2.1　Definitions (定义)

如果当 $|x|$ 无限增大时，函数 $f(x)$ 无限接近于一个确定的常数 L，则称 L 为 $f(x)$ 当 $x \to +\infty$ 时的极限，记为：$\lim\limits_{x \to \infty} f(x) = L$.

类似的，如果当 $x > 0$, $|x|$ 无限增大时，函数 $f(x)$ 无限接近于一个确定的常数 L，则称 L 为 $f(x)$ 当 $x \to +\infty$ 时的极限，记为：$\lim\limits_{x \to +\infty} f(x) = L$.

如果当 $x < 0$, $|x|$ 无限增大时，函数 $f(x)$ 无限接近于一个确定的常数 L，则称 L 为 $f(x)$ 当 $x \to -\infty$ 时的极限，记为：$\lim\limits_{x \to -\infty} f(x) = L$.

2.1.2.2　Theorem 2.2

$$\lim_{x \to \infty} f(x) = L \Leftrightarrow \lim_{x \to -\infty} f(x) = L \text{ and } \lim_{x \to +\infty} f(x) = L.$$

Example 2.7

Find $\lim\limits_{x \to \infty} \dfrac{1}{x}$.

Solution

在图 2.6 中可以观察出：

$$\lim_{x \to -\infty} \frac{1}{x} = 0 \text{ 且 } \lim_{x \to +\infty} \frac{1}{x} = 0 \Rightarrow \lim_{x \to \infty} \frac{1}{x} = 0.$$

直线 $y = 0$ 是函数 $y = \dfrac{1}{x}$ 的水平渐近线.

一般地，如果 $\lim\limits_{x \to -\infty} f(x) = c$ 或 $\lim\limits_{x \to +\infty} f(x) = c$，则直线 $y = c$ 称为函数 $y = f(x)$ 的图像的**水平渐近线 (horizontal asymptote)**.

Example 2.8　根据图像求极限.

Use your calculator to find the limits:

a. $\lim\limits_{x \to +\infty} x^3$　　　b. $\lim\limits_{x \to -\infty} x^3$　　　c. $\lim\limits_{x \to +\infty} x^6$　　　d. $\lim\limits_{x \to -\infty} x^6$

Solution

a. $\lim\limits_{x \to +\infty} x^3 = +\infty$ b. $\lim\limits_{x \to -\infty} x^3 = -\infty$ c. $\lim\limits_{x \to +\infty} x^6 = +\infty$ d. $\lim\limits_{x \to -\infty} x^6 = +\infty$

2.2 The Precise Definition of a Limit(极限的严格定义)

虽然极限的严格定义已不属于 AP 微积分的考试范围,但是为了适应大学学习的要求,请大家尽量通过概念的几何解释对极限的严格定义加以理解.

2.2.1 x Approaches c(x 趋近于常数 c)

Let $f(x)$ be defined on an open interval about c, except possibly at c itself. We say that the **limit of $f(x)$ as x approaches c is the number L**, and write

$$\lim_{x \to c} f(x) = L$$

if, for every number there exists a corresponding number such that for all x,

$$0 < |x - c| < \delta \Rightarrow |f(x) - L| < \varepsilon$$

设函数 $f(x)$ 在点 c 的某一去心邻域内有定义. 如果存在常数 L, 对于任意给定的正数 ε(不论它多么小), 总存在正数 δ, 使得当 x 满足不等式 $0 < |x - c| < \delta$ 时, 对应的函数值 $f(x)$ 都满足不等式 $|f(x) - L| < \varepsilon$, 那么常数 L 就叫作当 $x \to c$ 时函数 $f(x)$ 的极限, 记为:

$$\lim_{x \to c} f(x) = L \text{ 或 } f(x) \to L (\text{当 } x \to c \text{ 时}).$$

定义的简单表述:

$$\lim_{x \to c} f(x) = L \Leftrightarrow \forall \varepsilon > 0, \exists \delta > 0, \text{当 } 0 < |x - c| < \delta \text{ 时}, |f(x) - L| < \varepsilon$$

$\lim\limits_{x \to c} f(x) = L$ 的几何解释:任意给定一个正数 ε, 作平行于 x 轴的两条直线 $y = L + \varepsilon$ 和 $y = L - \varepsilon$, 介于这两条直线之间是一横条区域. 根据定义, 对于给定的 ε, 存在着点 c 的邻域 $(c - \delta, c + \delta)$, 当 $y = f(x)$ 的图像上的点的横坐标 c 落在 $(c - \delta, c + \delta)$ 内, 但 $x \neq c$ 时, 这些点的纵坐标 $f(x)$ 满足不等式 $|f(x) - L| < \varepsilon$, 或 $L - \varepsilon < f(x) < L + \varepsilon$, 也就是这些点落在图 2.7 中的横条区域内.

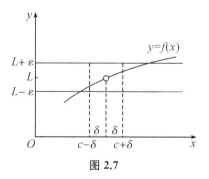

图 2.7

Example 2.9 由定义证明极限.

证明 $\lim\limits_{x \to 1}(2x - 1) = 1$.

分析:$|f(x) - L| = |(2x - 1) - 1| = 2|x - 1|$. $\forall \varepsilon > 0$, 要使 $|f(x) - L| < \varepsilon$, 只要 $|x - 1| < \dfrac{\varepsilon}{2}$.

证明：因为 $\forall \varepsilon > 0, \exists \delta = \dfrac{\varepsilon}{2}$，当 $0 < |x-1| < \delta$ 时，有
$$|f(x) - L| = |(2x-1) - 1| = 2|x-1| < \varepsilon,$$
所以 $\lim\limits_{x \to 1}(2x-1) = 1$.

Example 2.10 极限的严格定义.

Of the following choices of δ, which is the largest that could be used successfully with an arbitrary ε in an epsilon-delta proof of $\lim\limits_{x \to 2}(1-3x) = -5$?

(A) $\delta = 3\varepsilon$ (B) $\delta = \varepsilon$ (C) $\delta = \dfrac{\varepsilon}{2}$ (D) $\delta = \dfrac{\varepsilon}{4}$

Solution

D

Need to have $|(1-3x)-(-5)| < \varepsilon$ whenever $0 < |x-2| < \delta$.

$|(1-3x)-(-5)| = |6-3x| = 3|x-2| < \varepsilon$ if $|x-2| < \dfrac{\varepsilon}{3}$.

Thus we can use any $\delta < \dfrac{\varepsilon}{3}$.

Of the four choices, the largest satisfying this condition is $\delta = \dfrac{\varepsilon}{4}$.

2.2.2 x Approaches Infinity（x 趋近于无穷大）

设 $f(x)$ 当 $|x|$ 大于某一正数时有定义，如果存在常数 L，对于任意给定的正数 ε，总存在着正数 X，使得当 x 满足不等式 $|x| > X$ 时，对应的函数值 $f(x)$ 都满足不等式
$$|f(x) - L| < \varepsilon,$$
则常数 L 叫作函数 $f(x)$ 当 $x \to \infty$ 时的极限，记为 $\lim\limits_{x \to \infty} f(x) = L$ 或 $f(x) \to L (x \to \infty)$.

定义的简单表述：
$$\lim\limits_{x \to \infty} f(x) = L \Leftrightarrow \forall \varepsilon > 0, \exists X > 0, \text{当} |x| > X \text{ 时，} |f(x) - L| < \varepsilon.$$

$\lim\limits_{x \to \infty} f(x) = L$ 函数极限的几何解释：作直线 $y = L - \varepsilon$ 和 $y = L + \varepsilon$，则总有一个正数 X 存在，使得当 $|x| > X$ 时，函数 $y = f(x)$ 的图像位于这两条直线之间（图 2.8）.这时，直线 $y = L$ 是函数 $y = f(x)$ 的图像的**水平渐近线**（**horizontal asymptote**）.

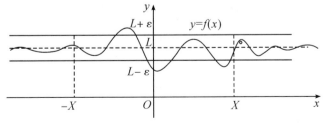

图 2.8

2.3 Theorems on Limits(极限的定理)

2.3.1 The Limit Laws(极限的运算定律)

极限的运算定律主要是两个函数极限的和、差、积、商、数乘以及乘方运算的极限法则，在下面的表述中，记号"lim"下面没有标明自变量的变化过程，实际上，下面的定理对任何自变量变化过程如：$x \to c, x \to c^-, x \to -\infty, x \to \infty, \cdots$ 都成立.

If L, M and k are real numbers and
$$\lim f(x) = L \text{ and } \lim g(x) = M, \text{then}$$

(1) Sum Rule: $\qquad \lim(f(x) + g(x)) = L + M$

和的极限等于极限的和.

(2) Difference Rule: $\qquad \lim(f(x) - g(x)) = L - M$

差的极限等于极限的差.

(3) Product Rule: $\qquad \lim(f(x) \cdot g(x)) = L \cdot M$

积的极限等于极限的积.

(4) Constant Multiple Rule: $\qquad \lim[k f(x)] = k \lim f(x) = k \cdot L$

函数数乘后的极限等于这个常数与函数极限的积，即求常数与函数之积的极限时，常数可以提出来.

(5) Quotient Rule: $\qquad \lim \dfrac{f(x)}{g(x)} = \dfrac{\lim f(x)}{\lim g(x)} = \dfrac{L}{M} (M \neq 0)$

分母不为零时，商的极限等于极限的商.

(6) Power Rule:

If r and s are integers with no common factor and $s \neq 0$ then
$$\lim(f(x))^{r/s} = (\lim f(x))^{r/s} = L^{r/s}$$
provided that $L^{r/s}$ is a real number (If s is even, we assume $L > 0$).

乘方的极限等于极限的乘方.

Example 2.11 利用运算律求极限.

a. $\lim\limits_{x \to 1}(2x - 1)$ \qquad b. $\lim\limits_{x \to 2}\dfrac{x^3 - 1}{x^2 - 5x + 3}$

Solution

a. $\lim\limits_{x \to 1}(2x-1) = \lim\limits_{x \to 1} 2x - \lim\limits_{x \to 1} 1 = 2\lim\limits_{x \to 1} x - 1 = 2 \times 1 - 1 = 1$.

b. $\lim\limits_{x \to 2}\dfrac{x^3 - 1}{x^2 - 5x + 3} = \dfrac{\lim\limits_{x \to 2}(x^3 - 1)}{\lim\limits_{x \to 2}(x^2 - 5x + 3)} = \dfrac{\lim\limits_{x \to 2} x^3 - \lim\limits_{x \to 2} 1}{\lim\limits_{x \to 2} x^2 - 5\lim\limits_{x \to 2} x + \lim\limits_{x \to 2} 3} = \dfrac{(\lim\limits_{x \to 2} x)^3 - 1}{(\lim\limits_{x \to 2} x)^2 - 5 - 2 + 3} = \dfrac{2^3 - 1}{2^2 - 10 + 3} = -\dfrac{7}{3}$.

Example 2.12 Limit of polynomial functions(多项式函数的极限).

若 $P(x) = a_0 x^n + a_1 x^{n-1} + \cdots + a_{n-1} x + a_n (a_0 \neq 0)$，则 $\lim\limits_{x \to x_a} P(x) = ?$

Solution

$$\lim_{x \to x_0} P(x) = \lim_{x \to x_0}(a_0 x^n) + \lim_{x \to x_0}(a_1 x^{n-1}) + \cdots + \lim_{x \to x_0}(a_{n-1} x) + \lim_{x \to x_0} a_n$$

$$= a_0 \lim_{x \to x_0}(x^n) + a_1 \lim_{x \to x_0}(x^{n-1}) + \cdots + a_{n-1} \lim_{x \to x_0} x + \lim_{x \to x_0} a_n$$

$$= a_0 (\lim_{x \to x_0} x)^n + a_1 (\lim_{x \to x_0} x)^{n-1} + \cdots + a_n$$

$$= a_0 x_0^n + a_1 x_0^{n-1} + \cdots + a_n$$

$$= P(x_0).$$

简而言之，多项式函数在常数点的极限等于该点处的函数值，即可以直接将自变量的值代入函数表达式计算．

2.3.2　The Squeeze or Sandwich Theorem 夹逼定理（又称三明治定理）

Suppose that $g(x) \leqslant f(x) \leqslant h(x)$ for all x in some open interval containing c, except possibly at itself. Suppose also that

$$\lim_{x \to c} g(x) = \lim_{x \to c} h(x) = L$$

Then $\lim_{x \to c} f(x) = L$.

如图 2.9 所示：

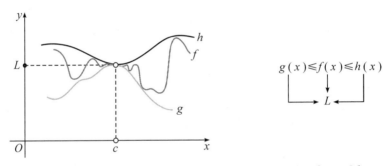

图 2.9　The graph of f is sandwiched between the graphs of g and h.

Example 2.13　Using the Sandwich Theorem.

证明一个重要极限：$\lim_{x \to 0} \dfrac{\sin x}{x} = 1$（$x$ 为弧度制）．

证明：首先注意到，函数 $\dfrac{\sin x}{x}$ 对于一切 $x \neq 0$ 都有定义．

如图 2.10，图中的圆为单位圆，$BC \perp OA$，$DA \perp OA$．

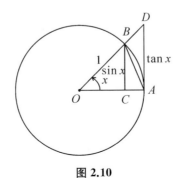

图 2.10

圆心角 $\angle AOB = x \left(0 < x < \dfrac{\pi}{2}\right)$.

显然 $\sin x = CB, x = \widehat{AB}, \tan x = AD$.

因为 $S_{\triangle AOB} < S_{\text{扇形}AOB} < S_{\triangle AOD}$,

所以 $\dfrac{1}{2}\sin x < \dfrac{1}{2}x < \dfrac{1}{2}\tan x$,

即 $\sin x < x < \tan x$.

不等号各边都除以 $\sin x$, 就有 $1 < \dfrac{x}{\sin x} < \dfrac{1}{\cos x}$,

或 $\cos x < \dfrac{\sin x}{x} < 1$.

注意, 此不等式当 $-\dfrac{\pi}{2} < x < 0$ 时也成立.

而 $\lim\limits_{x \to 0} \cos x = 1$, 根据三明治定理知: $\lim\limits_{x \to 0} \dfrac{\sin x}{x} = 1$.

2.4　Computing Limits(极限的计算)

2.4.1　Substitution in 代入法(适用于分母 $\neq 0$ 的情形)

Example 2.14　Find the Limits.

a. $\lim\limits_{x \to 8}(x^2 - 12x - 10)$　　　b. $\lim\limits_{x \to 5}\dfrac{x+3}{x^2-15}$　　　c. $\lim\limits_{x \to -1} x\cos(\pi x)$

Solution

a. $\lim\limits_{x \to 8}(x^2 - 12x - 10) = 8^2 - 12 \times 8 - 10 = -42$

b. $\lim\limits_{x \to 5}\dfrac{x+3}{x^2-15} = \dfrac{5+3}{5^2-15} = \dfrac{8}{10} = \dfrac{4}{5}$

c. $\lim\limits_{x \to -1} x\cos(\pi x) = (-1) \cdot \cos(-\pi) = (-1) \cdot (-1) = 1$

2.4.2　Denominator = 0 and Numerator $\neq$ 0(分母 = 0 但分子 $\neq$ 0 的情形)

定理:

若 $\lim f(x) = 0$, 且 $f(x) > 0$, 则 $\lim \dfrac{1}{f(x)} = +\infty$;

若 $\lim f(x) = 0$, 且 $f(x) < 0$, 则 $\lim \dfrac{1}{f(x)} = -\infty$.

Example 2.15

Find $\lim\limits_{x \to 1} \dfrac{e^x}{\ln x}$.

Solution

$\because \lim\limits_{x \to 1^-} \dfrac{e^x}{\ln x} = \dfrac{\lim\limits_{x \to 1^-} e^x}{\lim\limits_{x \to 1^-} \ln x} = -\infty$ and $\lim\limits_{x \to 1^+} \dfrac{e^x}{\ln x} = \dfrac{\lim\limits_{x \to 1^+} e^x}{\lim\limits_{x \to 1^+} \ln x} = \infty$

$$\lim_{x\to 1^-}\frac{e^x}{\ln x} \neq \lim_{x\to 1^+}\frac{e^x}{\ln x}$$

$\therefore \lim_{x\to 1}\frac{e^x}{\ln x}$ does not exist.

2.4.3　Denominator=0 and Numerator=0(分母=0且分子=0的情形)

Example 2.16　消去零因式法.

a. $\lim\limits_{x\to 2}\dfrac{x-2}{x^2-4}$　　　b. $\lim\limits_{h\to 0}\dfrac{(x+h)^2-x^2}{h}$　　　c. $\lim\limits_{x\to 1}\left(\dfrac{1}{1-x}-\dfrac{3}{1-x^3}\right)$

Solution

a. $\lim\limits_{x\to 2}\dfrac{x-2}{x^2-4}=\lim\limits_{x\to 2}\dfrac{x-2}{(x-2)(x+2)}=\lim\limits_{x\to 2}\dfrac{1}{x+2}=\dfrac{1}{4}$

b. $\lim\limits_{h\to 0}\dfrac{(x+h)^2-x^2}{h}=\lim\limits_{h\to 0}\dfrac{(x+h+x)(x+h-x)}{h}$

$\qquad\qquad\qquad\quad=\lim\limits_{h\to 0}\dfrac{2hx+h^2}{h}$

$\qquad\qquad\qquad\quad=\lim\limits_{h\to 0}(2x+h)=2x$

c. $\lim\limits_{x\to 1}\left(\dfrac{1}{1-x}-\dfrac{3}{1-x^3}\right)=\lim\limits_{x\to 1}\dfrac{1+x+x^2-3}{1-x^3}=\lim\limits_{x\to 1}\dfrac{(x-1)(x+2)}{(1-x)(1+x+x^2)}$

$\qquad\qquad\qquad\qquad\qquad\quad=-\lim\limits_{x\to 1}\dfrac{x+2}{1+x+x^2}=-1$

[注]

(1)消去零因式法主要用于当 $Q(x_0)=P(x_0)=0$ 时,求 $\lim\limits_{x\to x_0}\dfrac{P(x)}{Q(x)}$.其中 $Q(x_0)=P(x_0)=0$ 的原因是 $Q(x)=P(x)$ 都包含了零因式 $(x-x_0)$,因此要对分子、分母进行因式分解,分子、分母分解出零因式 $(x-x_0)$ 后,同时消去 $(x-x_0)$,再用代入法就可以计算出极限值.因式分解常用"十字相乘法"或公式法,其中公式法常用以下几个公式:

$$\boxed{a^2-b^2=(a+b)(a-b)}$$

$$\boxed{a^2\pm 2ab+b^2=(a\pm b)^2}$$

$$\boxed{a^3\pm b^3=(a\pm b)(a^2\mp ab+b^2)}$$

(2)对于 $\lim\limits_{h\to 0}\dfrac{(x+h)^2-x^2}{h}$,虽然函数表达式中包含了两个字母 x 和 h,但是由于 AP 微积分研究的是单变量函数,极限符号 lim 下方跟的是自变量的变化过程,所以此题中的自变量是 h,将 x 看成常数.

(3)用消去零因式法求极限的题都可以用后面将要学习的洛必达法则求解.

Example 2.17　共轭因式法.

a. $\lim\limits_{x\to 1}\dfrac{\sqrt{x}-1}{x-1}$　　　b. $\lim\limits_{x\to 2}\dfrac{\sqrt{2x+5}-\sqrt{x+7}}{x-2}$　　　c. $\lim\limits_{x\to -8}\dfrac{\sqrt{1-x}-3}{2+\sqrt[3]{x}}$

Solution

a. $\lim\limits_{x\to 1}\dfrac{\sqrt{x}-1}{x-1}=\lim\limits_{x\to 1}\dfrac{(\sqrt{x}-1)(\sqrt{x}+1)}{(x-1)(\sqrt{x}+1)}=\lim\limits_{x\to 1}\dfrac{x-1}{(x-1)(\sqrt{x}+1)}=\lim\limits_{x\to 1}\dfrac{1}{\sqrt{x}+1}=\dfrac{1}{2}$

b. $\lim\limits_{x\to 2}\dfrac{\sqrt{2x+5}-\sqrt{x+7}}{x-2}$

$=\lim\limits_{x\to 2}\dfrac{(\sqrt{2x+5}-\sqrt{x+7})(\sqrt{2x+5}+\sqrt{x+7})}{(x-2)(\sqrt{2x+5}+\sqrt{x+7})}$

$=\lim\limits_{x\to 2}\dfrac{(x-2)}{(x-2)(\sqrt{2x+5}+\sqrt{x+7})}$

$=\lim\limits_{x\to 2}\dfrac{1}{\sqrt{2x+5}+\sqrt{x+7}}=\dfrac{1}{6}$

c. $\lim\limits_{x\to -8}\dfrac{\sqrt{1-x}-3}{2+\sqrt[3]{x}}$

$=\lim\limits_{x\to -8}\dfrac{(\sqrt{1-x}-3)(\sqrt{1-x}+3)(4-2\sqrt[3]{x}+\sqrt[3]{x^2})}{(2+\sqrt[3]{x})(4-2\sqrt[3]{x}+\sqrt[3]{x^2})(\sqrt{1-x}+3)}$

$=\lim\limits_{x\to -8}\dfrac{(1-x-9)(4-2\sqrt[3]{x}+\sqrt[3]{x^2})}{(8+x)(\sqrt{1-x}+3)}$

$=-\lim\limits_{x\to -8}\dfrac{4-2\sqrt[3]{x}+\sqrt[3]{x^2}}{\sqrt{1-x}+3}$

$=-2$

共轭因式可以由常见的因式分解公式得到，比如：根据 $a^2-b^2=(a+b)(a-b)$ 可知，$\sqrt{x}-1$ 的共轭因式是 $\sqrt{x}+1$.

2.4.4 Limits of Rational Functions(有理函数的极限)

2.4.4.1 有理函数的极限 $\lim\limits_{x\to x_0}\dfrac{P(x)}{Q(x)}=\lim\limits_{x\to 0}\dfrac{a_0x^n+a_1x^{n-1}+\cdots+a_n}{b_0x^2+b_1x^2+\cdots+b_m}=?$

当 $Q(x_0)\neq 0$ 时，$\lim\limits_{x\to x_0}\dfrac{P(x)}{Q(x)}=\dfrac{P(x_0)}{Q(x_0)}$.

当 $Q(x_0)=0$ 且 $P(x_0)\neq 0$ 时，$\lim\limits_{x\to x_0}\dfrac{P(x)}{Q(x)}=\infty(+\infty \text{或} -\infty)$.

当 $Q(x_0)=P(x_0)=0$ 时，先将分子、分母的公因式 $(x-x_0)$ 约去.

2.4.4.2 The Rational Function Theorem(有理函数定理)

$$\lim\limits_{x\to\infty}\dfrac{a_0x^n+ax^{n-1}+\cdots+a_n}{b_0x^m+b_1x^{m-1}+\cdots+b_m}=\lim\limits_{x\to\infty}\dfrac{a_0}{b_0}=\begin{cases}0, & n<m\\ \dfrac{a_0}{b_0}, & n=m\\ \infty, & n>m\end{cases}(\lim\limits_{x\to\infty}\text{改为}\lim\limits_{x\to -\infty}\text{定理同样成立})$$

有理函数定理在国内的部分教材上被形象地称为"抓大头准则"，是因为在求有理函数的极限时，分子、分母同时除以分子、分母中 x 的最高次幂，由于其结果只与分子、分母的最高次幂有关，所以只需要抓住分子、分母的最高次幂这两个"大头"项，就可以快速求出答案．

Example 2.18 有理函数定理的应用．

a. $\lim\limits_{x\to\infty}\dfrac{x^4-5}{2x^4+25x+1}$ b. $\lim\limits_{n\to\infty}\dfrac{3n^2-1}{n^2+100n+30}$ c. $\lim\limits_{x\to\infty}\dfrac{\sqrt[3]{8x^3+6x^2+3x+1}}{2x-1}$

d. $\lim\limits_{x\to\infty}\dfrac{2x^2-3}{5-2x^3}$ e. $\lim\limits_{x\to-\infty}\dfrac{3x^5}{2+x^4}$ f. $\lim\limits_{x\to\infty}(\sqrt{x^2+x+1}-\sqrt{x^2-x+1})$

Solution

a. $\lim\limits_{x\to\infty}\dfrac{x^4-5}{2x^4+25x+1}=\lim\limits_{x\to\infty}\dfrac{1-\dfrac{5}{x^4}}{2+\dfrac{25x}{x^4}+\dfrac{1}{x^4}}=\dfrac{1-0}{2+0+0}=\dfrac{1}{2}$（分子、分母同时除以$x^4$）

或直接"抓"分子、分母的最高次项：

$\lim\limits_{x\to\infty}\dfrac{x^4-5}{2x^4+25x+1}=\lim\limits_{x\to\infty}\dfrac{x^4}{2x^4}=\dfrac{1}{2}$（"抓大头准则"）

b. $\lim\limits_{n\to\infty}\dfrac{3n^2-1}{n^2+100n+30}=\lim\limits_{n\to\infty}\dfrac{3-\dfrac{1}{n^2}}{1+\dfrac{100n}{n^2}+\dfrac{30}{n^2}}=\dfrac{3-0}{1+0+0}=3$（分子、分母同时除以$n^2$）

或$\lim\limits_{n\to\infty}\dfrac{3n^2-1}{n^2+100n+30}=\lim\limits_{n\to\infty}\dfrac{3n^2}{n^2}=3$（"抓大头准则"）

c. $\lim\limits_{x\to\infty}\dfrac{\sqrt[3]{8x^3+6x^2+3x+1}}{2x-1}$

$=\lim\limits_{x\to\infty}\dfrac{\sqrt[3]{8+\dfrac{6x^2}{x^3}+\dfrac{3x}{x^3}+\dfrac{1}{x^3}}}{2-\dfrac{1}{x}}=\dfrac{\sqrt[3]{8+0+0+0}}{2-0}=1$（分子、分母同时除以$x$）

或$\lim\limits_{x\to\infty}\dfrac{\sqrt[3]{8x^3+6x^2+3x+1}}{2x-1}=\lim\limits_{x\to\infty}\dfrac{\sqrt[3]{8x^3}}{2x}=\lim\limits_{x\to\infty}\dfrac{2x}{2x}=1$（"抓大头准则"）

d. $\lim\limits_{x\to\infty}\dfrac{2x^2-3}{5-2x^3}=\lim\limits_{x\to\infty}\dfrac{\dfrac{2x^2}{x^3}-\dfrac{3}{x^3}}{\dfrac{5}{x^3}-\dfrac{2x^3}{x^3}}=\dfrac{0}{-2}=0$（分子、分母同时除以$x^3$）或

$\lim\limits_{x\to\infty}\dfrac{2x^2-3}{5-2x^3}=\lim\limits_{x\to\infty}\dfrac{2x^2}{-2x^3}=\lim\limits_{x\to\infty}\dfrac{1}{-x}=0$（"抓大头准则"）

e. $\lim\limits_{x\to-\infty}\dfrac{3x^5}{2+x^4}=\lim\limits_{x\to-\infty}\dfrac{3}{\dfrac{2}{x^5}+\dfrac{x^4}{x^5}}=-\infty$（分子、分母同时除以$x^5$）或

$\lim\limits_{x\to-\infty}\dfrac{3x^5}{2+x^4}=\lim\limits_{x\to-\infty}\dfrac{3x^5}{x^4}=\lim\limits_{x\to-\infty}3x=-\infty$（"抓大头准则"）

f. $\lim\limits_{x\to\infty}(\sqrt{x^2+x+1}-\sqrt{x^2-x+1})$

$$=\lim_{x\to\infty}\frac{(\sqrt{x^2+x+1}-\sqrt{x^2-x+1})(\sqrt{x^2+x+1}+\sqrt{x^2-x+1})}{\sqrt{x^2+x+1}+\sqrt{x^2-x+1}}$$

$$=\lim_{x\to\infty}\frac{2x}{\sqrt{x^2+x+1}+\sqrt{x^2-x+1}}$$

$$=\lim_{x\to\infty}\frac{2x}{\sqrt{x^2}+\sqrt{x^2}}\text{（"抓大头准则"）}$$

$$=\lim_{x\to\infty}\frac{2x}{x+x}$$

$$=1$$

[注]

从 Example 2.18 的几道求极限的题目中可以看出，"抓大头准则"用起来比较方便，但是在使用前一定要满足两个条件：

(1) 变量过程只能是 $\lim\limits_{x\to\infty}$ 或 $\lim\limits_{x\to-\infty}$．

(2) 函数的形式应为 $\dfrac{P(x)}{Q(x)}$ 的形式，不能进行如下计算：

$$\lim_{x\to\infty}(\sqrt{x^2+x+1}-\sqrt{x^2-x+1})=\lim_{x\to\infty}(\sqrt{x^2}-\sqrt{x^2})=\lim_{x\to\infty}(x-x)=0\text{（错误）}$$

2.4.5 Two Important Limits（两个重要极限）

重要极限 1： $$\lim_{x\to 0}\frac{\sin x}{x}=1$$

公式可以推广到：在极限 $\lim\dfrac{\sin\alpha(x)}{\alpha(x)}$ 中，只要 $\lim\alpha(x)=0$，就有 $\lim\dfrac{\sin\alpha(x)}{\alpha(x)}=1$．

Example 2.19 **Evaluate the Following Limits.**

a. $\lim\limits_{x\to\frac{\pi}{2}}\dfrac{\sin\left(x-\dfrac{\pi}{2}\right)}{x-\dfrac{\pi}{2}}$
b. $\lim\limits_{x\to 0}\dfrac{\sin\dfrac{x}{2}}{x}$
c. $\lim\limits_{x\to 0}\dfrac{\tan 2x}{x}$
d. $\lim\limits_{x\to 0}\dfrac{\tan 3x}{4\sin x}$

e. $\lim\limits_{x\to 0}\dfrac{1-\cos x}{x^2}$
f. $\lim\limits_{x\to 0}\dfrac{\cos x-1}{x}$
g. $\lim\limits_{x\to 0}\dfrac{x-\sin 3x}{x+\sin 3x}$
h. $\lim\limits_{x\to\infty}x\cdot\sin\dfrac{1}{x}$

i. $\lim\limits_{x\to\infty}x\arcsin\dfrac{1}{x}$
j. $\lim\limits_{x\to\pi}\dfrac{\sin x}{1-\dfrac{x^2}{\pi^2}}$
k. $\lim\limits_{x\to 0}\dfrac{\sec x-\cos x}{x}$
l. $\lim\limits_{x\to 0}\dfrac{\cos\alpha x-\cos\beta x}{x^2}$

Solution

a. $\lim\limits_{x\to\frac{\pi}{2}}\dfrac{\sin\left(x-\dfrac{\pi}{2}\right)}{x-\dfrac{\pi}{2}}=1$

b. $\lim\limits_{x\to 0}\dfrac{\sin\dfrac{x}{2}}{x}=\lim\limits_{x\to 0}\dfrac{\sin\dfrac{x}{2}}{\dfrac{x}{2}\cdot 2}=\lim\limits_{x\to 0}\dfrac{\sin\dfrac{x}{2}}{\dfrac{x}{2}}\cdot\dfrac{1}{2}=\dfrac{1}{2}$

c. $\lim\limits_{x\to 0}\dfrac{\tan 2x}{x}=\lim\limits_{x\to 0}\dfrac{\dfrac{\sin 2x}{\cos 2x}}{x}=\lim\limits_{x\to 0}\dfrac{\sin 2x}{2x}\cdot\lim\limits_{x\to 0}\dfrac{2}{\cos 2x}=1\times 2=2$

d. $\lim\limits_{x\to 0}\dfrac{\tan 3x}{4\sin x}=\lim\limits_{x\to 0}\dfrac{\sin 3x}{4\sin x\cdot\cos 3x}=\lim\limits_{x\to 0}\dfrac{\sin 3x/3x}{\sin x/x}\cdot\dfrac{3}{4}\cdot\dfrac{1}{\lim\limits_{x\to 0}\cos 3x}=\dfrac{3}{4}$

e. $\lim\limits_{x\to 0}\dfrac{1-\cos x}{x^2}=\lim\limits_{x\to 0}\dfrac{2\sin^2\dfrac{x}{2}}{x^2}=\dfrac{1}{2}\lim\limits_{x\to 0}\dfrac{\sin^2\dfrac{x}{2}}{\left(\dfrac{x}{2}\right)^2}=\dfrac{1}{2}\lim\limits_{x\to 0}\left[\dfrac{\sin\dfrac{x}{2}}{\dfrac{x}{2}}\right]^2=\dfrac{1}{2}$

f. $\lim\limits_{x\to 0}\dfrac{\cos x-1}{x}=\lim\limits_{x\to 0}\dfrac{\cos x-1}{x}\cdot\dfrac{\cos x+1}{\cos x+1}=\lim\limits_{x\to 0}\dfrac{\cos^2 x-1}{x(\cos x+1)}$

$=\lim\limits_{x\to 0}\dfrac{-\sin^2 x}{x(\cos x+1)}=\lim\limits_{x\to 0}\dfrac{\sin x}{x}\cdot\dfrac{-\sin x}{\cos x+1}=1\times\dfrac{0}{1+1}=0$

或结合三角函数的倍角公式有：

$\lim\limits_{x\to 0}\dfrac{\cos x-1}{x}=\lim\limits_{x\to 0}\dfrac{-2\sin^2\dfrac{x}{2}}{x}=-\lim\limits_{x\to 0}\dfrac{\sin\dfrac{x}{2}}{\dfrac{x}{2}}\cdot\sin\dfrac{x}{2}=-\lim\limits_{x\to 0}\sin\dfrac{x}{2}=0$

g. $\lim\limits_{x\to 0}\dfrac{x-\sin 3x}{x+\sin 3x}=\lim\limits_{x\to 0}\dfrac{1-\dfrac{\sin 3x}{x}}{1+\dfrac{\sin 3x}{x}}=\dfrac{1-3\lim\limits_{x\to 0}\dfrac{\sin 3x}{3x}}{1+3\lim\limits_{x\to 0}\dfrac{\sin 3x}{3x}}=\dfrac{1-3}{1+3}=-\dfrac{1}{2}$

h. $\lim\limits_{x\to\infty}x\cdot\sin\dfrac{1}{x}=\lim\limits_{x\to\infty}\dfrac{\sin\dfrac{1}{x}}{\dfrac{1}{x}}=1$

i. $\lim\limits_{x\to\infty}x\operatorname{arc\,sin}\dfrac{1}{x}$ 中，令 $\operatorname{arc\,sin}\dfrac{1}{x}=t$，则 $\sin t=\dfrac{1}{x}$

即 $x=\dfrac{1}{\sin t}$

当 $x\to\infty$ 时，$t\to 0$，$\lim\limits_{x\to\infty}x\operatorname{arc\,sin}\dfrac{1}{x}=\lim\limits_{t\to 0}\dfrac{t}{\sin t}=\lim\limits_{t\to 0}\dfrac{1}{\dfrac{\sin t}{t}}=1$

j. $\lim\limits_{x\to\pi}\dfrac{\sin x}{1-\dfrac{x^2}{\pi^2}}=\lim\limits_{x\to\pi}\dfrac{\pi^2\sin(\pi-x)}{(\pi+x)(\pi-x)}=\lim\limits_{x\to\pi}\dfrac{\pi^2}{\pi+x}\cdot\lim\limits_{x\to\pi}\dfrac{\sin(\pi-x)}{\pi-x}=\dfrac{\pi}{2}$

k. $\lim\limits_{x\to 0}\dfrac{\sec x-\cos x}{x}=\lim\limits_{x\to 0}\dfrac{1-\cos^2 x}{x\cos x}=\lim\limits_{x\to 0}\dfrac{\sin^2 x}{x^2}\cdot\dfrac{x}{\cos x}=0$

l. $\lim\limits_{x\to 0}\dfrac{\cos\alpha x-\cos\beta x}{x^2}=\lim\limits_{x\to 0}\dfrac{(1-\cos\beta x)-(1-\cos\alpha x)}{x^2}$

$=\lim\limits_{x\to 0}\dfrac{2\left(\sin^2\dfrac{\beta x}{2}-\sin^2\dfrac{\alpha x}{2}\right)}{x^2}$

$$=\lim_{x\to 0}\frac{\beta^2}{2}\left(\frac{\sin\frac{\beta x^2}{2}}{\frac{\beta x}{2}}\right)^2-\lim_{x\to 0}\frac{\alpha^2}{2}\left(\frac{\sin\frac{\alpha x}{2}}{\frac{\pi x}{2}}\right)^2=\frac{\beta^2-\alpha^2}{2}$$

重要极限 2：
$$\lim_{x\to\infty}\left(1+\frac{1}{x}\right)^x=e$$

公式可以推广到：在极限 $\lim[1+\alpha(x)]^{\frac{1}{\alpha(x)}}$ 中，只要 $\lim\alpha(x)=0$，就有 $\lim[1+\alpha(x)]^{\frac{1}{\alpha(x)}}=e$.

e 是个无理数，它的值是 e＝2.718281828459045…，它是指数函数 $y=e^x$ 以及对数函数 $y=\ln x$ 的底数.

Example 2.20 **Evaluate the Following Limits.**

a. $\lim\limits_{x\to\infty}\left(1-\dfrac{1}{x}\right)^x$　　b. $\lim\limits_{x\to 0}(1+3x)^{\frac{1}{x}}$　　c. $\lim\limits_{x\to\infty}\left(\dfrac{x}{x+1}\right)^{x+2}$　　d. $\lim\limits_{x\to 0}\dfrac{\log_a(1+x)}{x}$

Solution

a. $\lim\limits_{x\to\infty}\left(1-\dfrac{1}{x}\right)^x=\lim\limits_{x\to\infty}\left(1-\dfrac{1}{x}\right)^{-x\cdot(-1)}=e^{-1}=\dfrac{1}{e}$

b. $\lim\limits_{x\to 0}(1+3x)^{\frac{1}{x}}=\lim\limits_{x\to 0}(1+3x)^{\frac{1}{3x}\cdot 3}=e^3$

c. $\lim\limits_{x\to\infty}\left(\dfrac{x}{x+1}\right)^{x+2}=\lim\limits_{x\to\infty}\left(\dfrac{x+1-1}{x+1}\right)^{x+2}=\lim\limits_{x\to\infty}\left(1-\dfrac{1}{x+2}\right)^{x+2}$

$\qquad=\lim\limits_{x\to\infty}\left(1-\dfrac{1}{x+1}\right)^{-(x+1)\cdot\left(-\frac{x+2}{x+1}\right)}=e^{-\lim\limits_{x\to\infty}\frac{x+2}{x+1}}=e^{-1}=\dfrac{1}{e}$

d. $\lim\limits_{x\to 0}\dfrac{\log_a(1+x)}{x}=\lim\limits_{x\to 0}\dfrac{1}{x}\log_a(1+x)=\lim\limits_{x\to 0}\log_a(1+x)^{\frac{1}{x}}=\log_a\left[\lim\limits_{x\to 0}(1+x)^{\frac{1}{x}}\right]$

$\qquad=\log_a e=\dfrac{\ln e}{\ln a}=\dfrac{1}{\ln a}$

2.4.6 无穷小乘以有界函数等于无穷小

如果函数 $y=f(x)$ 在某个自变量的变化过程中极限为零，那么称函数 $f(x)$ 为该极限过程的无穷小，无穷小与有界函数的积仍然是无穷小. AP 考试中常见的有界函数有 $y=\sin x$，$y=\cos x$ 和反三角函数.

Example 2.21 **Find the Following Limits.**

a. $\lim\limits_{x\to\infty}\dfrac{\sin x}{x}$　　b. $\lim\limits_{x\to 0}x\sin\dfrac{1}{x}$　　c. $\lim\limits_{x\to\infty}\dfrac{\arctan x}{x}$　　d. $\lim\limits_{x\to\infty}\dfrac{\cos x}{e^x+e^{-x}}$

Solution

a. $\lim\limits_{x\to\infty}\dfrac{\sin x}{x}=\lim\limits_{x\to\infty}\dfrac{1}{x}\cdot\sin x=0\left(\because\lim\limits_{x\to\infty}\dfrac{1}{x}=0,|\sin x|\leqslant 1\right)$

b. $\lim\limits_{x\to 0}x\sin\dfrac{1}{x}=0\left(\because\lim\limits_{x\to 0}x=0,\left|\sin\dfrac{1}{x}\right|\leqslant 1\right)$

c. $\lim\limits_{x\to\infty}\dfrac{\arctan x}{x}=0\left(\because\lim\limits_{x\to\infty}\dfrac{1}{x}=0,|\arctan x|<\dfrac{\pi}{2}\right)$

d. $\lim\limits_{x\to\infty}\dfrac{\cos x}{e^x+e^{-x}}=\lim\limits_{x\to\infty}\dfrac{1}{e^x+e^{-x}}\cdot\cos x=0(\because\lim\limits_{x\to\infty}\dfrac{1}{e^x+e^{-x}}=0,|\cos x|\leqslant 1)$

2.4.7 等价无穷小替换

两个无穷小的商的极限会有不同的结果,例如,当 $x\to 0$ 时,$3x,x^2,\sin x,x\sin\dfrac{1}{x}$ 都是无穷小,但 $\lim\limits_{x\to 0}\dfrac{3x}{x}=3,\lim\limits_{x\to 0}\dfrac{x^2}{x}=0,\lim\limits_{x\to 0}\dfrac{\sin x}{x}=1,\lim\limits_{x\to 0}\dfrac{x\sin\dfrac{1}{x}}{x}$ 不存在,$\lim\limits_{x\to 0}\dfrac{3x}{x^2}=\infty$,不同的结果反映了不同的无穷小趋近于零的速度快慢.

如何比较无穷小趋于零的速度的快慢呢?

2.4.7.1 定义

设 $f(x)$ 和 $g(x)$ 是同一自变量变化过程的无穷小,即
$$\lim f(x)=0,\quad \lim g(x)=0$$

若 $\lim\dfrac{f(x)}{g(x)}=0$,则称 $f(x)$ 是比 $g(x)$ 高阶的无穷小.此时 $f(x)$ 趋近于 0 的速度比 $g(x)$ 趋近于 0 的速度快一些,如:$\lim\limits_{x\to 0}\dfrac{x^5}{x^2}=0$,则 x^5 是 x^2 在 $x\to 0$ 时的高阶无穷小.

若 $\lim\dfrac{f(x)}{g(x)}=\infty$,则称 $f(x)$ 是比 $g(x)$ 低阶的无穷小.此时 $f(x)$ 趋近于 0 的速度比 $g(x)$ 趋近于 0 的速度慢一些,如:$\lim\limits_{x\to 0}\dfrac{x^3}{x^6}=\infty$.

若 $\lim\dfrac{f(x)}{g(x)}=c(c\neq 0$ 为常数$)$,则称 $f(x)$ 是 $g(x)$ 的同阶无穷小.

若 $\lim\dfrac{f(x)}{g(x)}=1$,则称 $f(x)$ 与 $g(x)$ 是等价无穷小,记作:$f(x)\sim g(x)$.

比如:$x\to 0$ 时,$2x^2$ 与 x^2 是同阶无穷小,$\sin x$ 与 x 是等价无穷小.

2.4.7.2 等价无穷小替换定理

假设 $f(x),g(x),m(x)$ 与 $n(x)$ 是对于同一自变量变化过程的无穷小,且 $f(x)\sim m(x),g(x)\sim n(x)$,则:
$$\lim\dfrac{f(x)}{g(x)}=\lim\dfrac{m(x)}{n(x)}$$

等价无穷小替换定理表明,在求两个无穷小的商的极限时,分子、分母可以用等价无穷小替代.

常用的等价无穷小:当 $x\to 0$ 时,有

$\sin x\sim x$ $\qquad\qquad$ $\tan x\sim x$ $\qquad\qquad$ $\arcsin x\sim x$

$\arctan x\sim x$ $\qquad\quad$ $\ln(1+x)\sim x$ $\qquad\quad$ $1-\cos x\sim\dfrac{1}{2}x^2$

Example 2.22 Find the Following Limits.

a. $\lim\limits_{x\to\frac{\pi}{2}}\dfrac{\sin\left(x-\dfrac{\pi}{2}\right)}{x-\dfrac{\pi}{2}}$ b. $\lim\limits_{x\to 0}\dfrac{\sin\dfrac{x}{2}}{x}$ c. $\lim\limits_{x\to 0}\dfrac{\tan 2x}{x}$ d. $\lim\limits_{x\to 0}\dfrac{\tan 3x}{4\sin x}$

e. $\lim\limits_{x\to 0}\dfrac{1-\cos x}{x^2}$ f. $\lim\limits_{x\to 0}\dfrac{\cos x-1}{x}$ g. $\lim\limits_{x\to 0}\dfrac{(x^2+2)\sin x}{\arcsin x}$

Solution

a. $\because \lim\limits_{x\to\frac{\pi}{2}}\left(x-\dfrac{\pi}{2}\right)=0 \qquad \sin x \sim x\,(x\to 0)$

$\therefore \sin\left(x-\dfrac{\pi}{2}\right)\sim\left(x-\dfrac{\pi}{2}\right)\left(x\to\dfrac{\pi}{2}\right)$

So $\lim\limits_{x\to\frac{\pi}{2}}\dfrac{\sin\left(x-\dfrac{\pi}{2}\right)}{x-\dfrac{\pi}{2}}=\lim\limits_{x\to\frac{\pi}{2}}\dfrac{x-\dfrac{\pi}{2}}{x-\dfrac{\pi}{2}}=1$

b. $\because \lim\limits_{x\to 0}\dfrac{x}{2}=0 \qquad \therefore \sin\dfrac{x}{2}\sim\dfrac{x}{2}\,(x\to 0)$

So $\lim\limits_{x\to 0}\dfrac{\sin\dfrac{x}{2}}{x}=\lim\limits_{x\to 0}\dfrac{\dfrac{x}{2}}{x}=\dfrac{1}{2}$

c. When $x\to 0$, $\tan 2x\sim 2x$

So $\lim\limits_{x\to 0}\dfrac{\tan 2x}{x}=\lim\limits_{x\to 0}\dfrac{2x}{x}=2$

d. When $x\to 0$, $\tan 3x\sim 3x$, $\sin x\sim x$

So $\lim\limits_{x\to 0}\dfrac{\tan 3x}{4\sin x}=\lim\limits_{x\to 0}\dfrac{3x}{4x}=\dfrac{3}{4}$

e. When $x\to 0$, $1-\cos x\sim \dfrac{1}{2}x^2$

So $\lim\limits_{x\to 0}\dfrac{1-\cos x}{x^2}=\lim\limits_{x\to 0}\dfrac{\dfrac{1}{2}x^2}{x^2}=\dfrac{1}{2}$

f. When $x\to 0$, $1-\cos x\sim \dfrac{1}{2}x^2$

So $\lim\limits_{x\to 0}\dfrac{\cos x-1}{x}=\lim\limits_{x\to 0}\dfrac{-\dfrac{1}{2}x^2}{x}=\lim\limits_{x\to 0}-\dfrac{1}{2}x=0$

g. When $x\to 0$, $\sin x\sim x$, $\arcsin x\sim x$

So $\lim\limits_{x\to 0}\dfrac{(x^2+2)\sin x}{\arcsin x}=\lim\limits_{x\to 0}\dfrac{(x^2+2)x}{x}=\lim\limits_{x\to 0}(x^2+2)=2$

[注]

(1)Example 2.22 中的 a～f 为 Example 2.19 中的例题,用等价无穷小替换法要比用重要极限公式求解简便快捷.

(2)等价无穷小替换在使用 L'Hôpital's Rule(洛必达法则)时可以简化计算.

(3)使用等价无穷小替换时一定要谨慎,最好只对分子或分母中的无穷小因子作等价无穷小替换,如:求 $\lim\limits_{x\to 0}\dfrac{\tan x-\sin x}{2x^3}$.

错误解法:

∵ When $x\to 0$, $\tan x\sim x$ and $\sin x\sim x$

∴ $\lim\limits_{x\to 0}\dfrac{\tan x-\sin x}{2x^3}=\lim\limits_{x\to 0}\dfrac{x-x}{2x^3}=\lim\limits_{x\to 0}\dfrac{0}{2x^3}=0$

正确解法:

∵ When $x\to 0$, $\tan x-\sin x=\tan x(1-\cos x)\sim x\cdot\dfrac{1}{2}x^2$

∴ $\lim\limits_{x\to 0}\dfrac{\tan x-\sin x}{2x^3}=\lim\limits_{x\to 0}\dfrac{\dfrac{1}{2}x^3}{2x^3}=\dfrac{1}{4}$

2.4.8 Rates of Growth(增长率)

对于幂函数 $y=x^n(n>0)$,当 $x\to\infty$ 时,随着 n 的增大,函数值趋于 ∞ 的速度越来越快,比如: $\lim\limits_{x\to\infty}\dfrac{x^2}{x^3}=0$ 的原因是当 $x\to\infty$ 时,分母 x^3 趋于 ∞ 的速度比分子 x^2 快得多.

当 $x\to\infty$ 时,幂函数 $x^n(n>0)$、指数函数 $a^x(a>1)$ 和对数函数 $\log_a x(a>1)$ 的增长率的关系是:对数函数<幂函数<指数函数,若函数为不同类型的函数的和时,其总的增长率由增长率较大的函数决定.

Example 2.23 比较增长率求极限.

a. $\lim\limits_{x\to\infty}\dfrac{e^x}{\ln x}$ b. $\lim\limits_{x\to\infty}\dfrac{x^5}{e^x}$ c. $\lim\limits_{x\to\infty}\dfrac{\sin x+\ln x}{e^x}$

Solution

a. $\lim\limits_{x\to\infty}\dfrac{e^x}{\ln x}=\infty$ (因为当 $x\to\infty$ 时,$y=e^x$ 较 $y=\ln x$ 趋于 ∞ 的速度快得多).

b. $\lim\limits_{x\to\infty}\dfrac{x^5}{e^x}=0$ (因为当 $x\to\infty$ 时,$y=e^x$ 较 $y=x^5$ 趋于 ∞ 的速度快得多).

c. $\lim\limits_{x\to\infty}\dfrac{\sin x+\ln x}{e^x}=\lim\limits_{x\to\infty}\dfrac{\ln x}{e^x}=0$ (因为 $y=\sin x$ 为有界函数,当 $x\to\infty$ 时,分子趋于 ∞ 的速度由 $y=\ln x$ 决定,又 $y=e^x$ 较 $y=\ln x$ 趋于 ∞ 的速度快得多,所以极限等于 0).

2.5 Asymptotes(渐近线)

不管是用图形计算器还是用数学软件绘制函数图像,都只能显示有限视窗下的函数图像,也就是自变量 x 和因变量 y 取有限值时函数的图像.那么,当函数中 x 或 y 无限增大时,函数图像的性质如何确定呢?通过对函数的渐近线进行研究,可以说明函数图像在 x 或 y

无限增大时的变化特征.

2.5.1 Horizontal Asymptotes(水平渐近线)

A line $y=b$ is a **horizontal asymptote** of the graph of a function $y=f(x)$ if either
$$\lim_{x\to\infty}f(x)=b \quad \text{or} \quad \lim_{x\to-\infty}f(x)=b$$
若 $\lim\limits_{x\to\infty}f(x)=b$ 或 $\lim\limits_{x\to-\infty}f(x)=b$，则直线 $y=b$ 是 $y=f(x)$ 的水平渐进线.

2.5.2 Vertical Asymptote(垂直渐近线)

A line $x=a$ is a vertical asymptote of the graph of a function $y=f(x)$ if either
$$\lim_{x\to a^-}f(x)=\pm\infty \quad \text{or} \quad \lim_{x\to a^+}f(x)=\pm\infty$$
由于 $x=a$ 是一条垂直于 x 轴的直线，所以称为垂直渐近线.

[注]

(1)水平渐近线和垂直渐近线描述了当 x 和 y 一个趋于无穷大，另一个趋于常数时的变化情况.

(2)斜渐近线(**oblique or slanted asymptote**)在 AP 考试中不作要求.

Example 2.24 Looking for Asymptotes.

求函数 $y=\dfrac{x^2+5x+6}{x^2-9}$ 的渐近线.

Solution

$$y=\frac{x^2+5x+6}{x^2-9}=\frac{(x+2)(x+3)}{(x+3)(x-3)}$$

若求水平渐进线 $y=b$，则考察 $x\to\infty$ 和 $x\to-\infty$ 时，函数值 y 是否趋于常数.

$\because \lim\limits_{x\to\infty}\dfrac{x^2+5x+6}{x^2-9}=1 \qquad \therefore y=1$ 为水平渐近线.

$\because \lim\limits_{x\to-\infty}\dfrac{x^2+5x+6}{x^2-9}=1 \qquad \therefore y=1$ 为水平渐近线.

说明函数 $y=\dfrac{x^2+5x+6}{x^2-9}$ 的水平渐近线只有 $y=1$.

若求函数 $y=\dfrac{x^2+5x+6}{x^2-9}=\dfrac{(x+2)(x+3)}{(x+3)(x-3)}$ 的垂直渐进线，需要考察 $x\to 3^-$, $x\to 3^+$, $x\to -3^-$, $x\to -3^+$ 时 y 是否趋于无穷大.

$\because \lim\limits_{x\to 3^-}\dfrac{x^2+5x+6}{x^2-9}=\lim\limits_{x\to 3^-}\dfrac{(x+2)(x+3)}{(x+3)(x-3)}=\lim\limits_{x\to 3^-}\dfrac{5\cdot 6}{6(x-3)}=-\infty$

$\therefore x=3$ 为垂直渐近线.

$\because \lim\limits_{x\to 3^+}\dfrac{x^2+5x+6}{x^2-9}=\lim\limits_{x\to 3^+}\dfrac{(x+2)(x+3)}{(x+3)(x-3)}=\lim\limits_{x\to 3^+}\dfrac{5\cdot 6}{6(x-3)}=\infty$

$\therefore x=3$ 为垂直渐近线.

$\because \lim\limits_{x\to -3^-}\dfrac{x^2+5x+6}{x^2-9}=\lim\limits_{x\to -3^-}\dfrac{x+2}{x-3}=\dfrac{1}{6}=\lim\limits_{x\to -3^+}\dfrac{x^2+5x+6}{x^2-9}$

$\therefore x=-3$ 不是渐近线.

综上所述，函数 $y=\dfrac{x^2+5x+6}{x^2-9}$ 的渐近线有 $x=3$ 和 $y=1$ 两条(图 2.11).

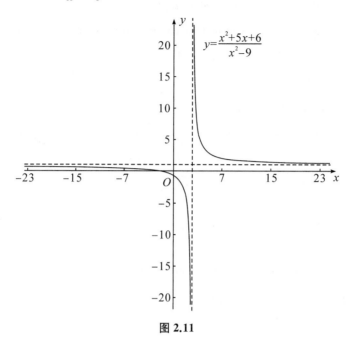

图 2.11

Example 2.25 求下列函数的渐近线.

a. $y=\mathrm{e}^x$ b. $y=\ln x$ c. $y=\dfrac{2x}{x+1}$ d. $y=\dfrac{x^2}{x-2}$

e. 参数方程: $\begin{cases} x=\dfrac{1}{t} \\ y=\dfrac{t}{t+1} \end{cases}$

Solution

a. $\because \lim\limits_{x\to\infty}\mathrm{e}^x=\infty,\therefore x\to\infty$ 时，$y=\mathrm{e}^x$ 没有渐近线.

$\because \lim\limits_{x\to-\infty}\mathrm{e}^x=\lim\limits_{x\to\infty}\dfrac{1}{\mathrm{e}^x}=0,\therefore x\to-\infty$ 时，$y=\mathrm{e}^x$ 有水平渐近线 $y=0$.

综上，函数 $y=\mathrm{e}^x$ 只有一条水平渐近线 $y=0$.

b. $\because \lim\limits_{x\to 0^+}\ln x=-\infty,\therefore x=0$ 为 $y=\ln x$ 的垂直渐近线.

c. $\because \lim\limits_{x\to -1^-}\dfrac{2x}{x+1}=+\infty$ or $\because \lim\limits_{x\to -1^+}\dfrac{2x}{x+1}=-\infty,\therefore x=-1$ 为 $y=\dfrac{2x}{x+1}$ 的垂直渐近线.

$\because \lim\limits_{x\to -\infty}\dfrac{2x}{x+1}=2$ or $\because \lim\limits_{x\to +\infty}\dfrac{2x}{x+1}=2,\therefore y=2$ 为 $y=\dfrac{2x}{x+1}$ 的水平渐近线.

综上，函数 $y=\dfrac{2x}{x+1}$ 有 $x=-1$ 和 $y=2$ 两条渐近线.

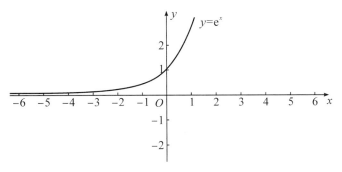

图 2.12

d. $\because \lim\limits_{x\to-\infty}\dfrac{x^2}{x-2}=-\infty$, and $\lim\limits_{x\to+\infty}\dfrac{x^2}{x-2}=+\infty$

$\therefore y=\dfrac{x^2}{x-2}$ 没有水平渐进线.

$\because \lim\limits_{x\to 2^-}\dfrac{x^2}{x-2}=-\infty$ or $\lim\limits_{x\to 2^+}\dfrac{x^2}{x-2}=+\infty$

$\therefore x=2$ 为 $y=\dfrac{x^2}{x-2}$ 的垂直渐进线.

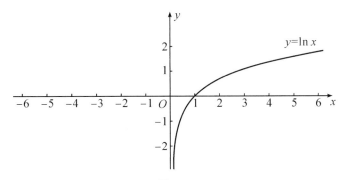

图 2.13

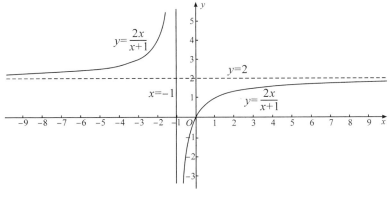

图 2.14

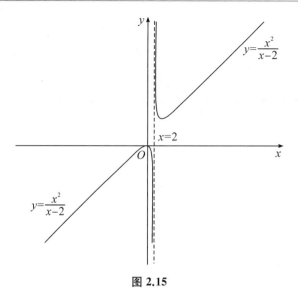

图 2.15

e.参数方程：$\begin{cases} x = \dfrac{1}{t} \\ y = \dfrac{t}{t+1} \end{cases}$

解法 1：将参数方程化为直角坐标方程

$\begin{cases} x = \dfrac{1}{t} \\ y = \dfrac{t}{t+1} \end{cases} \Rightarrow y = \dfrac{1}{1+\dfrac{1}{t}} = \dfrac{1}{1+x}(x \neq 0)$

$\because \lim\limits_{x \to -\infty} \dfrac{1}{1+x} = \lim\limits_{x \to +\infty} \dfrac{1}{1+x} = 0, \therefore y = 0$ 为其水平渐近线.

$\because \lim\limits_{x \to -1^-} \dfrac{1}{1+x} = -\infty \text{ or } \lim\limits_{x \to -1^+} \dfrac{1}{1+x} = +\infty, \therefore x = -1$ 为其垂直渐近线.

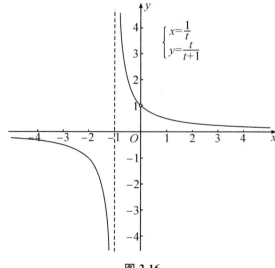

图 2.16

解法 2：根据渐进线的定义

渐进线的本质是描述 $x\to\pm\infty$ 或 $y\to\pm\infty$ 时函数图像的变化特点，所以分别考察：$x\to\pm\infty$ 时，y 是否趋近于一个常数；或 $y\to\pm\infty$ 时，x 是否趋近于一个常数.

For horizontal asymptotes consider the limit as $x\to\pm\infty$: $t\to 0 \Rightarrow y=0$ is an asymptote.

For vertical asymptotes consider the limit as $y\to\pm\infty$: $t\to -1 \Rightarrow x=-1$ is an asymptote.

2.6　Continuity(连续性)

2.6.1　Definition(定义)

2.6.1.1　Continuity at a Fixed Point(函数在某点连续的定义)

(1)函数在点 $x=c$ 连续.

A function $f(x)$ is continuous at a fixed point c if the limit of $f(x)$ at c is the same as the value of the function $f(x)$ at c. That is to say, function $f(x)$ is continuous at c if the three followings are met:

① $f(x)$ is defined;
② $\lim\limits_{x\to c} f(x)$ exists;
③ $\lim\limits_{x\to c} f(x) = f(c)$.

Any of the three conditions fails, then the function $f(x)$ is not continuous at c, or we say the function $f(x)$ is discontinuous at c, and $x=c$ is a **discontinuity**(不连续点或间断点) of $f(x)$.

若函数 $f(x)$ 在 $x=c$ 点的极限值等于该点的函数值，则称函数 $f(x)$ 在 $x=c$ 连续. $x=c$ 称为函数 $f(x)$ 的连续点，即函数 $f(x)$ 在 $x=c$ 连续必须满足以下三个条件：

① $f(x)$ 在 $x=c$ 有定义；
② 极限 $\lim\limits_{x\to c} f(x)$ 存在；
③ 函数 $f(x)$ 在 $x=c$ 点的极限值等于该点的函数值，即 $\lim\limits_{x\to c} f(x) = f(c)$.

若以上三个条件只要有一条不满足，则 $f(x)$ 在 $x=c$ 不连续，此时 $x=c$ 称为函数 $f(x)$ 的不连续点或间断点.

直观来讲，连续函数的图像是可以"一笔画成的"，间断点即为图像中曲线断开的点.

(2)函数在点 $x=c$ 单侧连续.

A function f is **left-continuous(or continuous from the left)** at c if $\lim\limits_{x\to c^-} f(x) = f(c)$.

如果 $\lim\limits_{x\to c^-} f(x) = f(c)$，则称 $y=f(x)$ 在点 $x=c$ 处左连续.

A function f is **right-continuous(or continuous from the right)** at c if $\lim\limits_{x\to c^+} f(x) = f(c)$.

如果 $\lim\limits_{x\to c^+} f(x) = f(c)$，则称 $y=f(x)$ 在点 $x=c$ 处右连续.

左连续和右连续又称单侧连续.

(3)左、右连续与连续的关系.

定理:函数 $y=f(x)$ 在点 $x=c$ 处连续 $\Leftrightarrow$ 函数 $y=f(x)$ 在点 $x=c$ 处既左连续又右连续,即 $\lim\limits_{x\to c}f(x)=f(c) \Leftrightarrow \lim\limits_{x\to c^+}f(x)=f(c)=\lim\limits_{x\to c^-}f(x)$.

2.6.1.2 Continuity on an Interval(函数在某区间连续的定义)

A function $y=f(x)$ is continuous on an open interval (a,b) if it is continuous at each point in the open interval.

若函数 $y=f(x)$ 在开区间 (a,b) 的每一点连续,则称函数 $y=f(x)$ 在开区间 (a,b) 连续.

A function $y=f(x)$ is continuous on a closed interval $[a,b]$ if it is continuous on the open interval (a,b), and also continuous from the right at the left endpoint a, $\lim\limits_{x\to a^+}f(x)=f(a)$, and continuous from the left at the right endpoint b, $\lim\limits_{x\to b^-}f(x)=f(b)$.

若函数 $y=f(x)$ 在开区间 (a,b) 的每一点连续,且在 $[a,b]$ 的左端点右连续,右端点左连续,即 $\lim\limits_{x\to a^+}f(x)=f(a)$ 且 $\lim\limits_{x\to b^-}f(x)=f(b)$,则称 $y=f(x)$ 在闭区间 $[a,b]$ 连续.

类似地,函数 $y=f(x)$ 在开区间 (a,b) 的每一点连续,且在 $[a,b)$ 的左端点右连续[即 $\lim\limits_{x\to a^+}f(x)=f(a)$],则称 $y=f(x)$ 在区间 $[a,b)$ 连续.

函数 $y=f(x)$ 在开区间 (a,b) 的每一点连续,且在 $(a,b]$ 的右端点左连续[即 $\lim\limits_{x\to b^-}f(x)=f(b)$],则称 $y=f(x)$ 在区间 $(a,b]$ 连续.

Example 2.26 由函数图像求连续区间.

若 $y=f(x)$ 的函数图像如图 2.17 所示,其中 $y=0,y=3$ 为其水平渐近线,根据图像求 $y=f(x)$ 的连续区间.

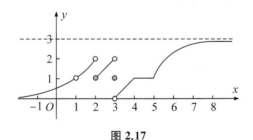

图 2.17

分析:从图像上观察,$y=f(x)$ 在整个区间 $(-\infty,\infty)$ 不能一笔画成,所以 $y=f(x)$ 在 $(-\infty,\infty)$ 不连续.但是 $y=f(x)$ 的图像在 $(-\infty,1),(1,2),(2,3),(3,\infty)$ 是可以一笔画成的,所以 $y=f(x)$ 在开区间 $(-\infty,1),(1,2),(2,3),(3,\infty)$ 分别连续.

从图中还可以观察出 $\lim\limits_{x\to 2^+}f(x)=f(2)$,所以 $y=f(x)$ 在 $x=2$ 右连续.

Solution

$f(x)$ is continuous on $(-\infty,1)$, $(1,2)$, $[2,3)$ and $(3,\infty)$. There are three discontinuities $x=1, x=2$ and $x=3$.

2.6.2 Discontinuity Types(间断点的类型)

函数 $f(x)$ 在 $x=c$ 连续必须满足以下三个条件:

(1) $f(x)$ 在 $x=c$ 有定义.

(2)极限$\lim\limits_{x \to c}f(x)$存在.

(3)函数 $f(x)$ 在 $x=c$ 的极限值等于该点的函数值,即$\lim\limits_{x \to c}f(x)=f(c)$.

若以上三个条件只要有一条不满足,则 $f(x)$ 在 $x=c$ 不连续,c 称为函数 $f(x)$ 的间断点.

2.6.2.1 Removable Discontinuity or Point Discontinuity(可去间断点)

若$\lim\limits_{x \to c^-}f(x) = \lim\limits_{x \to c^+}f(x)$[即$\lim\limits_{x \to c}f(x)$的极限存在],但 $f(x)$ 在 $x=c$ 没有定义[见图 2.18(b)]或$\lim\limits_{x \to c}f(x) \neq f(c)$[见图 2.18(c)],则这样的间断点称为可去间断点.

可去间断点可以通过改变定义或补充定义变成连续点.比如,在图 2.18(b)和图 2.18(c)中,我们让 $y=\begin{cases}f(x) & (x \neq 0)\\ 1 & (x=0)\end{cases}$[见图 2.18(a)],这样 $x=0$ 变为连续点,这就是为什么将此类间断点称为可去间断点.

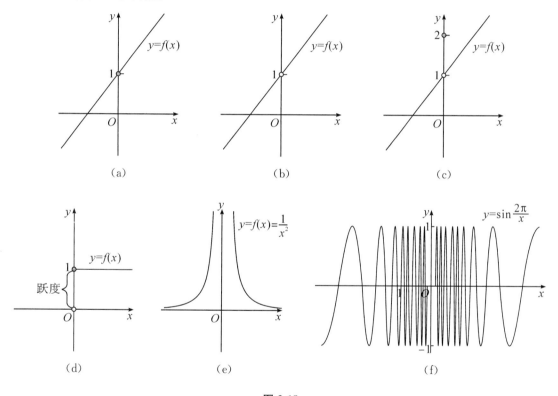

图 **2.18**

2.6.2.2 Jump Discontinuity(跳跃间断点)

若 $f(x)$ 在 $x=c$ 的左、右极限均存在,但是$\lim\limits_{x \to c^-}f(x) \neq \lim\limits_{x \to c^+}f(x)$,此时 $x=c$ 称为 $f(x)$ 的跳跃间断点.

由于在跳跃间断点处$\lim\limits_{x \to c^-}f(x) \neq \lim\limits_{x \to c^+}f(x)$,所以图像表现为 $f(x)$ 在点 $x=c$ 左、右两边的极限值有差异,这个差异称为"跃度"[见图 2.18(d)],因此这样的间断点称为跳跃间断点.

2.6.2.3 Infinite Discontinuity(无穷间断点)

若$\lim\limits_{x \to c^-}f(x)=\pm\infty$ or $\lim\limits_{x \to c^+}f(x)=\pm\infty$[即 $f(x)$ 在 $x=c$ 处有垂直渐近线],则 $x=c$ 称

为 $f(x)$ 的无穷间断点,如 $x=0$ 是 $y=\dfrac{1}{x^2}$ 的无穷间断点[见图 2.18(e)].

2.6.2.4 Oscillating Discontinuity[振荡间断点(AP 考试不要求)]

函数 $y=\sin\dfrac{2\pi}{x}$ 在点 $x=0$ 没有定义,所以点 $x=0$ 是函数 $y=\sin\dfrac{2\pi}{x}$ 的间断点.

当 $x\to 0$ 时,函数值在 -1 与 $+1$ 之间来回振荡,所以点 $x=0$ 称为函数 $y=\sin\dfrac{2\pi}{x}$ 的振荡间断点[见图 2.18(f)].

Example 2.27 由函数图像判断间断点的类型.

根据图 2.19 中 $f(x)$ 和 $g(x)$ 的图像,判断其间断点并判断其类型.

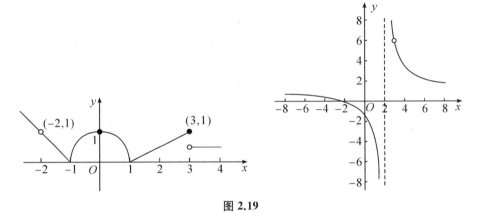

图 2.19

Solution

从 $f(x)$ 图像可见: $x=-2$ 为 $f(x)$ 的可去间断点, $x=3$ 为 $f(x)$ 的跳跃间断点.

从 $g(x)$ 图像可见: $x=2$ 为 $f(x)$ 的无穷间断点, $x=3$ 为 $f(x)$ 的可去间断点.

Example 2.28 由函数表达式判断间断点的类型.

判断下列函数的间断点所属的类型:

a. $y=\dfrac{1}{(x+1)^2}, x=-1$

b. $f(x)=\begin{cases} x^2+1, & x\leqslant 2 \\ 3, & x>2 \end{cases}, x=2$

c. $y=\dfrac{x^2-1}{x^2-3x+2}, x=1, x=2$

d. $y=\dfrac{1}{x}\ln(1+x), x=0$

e. $y=\cos\dfrac{1}{x}, x=0$

Solution

a. $\because \lim\limits_{x\to -1}\dfrac{1}{(x+1)^2}=\infty, \therefore x=-1$ is a infinite discontinuity.

b. $\because \lim\limits_{x\to 2^-}f(x)=\lim\limits_{x\to 2^-}(x^2+1)=5\neq 3=\lim\limits_{x\to 2^+}f(x), \therefore x=2$ is a jump discontinuity.

c. $\because \lim\limits_{x\to 1}\dfrac{x^2-1}{x^2-3x+2}=\lim\limits_{x\to 1}\dfrac{x+1}{x-2}=-2, \therefore x=1$ is a removable discontinuity.

$\because \lim\limits_{x\to 2^+}\dfrac{x^2-1}{x^2-3x+2}=\infty, \therefore x=2$ is a infinite discontinuity.

d. $\because \lim\limits_{x\to 0}\dfrac{1}{x}\ln(1+x)=\ln(\lim\limits_{x\to 0}(1+x)^{\frac{1}{x}})=\ln e=1, \therefore x=0$ is a removable discontinuity.

e. $\because x\to 0$ 时, $y=\cos\dfrac{1}{x}$ 在 -1 和 1 之间来回振荡, $\lim\limits_{x\to 0}\cos\dfrac{1}{x}$ 不存在.

$\therefore x=0$ 为 $y=\cos\dfrac{1}{x}$ 的振荡间断点.

2.7 连续函数定理

Theorem 2.3 连续函数的算术运算

If the functions f and g are continuous at c, then

(1) $f+g$ is continuous at c.

(2) $f-g$ is continuous at c.

(3) fg is continuous at c.

(4) f/g is continuous at c if $g(c)\neq 0$ and has a discontinuity at c if $g(c)=0$.

简而言之:连续函数的和、差、积、商仍为连续函数.

Theorem 2.4 反函数的连续性

If f is a one-to-one function that is continuous at each point of its domain, then f^{-1} is continuous at each point of its domain; that is, f^{-1} is continuous at each point in the range of f.

若原函数在其定义域内连续,则其反函数也连续.

Theorem 2.5 复合函数的连续性

If f is continuous at c and g is continuous at c, then the composite is $g\circ f$ continuous at c.

两个连续函数的复合函数连续.

Theorem 2.6 基本初等函数的连续性

Basic elementary functions are continuous at each point in their domains.

基本初等函数在它们的定义域内都是连续的.

[注]

(1)基本初等函数为对数函数、反三角函数、幂函数、指数函数以及常数函数,有国外的参考书用 LIATE (Logarithmic, Inverse trigonometric, Algebraic, Trigonometric,

Exponential)帮助学生记住这些函数.

(2)由以上定理可以推出:初等函数在其定义区间内连续,其中定义区间为包含定义域在内的区间.

Example 2.29 *由连续的定义求未知参数.*

设 $f(x)=\begin{cases} a+x^3, & x<0 \\ 1, & x=0 \\ \ln(b+2x), & x>0 \end{cases}$ 在 $(-\infty,\infty)$ 连续,求 a 和 b 的值.

Solution

由于分段函数 $f(x)$ 的各段函数均为初等函数,所以 $f(x)=a+x^3$ 在区间 $(-\infty,0)$ 连续,$f(x)=\ln(b+2x)$ 在区间 $(0,\infty)$ 连续.若 $f(x)$ 在 $(-\infty,\infty)$ 连续,则 $f(x)$ 在分段点 $x=0$ 处连续.

$\lim\limits_{x\to 0^-}f(x)=\lim\limits_{x\to 0^-}(a+x^3)=a$,$\lim\limits_{x\to 0^+}f(x)=\lim\limits_{x\to 0^+}\ln(b+2x)=\ln b$, $f(0)=1$

∵ $f(x)$ 在 $x=0$ 处连续,∴ $\lim\limits_{x\to 0^-}f(x)=\lim\limits_{x\to 0^+}f(x)=f(0)$.

即 $a=\ln b=1$,从而 $a=1, b=e$.

Theorem 2.7 The Extreme Value Theorem(最值定理)

The continuous function defined on a closed interval must have global maximum and global minimum.

闭区间连续的函数在该区间必有最大值和最小值.

Theorem 2.8 The Intermediate Value Theorem(介值定理)

A function $y=f(x)$ that is continuous on a closed interval $[a,b]$ takes on every value between $f(a)$ and $f(b)$. That is, if is any value y_0 between $f(a)$ and $f(b)$, then $y_0=f(c)$ for some c in $[a,b]$.

设函数 $y=f(x)$ 在闭区间 $[a,b]$ 上连续,数值 y_0 介于 $f(a)$ 和 $f(b)$ 之间,则存在某个数 $c\in[a,b]$,使得 $y_0=f(c)$,即若函数 $y=f(x)$ 在闭区间 $[a,b]$ 上连续,则 $y=f(x)$ 必能取到介于 $f(a)$ 和 $f(b)$ 之间的任意数值,因此定理称为"介值定理".

几何意义:连续曲线弧 $y=f(x)$ 与水平直线 $y=y_0$ 至少交于一点(如图2.20).

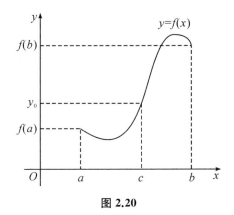

图 2.20

介值定理的特殊情形(零点定理)

If $y=f(x)$ is continuous on the closed interval $[a,b]$, and $f(a)$ and $f(b)$ have opposite signs, then $y=f(x)$ has at least an zero in that interval, that is, there is a value c, in (a,b) where $f(c)=0$.

设函数 $y=f(x)$ 在闭区间 $[a,b]$ 上连续,且 $f(a)$ 与 $f(b)$ 异号,即 $f(a) \cdot f(b)<0$,那么在开区间 (a,b) 内至少有一点 c,使 $f(c)=0$。由于 $x=c$ 是使函数值为零的点,所以该定理称为零点定理,在 AP 考试中都写成 "by IVT"(根据介值定理).

Example 2.30 闭区间连续定理综合.

Let f be a continuous function on the closed interval $[0,1]$. Let $f(0)=1$ and $f(1)=-1$. Which of the following is NOT necessarily true(不一定正确的是哪一个)?

(A) There exists a number h in $[0,1]$ such that $f(x) \leqslant f(h)$ for all x in $[0,1]$.

(B) There exists a number h in $[0,1]$ such that $f(h)=0$.

(C) There exists a number h in $[0,1]$ such that $f(h)=\dfrac{3}{2}$.

(D) For all h in the open interval $(0,1)$, $\lim\limits_{x \to h}f(x)=f(h)$.

Solution

(C)

根据最值定理(The Extreme Value Theorem),函数 $f(x)$ 在 $[0,1]$ 有最大值,即存在某个数 h,使得对所有的 $x \in [0,1]$,有 $f(x) \leqslant f(h)$,故选项 A 正确.

根据介值定理(The Intermediate Value Theorem),函数 $f(x)$ 在 $[0,1]$ 可以取得任何介于 1 和 -1 之间的值,因此 B 正确,而 $\dfrac{3}{2}>1$,所以 C 不一定正确.

选项 D 用数学语言描述了函数 $f(x)$ 在区间 $(0,1)$ 连续,所以选项 D 正确.

综上,本题答案为 C.

Example 2.31 介值定理

Train A runs back and forth on an east-west section of railroad track. Train A's velocity is given by a continuous function $v(t)$. Selected values for $v(t)$ are given in the table below.

t(minutes)	0	2	4	9	13	15
$v(t)$(meters/minutes)	0	30	60	100	-110	-140

Do the data in the table support the conclusion that train A's velocity is -100 meters per minute at some time t with $2<t<13$? Give a reason for your answer.

Solution

$v(t)$ is continuous on the interval $[2,13]$, and $v(13)=-110<-100<30=v(2)$.

Therefore, by the Intermediate Value Theorem(可简写不 by IVT), there is a time t, $2<t<13$, such that $v(t)=-100$.

Practice Exercises (习题)

1. The statement "$\lim_{x \to c} f(x) = L$" means that for each $\varepsilon > 0$, there exists a $\delta > 0$ such that

   (A) if $0 < |x - c| < \varepsilon$, then $|f(x) - L| < \delta$

   (B) if $0 < |f(x) - L| < \varepsilon$, then $|x - c| < \delta$

   (C) if $|f(x) - L| < \delta$, then $0 < |x - c| < \varepsilon$

   (D) if $0 < |x - c| < \delta$, then $|f(x) - L| < \varepsilon$

2. $\lim_{x \to 1} \dfrac{x}{\ln x}$ is

   (A) 0 (B) $\dfrac{1}{e}$ (C) e (D) nonexistent

3. $\lim_{x \to \frac{\pi}{7}} \dfrac{\sin\left(x - \frac{\pi}{7}\right)}{x - \frac{\pi}{7}}$ is

   (A) 0 (B) $\dfrac{1}{\sqrt{2}}$ (C) 1 (D) nonexistent

4. If $f(x) = 2x^2 + 1$, then $\lim_{x \to 0} \dfrac{f(x) - f(0)}{x^2}$ is

   (A) 0 (B) 1 (C) 2 (D) 4

5. If $f(x) = \begin{cases} \ln x & \text{for } 0 < x \leq 2 \\ x^2 \ln 2 & \text{for } 2 < x \leq 4 \end{cases}$, then $\lim_{x \to 2} f(x)$ is

   (A) ln2 (B) ln8 (C) 4 (D) nonexistent

6. Determine $\lim_{x \to \infty} \left[4x^2 \left(\sin^2 \dfrac{2}{x} \right) \right]$

   (A) 16 (B) 2 (C) 32 (D) nonexistent

7. The limit of the sequence $x_n = \dfrac{1 + cn^2}{(2n + 5 + 2\sin x)^2}$ as n approaches ∞ is -3. What is the value of c?

   (A) $-\dfrac{3}{4}$ (B) $-\dfrac{2}{3}$ (C) -9 (D) -12

8. $\lim_{x \to 0} \dfrac{\sin x \cos x}{x}$ is

   (A) -1 (B) 0 (C) 1 (D) nonexistent

9.

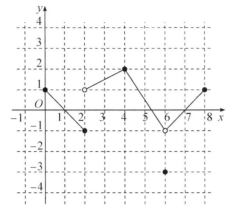

The figure above shows the graph of the function f. Which of the following statements are true?

Ⅰ. $\lim\limits_{x \to 2^-} f(x) = f(2)$

Ⅱ. $\lim\limits_{x \to 6^-} f(x) = \lim\limits_{x \to 6^+} f(x)$

Ⅲ. $\lim\limits_{x \to 6} f(x) = f(6)$

(A) Ⅱ only (B) Ⅲ only (C) Ⅰ and Ⅱ (D) Ⅰ, Ⅱ and Ⅲ

10. Let f be the function defined by $f(x) = \begin{cases} x^2 + 2 & \text{for } x \leq 3 \\ 6x + k & \text{for } x > 3 \end{cases}$, if f is continuous at $x = 3$, what is the value of k?

(A) -7 (B) 2

(C) 3 (D) There is no such value of k

11.

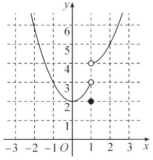

The figure above shows the graph of the function f. The value of $\lim\limits_{x \to 0} f(1-x^2)$ is

(A) 0 (B) 1 (C) 2 (D) 3

12. What are the equations of the horizontal asymptotes of the graph of $y = \dfrac{2x}{\sqrt{x^2-1}}$?

(A) $y = 0$ only (B) $y = 1$ only

(C) $y = -2$ and $y = 2$ (D) $y = -1$ and $y = 1$

13. The line $y = 4$ is a horizontal asymptote to the graph of which of the following functions?

(A) $y = \dfrac{\sin(4x)}{x}$ (B) $y = \dfrac{20x^2 + 3x}{5 + 5x^2}$

(C) $y = 4x$ (D) $y = \dfrac{1}{x-4}$

14. Let f be the function defined by the following.
$$f(x)=\begin{cases} \sin x, & x<0 \\ x^2, & 0\leqslant x<1 \\ 2-x, & 1\leqslant x<2 \\ x-3, & x\geqslant 2 \end{cases}$$

For what values of x is f NOT continuous?
(A) 0 only   (B) 0 and 2 only   (C) 2 only   (D) 0, 1 and 2 only

15. 证明：$\lim\limits_{x\to 1}\dfrac{x^2-1}{x-1}=2.$

16. 求 $\lim\limits_{x\to +\infty}\dfrac{\sin x}{e^x}.$

17. 求极限 $\lim\limits_{x\to 0}(1-2x^2)\lim\limits_{x\to 0}(1-2x^2)^{\frac{1}{x^2}}.$

18. 求极限 $\lim\limits_{x\to 1^+}\dfrac{\ln(1+\sqrt{x-1})}{\arcsin(2\sqrt{x-1})}.$

19. 求极限 $\lim\limits_{x\to -\infty} 2x e^x.$

习题参考答案

1.D 2.D 3.C 4.C 5.D 6.A 7.D 8.C 9.C 10.A 11.D 12.C 13.B 14.C

15. 分析：注意函数在 $x=1$ 是没有定义的，但这与函数在该点是否有极限并无关系.

当 $x\neq 1$ 时，$|f(x)-L|=|\dfrac{x^2-1}{x-1}-2|=|x-1|.$ $\forall \varepsilon>0$，要使 $|f(x)-L|<\varepsilon$，只要 $|x-1|<\varepsilon.$

证明：$\because \forall \varepsilon>0, \exists \delta=\varepsilon$，当 $0<|x-1|<\delta$ 时，有
$$|f(x)-A|=|\dfrac{x^2-1}{x-1}-2|=|x-1|<\varepsilon,$$

$\therefore \lim\limits_{x\to 1}\dfrac{x^2-1}{x-1}=2.$

16. 解：$\because$ 当 $x\to +\infty$ 时，$e^x\to +\infty, \dfrac{1}{e^x}\to 0$，而 $|\sin x|\leqslant 1.$

$\therefore \lim\limits_{x\to +\infty}\dfrac{\sin x}{e^x}=0.$

17. 解：原式 $=\lim\limits_{x\to 0}[(1-2x^2)^{-\frac{1}{2x^2}}]^{-2}=e^{-2}=\dfrac{1}{e^2}.$

18. 解：因为当 $x\to 1^+$ 时，$\sqrt{x-1}\to 0$
$\ln(1+\sqrt{x-1})\sim \sqrt{x-1}$ $\arcsin(2\sqrt{x-1})\sim 2\sqrt{x-1}$

故原式 $=\lim\limits_{x\to 1^+}\dfrac{\sqrt{x-1}}{2\sqrt{x-1}}=\dfrac{1}{2}.$

19. 解：$\lim\limits_{x\to -\infty} 2x e^x = \dfrac{\lim\limits_{x\to -\infty} 2x}{\lim\limits_{x\to -\infty} e^{-x}}=\dfrac{\lim\limits_{x\to -\infty} 2x}{\lim\limits_{x\to +\infty} e^x}=0$（比较变化率）

Chapter 3　Definition of Derivative(导数的定义)

3.1　Definition of Derivative(导数的定义)

3.1.1　引例

Example 3.1　几何问题：求切线的斜率.

设 $M(x_0, y_0)$ 为曲线 $y=f(x)$ 上的一点,在曲线上另取一点 $N(x,y)$,作 **secant line**(割线)MN.当点 N 沿曲线 $y=f(x)$ 趋近于点 $M(x_0,y_0)$ 时,割线 MN 绕点 M 旋转且越来越接近于 MT,直线 MT 称为曲线 $y=f(x)$ 在点 $M(x_0,y_0)$ 处的切线.

如图 3.1,割线 MN 的斜率为：

$$\tan\varphi = \frac{y-y_0}{x-x_0} = \frac{f(x)-f(x_0)}{x-x_0} \quad \text{(The slope of secant line)割线的斜率}$$

其中 φ 为割线 MN 的倾角.

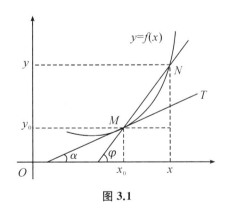

图 3.1

当点 N 沿曲线 $y=f(x)$ 趋于点 M(即 $x \to x_0$)时,若上式的极限存在,设为 k.

$$k = \lim_{x \to x_0} \frac{f(x)-f(x_0)}{x-x_0} \quad \text{(The slope of tangent line,切线的斜率)}$$

则此极限 k 为切线的斜率.这里 $k=\tan\alpha$,其中 α 是切线 MT 的倾角.于是,通过点 $M(x_0, f(x_0))$ 且以 k 为斜率的直线 MT 便是曲线 $y=f(x)$ 在点 M 处的切线.

切线 MT 的方程为： $\qquad y = k(x - x_0) + y_0$

令 $\Delta x = x - x_0$ 表示自变量 x 的增加量，相应的函数值的增加量表示为：

$$\Delta y = f(x_0 + \Delta x) - f(x_0) = f(x) - f(x_0)$$

$$k = \lim_{x \to x_0} \frac{f(x) - f(x_0)}{x - x_0} = \lim_{\Delta x \to 0} \frac{\Delta y}{\Delta x}$$

Example 3.2 物理问题：变速直线运动的瞬时速度.

设一质点在 x 轴上做变速运动，t 时刻质点的位置为 s，**位置函数**（position function）s 是 t 的函数，求质点在 t_0 时刻的瞬时速度.

考虑质点在 t 与 t_0 时刻的 **Average Velocity**（平均速度）：

$$\text{average velocity} = \frac{s(t) - s(t_0)}{t - t_0} = \frac{\Delta s}{\Delta t}$$

如果时间间隔非常短，平均速度在实践中也可用来说明动点在时刻 t_0 的速度. 但这样做是不精确的，更精确的应当这样：

当 $\Delta t \to 0$ 时，取比值 $\dfrac{s(t) - s(t_0)}{t - t_0}$ 的极限，如果这个极限存在，就把这个极限值称为质点在时刻 t_0 的 **Instantaneous Velocity**（瞬时速度）.

$$\text{instantaneous velocity} = \lim_{t \to t_0} \frac{s(t) - s(t_0)}{t - t_0} = \lim_{\Delta t \to 0} \frac{\Delta s}{\Delta t}$$

3.1.2 Derivative at a Point（点导数）

以上两个引例虽然属于两个不同的领域，一个是物理问题，一个是数学问题，但解决问题的数学模型却一样，都是用函数的增加量除以自变量的增加量趋于零过程中的极限值，抽象成数学概念便是函数的 **derivative**（导数）.

The derivative of $y = f(x)$ at $x = x_0$, denoted by $f'(x_0)$:

$$f'(x_0) = \lim_{\Delta x \to 0} \frac{\Delta y}{\Delta x} = \lim_{\Delta x \to 0} \frac{f(x_0 + \Delta x) - f(x_0)}{\Delta x}$$

if this limit exists.

函数 $y = f(x)$ 在点 x_0 处的导数 $f'(x_0)$ 也可记为 $y'\big|_{x=x_0}$，$\dfrac{dy}{dx}\bigg|_{x=x_0}$ 或 $\dfrac{df(x)}{dx}\bigg|_{x=x_0}$.

函数 $f(x)$ 在点 x_0 处可导有时也说成 $f(x)$ 在点 x_0 具有导数或导数存在.

[注]

(1) 导数的定义式有多种不同的形式，常见的有：

$$f'(x_0) = \lim_{\Delta x \to 0} \frac{\Delta y}{\Delta x} = \lim_{\Delta x \to 0} \frac{f(x_0 + \Delta x) - f(x_0)}{\Delta x}$$

$$f'(x_0) = \lim_{h \to 0} \frac{f(x_0 + h) - f(x_0)}{h}$$

$$f'(x_0) = \lim_{x \to x_0} \frac{f(x) - f(x_0)}{x - x_0}.$$

(2) The fraction $\dfrac{\Delta y}{\Delta x} = \dfrac{f(x) - f(x_0)}{x - x_0}$ is called **the difference quotient**（差商）for f at x_0

and represents **the average rate of change(平均变化率)** of f from x_0 to x.

分数 $\dfrac{\Delta y}{\Delta x} = \dfrac{f(x)-f(x_0)}{x-x_0}$ 称为差商,函数 $y=f(x)$ 在 $x=x_0$ 的差商表示函数 $f(x)$ 从 x_0 到 x 的平均变化率.

(3) $\mathrm{d}y$ 称为 y 的微分,$\mathrm{d}x$ 称为 x 的微分,导数 $f'(x_0) = \left.\dfrac{\mathrm{d}y}{\mathrm{d}x}\right|_{x=x_0}$ 又称为 $y=f(x)$ 在 $x=x_0$ 的微商,表示函数 $y=f(x)$ 在 $x=x_0$ 的瞬时变化率(**instantaneous rate of change**),几何上表示曲线 $y=f(x)$ 过切点 $(x_0, f(x_0))$ 的切线的斜率.

3.1.3 Derivative(导函数)

如果函数 $y=f(x)$ 在开区间 I 内的每一点处都可导,就称函数 $f(x)$ 在开区间 I 内可导. 这时,对于任意 $x \in I$,都对应着 $f(x)$ 的一个确定的导数值,这样就构成了一个新的函数,这个函数叫作原来函数 $y=f(x)$ 的导函数,记作 y',$f'(x)$,$\dfrac{\mathrm{d}y}{\mathrm{d}x}$ 或 $\dfrac{\mathrm{d}f(x)}{\mathrm{d}x}$.

3.1.3.1 导函数的定义式

$$y' = \lim_{\Delta x \to 0} \frac{f(x+\Delta x)-f(x)}{\Delta x} = \lim_{h \to 0} \frac{f(x+h)-f(x)}{h}$$

3.1.3.2 点导数与导函数之间的关系

函数 $f(x)$ 在点 $x=x_0$ 处的导数 $f'(x_0)$ 就是导函数 $f'(x)$ 在点 $x=x_0$ 处的函数值,即

$$f'(x_0) = f'(x)|_{x=x_0}$$

导函数 $y=f'(x)$ 简称导数,而 $y=f'(x_0)$ 是导数 $y=f'(x)$ 在 x_0 处的值,也就是将 $x=x_0$ 代入 $y=f'(x)$,就可以求点导数 $f'(x_0)$.

Example 3.3 常数函数的导数.

求函数 $f(x)=C$(C 为常数)的导数.

Solution

$$f'(x) = \lim_{h \to 0} \frac{f(x+h)-f(x)}{h} = \lim_{h \to 0} \frac{C-C}{h} = 0.$$

即 $(C)' = 0$.

Example 3.4

求 $f(x) = \dfrac{1}{x}$ 的导数.

Solution

$$f'(x) = \lim_{h \to 0} \frac{f(x+h)-f(x)}{h}$$

$$= \lim_{h \to 0} \frac{\dfrac{1}{x+h} - \dfrac{1}{x}}{h}$$

$$= \lim_{h \to 0} \frac{-h}{h(x+h)x}$$

$$= -\lim_{h \to 0} \frac{1}{(x+h)x}$$

$$= -\frac{1}{x^2}$$

Example 3.5

求 $f(x) = \sqrt{x}$ 的导数.

Solution

$$f'(x) = \lim_{h \to 0} \frac{f(x+h) - f(x)}{h}$$

$$= \lim_{h \to 0} \frac{\sqrt{x+h} - \sqrt{x}}{h}$$

$$= \lim_{h \to 0} \frac{(\sqrt{x+h} - \sqrt{x})(\sqrt{x+h} + \sqrt{x})}{h(\sqrt{x+h} + \sqrt{x})}$$

$$= \lim_{h \to 0} \frac{h}{h(\sqrt{x+h} + \sqrt{x})}$$

$$= \lim_{h \to 0} \frac{1}{\sqrt{x+h} + \sqrt{x}}$$

$$= \frac{1}{2\sqrt{x}}$$

Example 3.6

求函数 $f(x) = x^n$ (n 为正整数) 在 $x = a$ 处的导数.

Solution

$$f'(a) = \lim_{x \to a} \frac{f(x) - f(a)}{x - a}$$

$$= \lim_{x \to a} \frac{x^n - a^n}{x - a}$$

$$= \lim_{x \to a} a(x^{n-1} + ax^{n-2} + \cdots + a^{n-1})$$

$$= na^{n-1}$$

把以上结果中的 a 换成 x, 得 $f'(x) = nx^{n-1}$, 即

$$(x^n)' = nx^{n-1}$$

更一般地, 有

$$(x^\mu)' = \mu x^{\mu-1}, \text{其中 } \mu \text{ 为常数}$$

Example 3.7

求函数 $f(x) = \sin x$ 的导数.

Solution

$$f'(x) = \lim_{h \to 0} \frac{f(x+h) - f(x)}{h}$$

$$=\lim_{h\to 0}\frac{\sin(x+h)-\sin x}{h}$$

$$=\lim_{h\to 0}\frac{1}{h}\cdot 2\cos\left(x+\frac{h}{2}\right)\sin\frac{h}{2}$$

$$=\lim_{h\to 0}\cos\left(x+\frac{h}{2}\right)\cdot\frac{\sin\frac{h}{2}}{\frac{h}{2}}$$

$$=\cos x$$

即 $(\sin x)' = \cos x$.

用类似的方法,可求得 $(\cos x)' = -\sin x$.

Example 3.8

求函数 $f(x)=a^x(a>0, a\neq 1)$ 的导数.

Solution

$$f'(x)=\lim_{h\to 0}\frac{f(x+h)-f(x)}{h}$$

$$=\lim_{h\to 0}\frac{a^{x+h}-a^x}{h}$$

$$=a^x\lim_{h\to 0}\frac{a^h-1}{h}$$

令 $a^h-1=t$,则 $h=\log_a(1+t)$ $h\to 0$ 时,$t\to 0$

$$原式=a^x\lim_{t\to 0}\frac{t}{\log_a(1+t)}$$

$$=a^x\lim_{t\to 0}\frac{1}{\frac{1}{t}\log_a(1+t)}$$

$$=a^x\lim_{t\to 0}\frac{1}{\log_a(1+t)^{\frac{1}{t}}}$$

$$=\frac{a^x}{\log_a e}$$

$$=a^x\ln a$$

特别地,有 $(e^x)'=e^x$.

Example 3.9

求函数 $f(x)=\log_a x(a>0, a\neq 1)$ 的导数.

Solution

$$f'(x)=\lim_{h\to 0}\frac{f(x+h)-f(x)}{h}$$

$$=\lim_{h\to 0}\frac{\log_a(x+h)-\log_a x}{h}$$

$$= \lim_{h \to 0} \frac{1}{h} \log_a \left(\frac{x+h}{x} \right)$$

$$= \frac{1}{x} \lim_{h \to 0} \frac{x}{h} \log_a \left(1 + \frac{h}{x} \right)$$

$$= \frac{1}{x} \lim_{h \to 0} \log_a \left(1 + \frac{h}{x} \right)^{\frac{x}{h}}$$

$$= \frac{1}{x} \log_a e$$

$$= \frac{1}{x \ln a}.$$

即 $(\log_a x)' = \dfrac{1}{x \ln a}.$

特殊地, $(\ln x)' = \dfrac{1}{x}.$

Example 3.10 利用导数的定义求极限.

求下列极限的值：

a. $\lim\limits_{h \to 0} \dfrac{5\left(\frac{1}{2}+h\right)^5 - 5\left(\frac{1}{2}\right)^5}{h}$

b. $\lim\limits_{h \to 0} \dfrac{\sin(x+h) - \sin x}{h}$

c. $\lim\limits_{h \to 0} \dfrac{e^h - 1}{h}$

Solution

$$f'(x_0) = \lim_{h \to 0} \frac{f(x_0 + h) - f(x_0)}{h}$$

a. 这里 $f(x) = 5x^5, x_0 = \dfrac{1}{2}$

$$\lim_{h \to 0} \frac{5\left(\frac{1}{2}+h\right)^5 - 5\left(\frac{1}{2}\right)^5}{h} = (5x^5)' \big|_{x=\frac{1}{2}} = 25x^4 \big|_{x=\frac{1}{2}} = \frac{25}{16}$$

b. 由 $f'(x) = \lim\limits_{h \to 0} \dfrac{f(x+h) - f(x)}{h}$ 得：

$$\lim_{h \to 0} \frac{\sin(x+h) - \sin x}{h} = (\sin x)' = \cos x$$

c. 这里 $f(x) = e^x, x_0 = 0$

$$\lim_{h \to 0} \frac{e^h - 1}{h} = (e^x)' \bigg|_{x=0} = e^0 = 1$$

3.1.4 One-sided Derivative(单侧导数)

我们用极限的定义引出了左极限和右极限的定义，因此可以根据点导数的定义式来定义左导数和右导数.

(1)**Left-hand derivative at x_0(左导数)**:
$$f'_-(x_0) = \lim_{x \to x_0^-} \frac{f(x) - f(x_0)}{x - x_0}$$

(2)**Right-hand derivative at x_0(右导数)**:
$$f'_+(x_0) = \lim_{x \to x_0^+} \frac{f(x) - f(x_0)}{x - x_0}$$

(3)**导数与单侧导数的关系**:
$$f'(x_0) = A \Leftrightarrow f'_-(x_0) = f'_+(x_0) = A$$

有了单侧导数的定义后,我们可以定义闭区间可导:如果函数 $f(x)$ 在开区间 (a,b) 内可导,且右导数 $f'_+(a)$ 和左导数 $f'_-(b)$ 都存在,就说 $f(x)$ 在闭区间 $[a,b]$ 上可导.

Example 3.11 $y = |x|$ **Is Not Differentiable at the Origin.**
绝对值函数 $y = |x|$ 在 $x = 0$ 处不可导.

Show that the function $y = |x|$ is differentiable on $(-\infty, 0)$ and $(0, \infty)$ but has no derivative at $x = 0$.

Solution

To the right of the origin: $(|x|)' = (x)' = 1$

To the left of the origin: $(|x|)' = (-x)' = 1.$

There can be no derivative at the origin because the one-sided derivatives differ there:

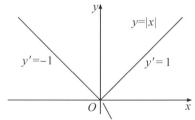

y' not defined at $x=0$: right-hand derivative $\neq$ left-hand derivative

图 3.2

Right-hand derivative of $|x|$ at zero $= \lim\limits_{h \to 0^+} \frac{|0+h| - |0|}{h} = \lim\limits_{h \to 0^+} \frac{h}{h} = \lim\limits_{h \to 0^+} 1 = 1$

Left-hand derivative of $|x|$ at zero $= \lim\limits_{h \to 0^-} \frac{|0+h| - |0|}{h} = \lim\limits_{h \to 0^-} \frac{-h}{h} = \lim\limits_{h \to 0^-} -1 = -1$

3.2 高阶导数

一般地,函数 $y = f(x)$ 的导数 $y' = f'(x)$ 仍然是 x 的函数,我们把 $y' = f'(x)$ 的导数叫作函数 $y = f(x)$ 的二阶导数,记作:

$$y'', f''(x) \text{ 或 } \frac{\mathrm{d}^2 y}{\mathrm{d} x^2},$$

即 $y''=(y')', f''(x)=[f'(x)]',$
$$\frac{d^2y}{dx^2}=\frac{d}{dx}\left(\frac{dy}{dx}\right).$$

相应地,把 $y=f(x)$ 的导数 $f'(x)$ 叫作函数 $y=f(x)$ 的一阶导数.

类似地,二阶导数的导数叫作三阶导数,三阶导数的导数叫作四阶导数,…,$(n-1)$ 阶导数的导数叫作 n 阶导数,分别记作:

$$y''', y^{(4)}, \cdots, y^{(n)} \text{ 或 } \frac{d^3y}{dx^3}, \frac{d^4y}{dx^4}, \cdots, \frac{d^ny}{dx^n}.$$

函数 $f(x)$ 具有 n 阶导数,也常说成函数 $f(x)$ 为 n 阶可导.如果函数 $f(x)$ 在点 x_0 处具有 n 阶导数,那么函数 $f(x)$ 在点 x_0 的某一邻域内必定具有一切低于 n 阶的导数.

二阶及二阶以上的导数统称高阶导数,y' 称为一阶导数,$y'', y''', y^{(4)}, \cdots, y^{(n)}$ 都称为高阶导数.

Example 3.12 求函数 $y=x^n$ 的 n 阶导数.
Solution
$$y'=(x^n)'=nx^{n-1}, y''=(nx^{n-1})'=n(n-1)x^{n-2}, \cdots$$

一般地,可得:
$$y^{(n)}=(x^n)^{(n)}=n!$$
$$y^{(n+1)}=(n!)'=0$$

Example 3.13 求函数 $y=e^x$ 的 n 阶导数.
Solution
$$y'=e^x, y''=e^x, y'''=e^x, y^{(4)}=e^x, \cdots$$

一般地,可得:
$$y^{(n)}=e^x,$$
即
$$(e^x)^{(n)}=e^x.$$

3.3 The Relationship between Differentiability and Continuity(可导与连续的关系)

Theorem Differentiability Implies Continuity(可导必然连续):
If a function $f(x)$ is differentiable at x_0, then $f(x)$ is continuous at x_0.
如果函数 $y=f(x)$ 在点 x_0 处可导,那么函数在该点必连续(可导 $\Rightarrow$ 连续).
反之不一定成立,一个函数在某点连续却不一定在该点可导.

Example 3.14 由函数图像作导函数的图像.

The graph of the function $y=f(x)$ shown here(图 3.3) is made of line segments joined end to end.

a.Graph the function's derivative.

b.At what values of x between $x=-4$ and $x=6$ is the function not differentiable?

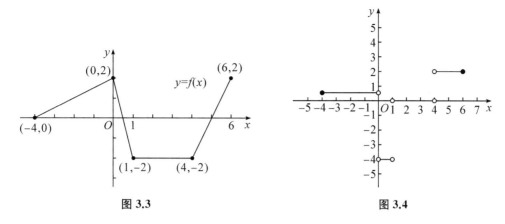

图 3.3　　　　　　　　　图 3.4

Solution

a. The slope from $x=-4$ to $x=0$ is $\dfrac{2-0}{0-(-4)}=\dfrac{1}{2}$.

The slope from $x=0$ to $x=1$ is $\dfrac{-2-2}{1-0}=-4$.

The slope from $x=1$ to $x=4$ is $\dfrac{-2-(-2)}{4-1}=0$.

The slope from $x=1$ to $x=4$ is $\dfrac{2-(-2)}{6-(-4)}=2$.

The graph of the derivative is shown (figure 3.4).

b. The derivative is undefined at $x=0, x=1,$ and $x=4$(在这些地方左导数$\neq$右导数).

The function is differentiable at $x=-4$ and $x=6$ because these are endpoints of the domain and the one-sided derivatives exist.

Example 3.15 判断函数的可导性.

a. 判断分段函数 $f(x)=\begin{cases}x^2+x, & x\leqslant 1\\ 3x-2, & x>1\end{cases}$ 在分段点 $x=1$ 处的可导性.

b. 判断分段函数 $f(x)=\begin{cases}x^2, x\leqslant 3\\ 6x-9, x>3\end{cases}$ 在分段点 $x=3$ 处的可导性.

Solution

a. 首先判断该分段函数的连续性：

$\because \lim\limits_{x\to 1^-}f(x)=2, \lim\limits_{x\to 1^+}f(x)=1$

$\lim\limits_{x\to 1^-}f(x)\neq \lim\limits_{x\to 1^+}f(x)$

$\therefore y=f(x)$ 在 $x=1$ 处不连续，进而不可导.

b. 首先判断该分段函数的连续性：

$\because \lim\limits_{x\to 3^-}f(x)=\lim\limits_{x\to 3^-}x^2=9, \lim\limits_{x\to 3^+}f(x)=\lim\limits_{x\to 3^+}(6x-9)=9$

$\therefore y=f(x)$ 在 $x=3$ 处连续.

由 $f(x)=\begin{cases}x^2, x\leqslant 3\\ 6x-9, x>3\end{cases}$ 得：

$$f'(x)=\begin{cases}2x, x<3\\ 6, x>3\end{cases}$$

∵ $f'_-(3)=6=f'_+(3)$

∴ $y=f(x)$ 在 $x=3$ 处可导且 $f'(3)=6$.

Example 3.16 由函数的可导性求未知参数的值.

设 $f(x)=\begin{cases}x^2, x\leq 1\\ ax+b, x>1\end{cases}$，为使 $f(x)$ 在 $x=1$ 处可导，应如何选择 a,b?

Solution

因为 $f(x)$ 在 $x=1$ 处可导，所以 $f(x)$ 在 $x=1$ 处连续，

则有 $\lim\limits_{x\to 1^-}f(x)=f(1)=\lim\limits_{x\to 1^+}f(x)$, $\lim\limits_{x\to 1^-}f(x)=\lim\limits_{x\to 1^-}x^2=f(1)=1$

$\lim\limits_{x\to 1^+}f(x)=\lim\limits_{x\to 1^+}(ax+b)=a+b$

所以 $a+b=1$，即 $b=1-a$

又 $f(x)$ 在 $x=1$ 处可导，从而有 $f'_+(1)=f'_-(1)$

$f'_-(1)=\lim\limits_{x\to 1^-}\dfrac{f(x)-f(1)}{x-1}=\lim\limits_{x\to 1^-}\dfrac{x^2-1}{x-1}=\lim\limits_{x\to 1^-}(x+1)=2$

$f'_+(1)=\lim\limits_{x\to 1^+}\dfrac{f(x)-f(1)}{x-1}=\lim\limits_{x\to 1^+}\dfrac{ax+b-1}{x-1}=\lim\limits_{x\to 1^+}\dfrac{ax-a}{x-1}=a=2$

故 $a=2, b=-1$.

3.4 不可导点的类型

3.4.1 A Discontinuity（不连续点）

根据可导与连续的关系，有：可导⇒连续. 其逆否命题：不连续⇒不可导. 所以不连续点一定是不可导点.

3.4.2 A Corner（角点）

A corner, where the one-sided derivatives differ.

左、右导数存在，但是不相等的点称为角点. 例如：绝对值函数 $y=|x|$，在 $x=0$ 处的左、右导数存在，但是不相等. $x=0$ 是函数 $y=|x|$ 的角点，不可导（$y=|x|$ 的图像见图 3.5）.

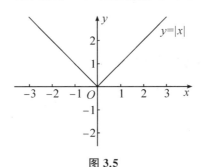

图 3.5

3.4.3 A Cusp 尖点

A cusp, where the slope of secant line approaches ∞ from one side and −∞ from the other.

尖点两侧的割线的斜率分别趋近于 ∞ 和 −∞，如：$y = x^{\frac{2}{3}}$ 在 $x = 0$ 处为尖点，不可导（见图 3.6）.

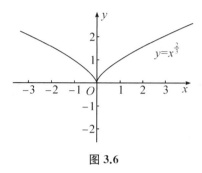

图 3.6

3.4.4 A Vertical Tangent (有垂直切线的点)

A vertical tangent, where the slopes of the secant lines approach either ∞ or −∞ from both sides.

有垂直切线的点的左、右两侧的切线的斜率同时趋近于 ∞ 或同时趋近于 −∞，如：$f(x) = \sqrt[3]{x}$ 在 $x = 0$ 处有垂直切线，所以不可导（见图 3.7）.

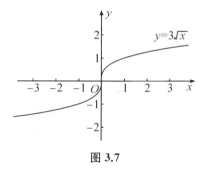

图 3.7

Example 3.17 根据图像判断函数的可导性.

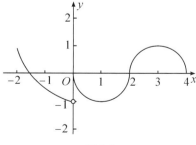

图 3.8

The graph of the function f shown in the figure (图 3.8) above has a vertical tangent at the point (2,0) and horizontal tangents at the points (1,−1) and (3,1). For what values of x, $2<x<4$, is f not differentiable?

 (A) 0 only   (B) 0 and 2 only   (C) 1 and 3 only   (D) 0, 1 and 3 only

Solution

(B)

当 $2<x<4$ 时，从图中观察，有：

在 $x=0$ 处为不连续点，不可导．

在 $x=2$ 处有垂直切线，不可导．

Practice Exercises（习题）

1. $\lim\limits_{h\to 0}\dfrac{e^h-1}{2h}$ is

 (A) 0   (B) $\dfrac{1}{2}$ (C) e   (D) nonexistent

2. $\lim\limits_{h\to 0}\dfrac{\ln(e+h)-1}{h}$ is

 (A) $f'(e)$, where $f(x)=\ln x$ (B) $f'(e)$, where $f(x)=\dfrac{\ln x}{x}$

 (C) $f'(1)$, where $f(x)=\ln x$ (D) $f'(1)$, where $f(x)=\ln(x+e)$

3. If $f(x)=2+|x-3|$ for all x, then the value of the derivative $f'(x)$ at $x=3$ is

 (A) 0   (B) 1   (C) 2   (D) nonexistent

4.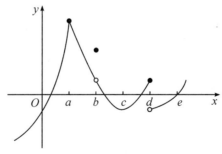

The graph of a function f is shown above. At which value of x is f continuous, but not differentiable?

 (A) a (B) b (C) c (D) d

5.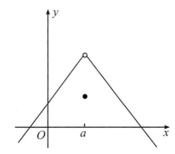

The graph of a function f is shown above. Which of the following statements must be false?

(A) $f(a)$ exists

(B) $f(x)$ is defined for $0 < x < a$

(C) f is not continuous at $x = a$

(D) $\lim\limits_{x \to a} f'(x)$ exists

6. Let f be the function defined below. Which of the following statements about f are true?

$$f(x) = \begin{cases} \dfrac{x^2 - 4}{x - 2}, & \text{if } x \neq 2 \\ 1, & \text{if } x = 2 \end{cases}$$

Ⅰ. f has a limit at $x = 2$

Ⅱ. f is continuous at $x = 2$

Ⅲ. f is differentiable at $x = 2$

(A) Ⅰ only   (B) Ⅱ only   (C) Ⅰ and Ⅱ   (D) Ⅰ, Ⅱ and Ⅲ

7. Let f be the function defined by $f(x) = \sqrt{|x-2|}$ for all x. Which of the following statements is true?

(A) f is continuous but not differentiable at $x = 2$

(B) f is differentiable at $x = 2$

(C) f is not continuous at $x = 2$

(D) $x = 2$ is a vertical asymptote of the graph of f.

8. Let $f(x) = \begin{cases} x^2 - 1, & x < 0 \\ 2x - 1, & x \geq 0 \end{cases}$. Which of the following is equal to the right-hand derivative of f at $x = 0$?

(A) 0   (B) -2 (C) 1   (D) 2

习题参考答案

1.B   2.A   3.D   4.A   5.D   6.A   7.A   8.D

Chapter 4 Computation of Derivative(导数的计算)

4.1 Arithmetic Operations on Derivative(导数的代数运算)

If u and v are differentiable functions of x, and C is a real number, then the following results are also differentiable and they follow the rules below:

设 $u(x), v(x)$ 可导，C 为常数，则下列函数也可导并且满足相应的规则.

4.1.1 Derivative of a Constant Function(常数的导数)

If f has the constant value $f(x) = C$ then
$$\frac{\mathrm{d}f}{\mathrm{d}x} = \frac{\mathrm{d}(C)}{\mathrm{d}x} = \frac{\mathrm{d}}{\mathrm{d}x}(C) = 0 \quad \text{or} \quad C' = 0$$

4.1.2 Power Rule(乘方法则)

If n is a positive integer, then
$$\frac{\mathrm{d}(x^n)}{\mathrm{d}x} = \frac{\mathrm{d}}{\mathrm{d}x}(x^n) = nx^{n-1} \quad \text{or} \quad (x^n)' = nx^{n-1}$$

常用的公式有：

$$x' = 1 \qquad \left(\frac{1}{x}\right)' = -\frac{1}{x^2} \qquad (\sqrt{x})' = \frac{1}{2\sqrt{x}} \qquad \left(\frac{1}{\sqrt{x}}\right)' = -\frac{1}{2\sqrt{x^3}}$$

4.1.3 Constant Multiple Rule(数乘法则)

$$\frac{\mathrm{d}(Cu(x))}{\mathrm{d}x} = \frac{\mathrm{d}}{\mathrm{d}x}(Cu(x)) = C\frac{\mathrm{d}(u(x))}{\mathrm{d}x} \quad \text{or} \quad (Cu)' = Cu' \quad (C \text{ is a constant})$$

4.1.4 Sum and Difference Rule(和、差法则)

$$\frac{\mathrm{d}(u(x) \pm v(x))}{\mathrm{d}x} = \frac{\mathrm{d}(u(x))}{\mathrm{d}x} \pm \frac{\mathrm{d}(v(x))}{\mathrm{d}x} \quad \text{or} \quad (u \pm v)' = u' \pm v'$$

Proof

$$[u(x) \pm v(x)]' = \lim_{h \to 0} \frac{[u(x+h) \pm v(x+h)] - [u(x) \pm v(x)]}{h}$$

$$=\lim_{h\to 0}\left[\frac{u(x+h)-u(x)}{h}\pm\frac{v(x+h)-v(x)}{h}\right]$$
$$=u'(x)\pm v'(x)$$

上述法则可推广到任意有限个可导函数的情形.例如：

设 $u=u(x), v=v(x), w=w(x)$ 均可导,则有：
$$(u+v-w)'=u'+v'-w'$$

4.1.5　Product Rule(乘法法则)

$$\frac{\mathrm{d}(u(x)v(x))}{\mathrm{d}x}=\frac{\mathrm{d}(u(x))}{\mathrm{d}x}\cdot v(x)\pm u(x)\cdot\frac{\mathrm{d}(v(x))}{\mathrm{d}x} \quad \text{or} \quad (uv)'=u'v+uv'$$

Proof

$$[u(x)\cdot v(x)]'=\lim_{h\to 0}\frac{u(x+h)v(x+h)-u(x)v(x)}{h}$$
$$=\lim_{h\to 0}\frac{1}{h}[u(x+h)v(x+h)-u(x)v(x+h)+u(x)v(x+h)-u(x)v(x)]$$
$$=\lim_{h\to 0}\left[\frac{u(x+h)-u(x)}{h}v(x+h)+u(x)\frac{v(x+h)-v(x)}{h}\right]$$
$$=\lim_{h\to 0}\frac{u(x+h)-u(x)}{h}\cdot\lim_{h\to 0}v(x+h)+u(x)\cdot\lim_{h\to 0}\frac{v(x+h)-v(x)}{h}$$
$$=u'(x)v(x)+u(x)v'(x)$$

上述法则可推广到任意有限个可导函数的情形.例如,设 $u=u(x), v=v(x), w=w(x)$ 均可导,则有：
$$(uvw)'=[(uv)w]'=(uv)'w+(uv)w'=(u'v+uv')w+uvw'=u'vw+uv'w+uvw'$$
即 $(uvw)'=u'vw+uv'w+uvw'$.

如果 $v=C$（C 为常数）,则有：$(Cu)'=Cu'$,所以 Constant Multiple Rule 是 Product Rule 的特殊情形.

4.1.6　Quotient Rule(除法法则)

$$\frac{\mathrm{d}\left(\frac{u}{v}\right)}{\mathrm{d}x}=\frac{\frac{\mathrm{d}u}{\mathrm{d}x}\cdot(v-u)\cdot\frac{\mathrm{d}v}{\mathrm{d}x}}{v^2} \quad \text{or} \quad \left(\frac{u}{v}\right)'=\frac{u'v-uv'}{v^2}$$

Proof

$$\left[\frac{u(x)}{v(x)}\right]'=\lim_{h\to 0}\frac{\frac{u(x+h)}{v(x+h)}-\frac{u(x)}{v(x)}}{h}=\lim_{h\to 0}\frac{u(x+h)v(x)-u(x)v(x+h)}{v(x+h)v(x)h}$$
$$=\lim_{h\to 0}\frac{[u(x+h)-u(x)]v(x)-u(x)[v(x+h)-v(x)]}{v(x+h)v(x)h}$$
$$=\lim_{h\to 0}\frac{\frac{u(x+h)-u(x)}{h}v(x)-u(x)\frac{v(x+h)-v(x)}{h}}{v(x+h)v(x)}$$
$$=\frac{u'(x)v(x)-u(x)v'(x)}{v^2(x)}$$

Example 4.1

$y = 2x^3 - 5x^2 + 3x - 7$,求 y'.

Solution

$$\begin{aligned}
y' &= (2x^3 - 5x^2 + 3x - 7)' \\
&= (2x^3)' - (5x^2)' + (3x)' - (7)' \\
&= 2(x^3)' - 5(x^2)' + 3(x)' \\
&= 2 \cdot 3x^2 - 5 \cdot 2x + 3 \\
&= 6x^2 - 10x + 3.
\end{aligned}$$

Example 4.2

$f(x) = x^3 + 4\cos x - \sin \dfrac{\pi}{2}$,求 $f'(x)$ 及 $f'\left(\dfrac{\pi}{2}\right)$.

Solution

$$f'(x) = (x^3)' + (4\cos x)' - \left(\sin \dfrac{\pi}{2}\right)' = 3x^2 - 4\sin x,$$

$$f'\left(\dfrac{\pi}{2}\right) = \dfrac{3}{4}\pi^2 - 4.$$

Example 4.3

$y = e^x(\sin x + \cos x)$,求 y'.

Solution

$$\begin{aligned}
y' &= (e^x)'(\sin x + \cos x) + e^x(\sin x + \cos x)' \\
&= e^x(\sin x + \cos x) + e^x(\cos x - \sin x) \\
&= 2e^x \cos x
\end{aligned}$$

Example 4.4

$y = 2^x + \sqrt{x} \ln x$,求 y'.

Solution

$$\begin{aligned}
y' &= (2^x)' + (\sqrt{x})' \ln x + \sqrt{x} (\ln x)' \\
&= 2^x \ln 2 + \dfrac{1}{2\sqrt{x}} \ln x + \dfrac{1}{\sqrt{x}} \cdot \dfrac{1}{x} \\
&= 2^x \ln 2 + \dfrac{\ln x \sqrt{x}}{2x} + \dfrac{\sqrt{x}}{x^2}
\end{aligned}$$

Example 4.5

$y = \tan x$,求 y'.

Solution

$$\begin{aligned}
y' &= (\tan x)' = \left(\dfrac{\sin x}{\cos x}\right)' = \dfrac{(\sin x)' \cos x - \sin x (\cos x)'}{\cos^2 x} \\
&= \dfrac{\cos^2 x + \sin^2 x}{\cos^2 x}
\end{aligned}$$

$$=\frac{1}{\cos^2 x}$$
$$=\sec^2 x$$

即 $(\tan x)' = \sec^2 x.$

Example 4.6

$y = \sec x$,求 y'.

Solution

$$y' = (\sec x)' = \left(\frac{1}{\cos x}\right)' = \frac{(1)'\cos x - 1 \cdot (\cos x)'}{\cos^2 x} = \frac{\sin x}{\cos^2 x} = \sec x \tan x.$$

即 $(\sec x)' = \sec x \tan x.$

用类似方法,还可求得余切函数及余割函数的导数公式:
$$(\cot x)' = -\csc^2 x,$$
$$(\csc x)' = -\csc x \cot x.$$

Example 4.7

$y = \dfrac{\sec x}{1 + \tan x}$,求 y'.

Solution

$$y' = \frac{(\sec x)'(1 + \tan x) - \sec x (1 + \tan x)'}{(1 + \tan x)^2}$$
$$= \frac{\sec x \tan x (1 + \tan x) - \sec x \sec^2 x}{(1 + \tan x)^2}$$
$$= \frac{\sec x (\tan x - 1)}{(1 + \tan x)^2}$$

4.2 Derivative of Inverse Function(反函数的导数)

如果函数 $x = f(y)$ 在某区间内单调、可导,且 $f'(y) \neq 0$,那么它的反函数 $y = f^{-1}(x)$ 在对应区间也可导,且

$$[f^{-1}(x)]' = \frac{1}{f'[f^{-1}(x)]} = \frac{1}{f'(y)} \text{ 或 } \frac{dy}{dx} = \frac{1}{\frac{dx}{dy}}$$

反函数求导法则比较难理解,在应用中关键是要注意到等式两边的自变量是不同的,比如:令 f 和 g 互为反函数,若点 (x_0, y_0) 在曲线 f 上,则点 (y_0, x_0) 在曲线 g 上(见图 4.1),则:

$$g'(y_0) = \frac{1}{f'(x_0)}$$

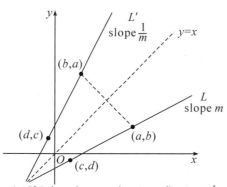
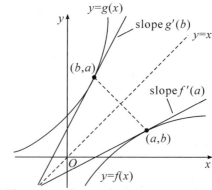

(a) If L has slope m, then its reflection L' has slope $1/m$.

(b) The tangent line to the inverse $y=g(x)$ is the reflection of the tangent line to $y=f(x)$.

图 4.1

Example 4.8

Let $f(x)=x^5+2x^3+x-1$.

a. Find $f(1)$ and $f'(1)$.

b. Find $f^{-1}(3)$ and $(f^{-1}(3))'$.

Solution

a. Note that $f'(x)=5x^4+6x^2+1$, thus $f(1)=3$ and $f'(1)=12$.

b. Since the graph of $y=f(x)$ include the point $(1,3)$ and the slope of the graph is 12 at this point, the graph of $y=f^{-1}(x)$ will include $(3,1)$ and the slope will be $\dfrac{1}{f'(1)}=\dfrac{1}{12}$.

Thus, $f^{-1}(3)=1$ and $(f^{-1}(3))'=\dfrac{1}{12}$.

Example 4.9 *反函数的导数*

$x=\sin y$ 的反函数是 $y=\arcsin x$,$x=\tan y$ 的反函数是 $y=\arctan x$,$x=a^y$ 的反函数是 $y=\log_a x$,利用反函数求导法则分别求 $(\arcsin x)'$,$(\arctan x)'$ 和 $(\log_a x)'$.

Solution

$$(\arcsin x)' = \frac{1}{(\sin y)'} = \frac{1}{\cos y} = \frac{1}{\sqrt{1-\sin^2 y}} = \frac{1}{\sqrt{1-x^2}}.$$

类似地有:$(\arccos x)' = -\dfrac{1}{\sqrt{1-x^2}}$

$$(\arctan x)' = \frac{1}{(\tan y)'} = \frac{1}{\sec^2 y} = \frac{1}{1+\tan^2 y} = \frac{1}{1+x^2}.$$

类似地有:$(\text{arccot } x)' = -\dfrac{1}{1+x^2}$

$$(\log_a x)' = \frac{1}{(a^y)'} = \frac{1}{a^y \ln a} = \frac{1}{x \ln a}$$

4.3 Essential Formulas(基本公式)

(1) $(c)' = 0$; (2) $x^\mu = \mu x^{\mu-1}$;

(3) $(\sin x)' = \cos x$; (4) $(\cos x)' = -\sin x$;

(5) $(\tan x)' = \sec^2 x$; (6) $(\cot x)' = -\csc^2 x$;

(7) $(\sec x)' = \sec x \cdot \tan x$; (8) $(\csc x)' = -\csc x \cdot \cot x$;

(9) $(a^x)' = a^x \ln a$; (10) $(e^x)' = e^x$;

(11) $(\log_a x)' = \dfrac{1}{x \ln a}$; (12) $(\ln x)' = \dfrac{1}{x}$;

(13) $(\arcsin x)' = \dfrac{1}{\sqrt{1-x^2}}$; (14) $(\arccos x)' = -\dfrac{1}{\sqrt{1-x^2}}$;

(15) $(\arctan x)' = \dfrac{1}{1+x^2}$; (16) $(\operatorname{arccot} x)' = -\dfrac{1}{1+x^2}$.

4.4 Chain Rule(链式法则)

4.4.1 The Chain Rule(链式法则)

If $f(u)$ is differentiable at the point $u = g(x)$ and $g(x)$ is differentiable at x, then the composite function $(f \circ g)(x) = f(g(x))$ is differentiable at x, and
$$(f \circ g)'(x) = f'(g(x)) \cdot g'(x)$$
In Leibniz's notation, if $y = f(u)$ and $u = g(x)$, then
$$\frac{dy}{dx} = \frac{dy}{du} \cdot \frac{du}{dx}$$
Where $\dfrac{dy}{du}$ is evaluated at $u = g(x)$.

对于复合函数 $y = f(g(x))$,若令 $y = f(u), u = g(x)$,则称 $y = f(u)$ 为复合函数的"外层函数", $u = g(x)$ 为复合函数的"内层函数".

链式法则 $(f \circ g)'(x) = f'(g(x)) \cdot g'(x)$ 表示一个复合函数的导数等于"外层函数"的导数与"内层函数"的导数之积.

$$y \xrightarrow{\frac{dy}{du}} u \xrightarrow{\frac{du}{dx}} x$$
（因变量）　（中间变量）　（自变量）

链式法则 $\dfrac{dy}{dx} = \dfrac{dy}{du} \cdot \dfrac{du}{dx}$ 就像链条一样"环环相扣",用各部分的导数的乘积的形式表现出来,既有简洁的美感,又有深刻的数学含义.

对于三层函数构成的复合函数 $y = f(u(v(x)))$,可令 $y = f(u), u = u(v(x)), v = v(x)$.

$y = f(u(v(x)))$ 的函数关系可以表示为:

$$y \xrightarrow{\frac{dy}{du}} u \xrightarrow{\frac{du}{dv}} v \xrightarrow{\frac{dv}{dx}} x$$
（因变量）　　（中间变量）　　（中间变量）　　（自变量）

则有链式法则：$\dfrac{dy}{dx} = \dfrac{dy}{du} \cdot \dfrac{du}{dv} \cdot \dfrac{dv}{dx}$.

用链式法则可以证明反函数求导法则.

根据直接函数与反函数的关系：$f[f^{-1}(x)] = x$

等式两边对 x 求导得：$f'[f^{-1}(x)] \cdot [f^{-1}(x)]' = 1$

即 $[f^{-1}(x)]' = \dfrac{1}{f'[f^{-1}(x)]} = \dfrac{1}{f'(y)}$

Example 4.10 *复合函数求导.*

a. $y = \ln\tan x$，求 $\dfrac{dy}{dx}$.

b. $y = \ln\cos(e^x)$，求 $\dfrac{dy}{dx}$.

c. $y = e^{\sin^2 \frac{1}{x}}$，求 $\dfrac{dy}{dx}$.

d. $y = \sin x$，求 $\dfrac{d^n y}{dx^n}$.

Solution

a. Let $y = \ln u, u = \tan x$

$\dfrac{dy}{dx} = \dfrac{dy}{du} \cdot \dfrac{du}{dx} = \dfrac{1}{u} \cdot \sec^2 x = \cot x \cdot \sec^2 x = \dfrac{1}{\sin x \cos x}$

b. let $y = \ln u, u = \cos v, v = e^x$

$\dfrac{dy}{dx} = \dfrac{dy}{du} \cdot \dfrac{du}{dv} \cdot \dfrac{dv}{dx} = \dfrac{1}{u}(-\sin v) \cdot e^x = -e^x \tan(e^x)$

省略设中间变量的过程，可简写为：

$\dfrac{dy}{dx} = \dfrac{1}{\cos(e^x)}(\cos e^x)' = \dfrac{1}{\cos(e^x)}(-\sin(e^x))(e^x)' = -e^x \tan(e^x)$

c. $\dfrac{dy}{dx} = (e^{\sin^2 \frac{1}{x}})' = e^{\sin^2 \frac{1}{x}} \left(\sin^2 \dfrac{1}{x}\right)' = e^{\sin^2 \frac{1}{x}} \cdot 2\sin \dfrac{1}{x} \cdot \left(\sin \dfrac{1}{x}\right)'$

$= e^{\sin^2 \frac{1}{x}} \cdot 2\sin \dfrac{1}{x} \cdot \cos \dfrac{1}{x} \cdot \left(\dfrac{1}{x}\right)' = e^{\sin^2 \frac{1}{x}} \cdot 2\sin \dfrac{1}{x} \cdot \cos \dfrac{1}{x} \cdot \left(-\dfrac{1}{x^2}\right)$

$= -\dfrac{1}{x^2} e^{\sin^2 \frac{1}{x}} \cdot \sin \dfrac{2}{x}$

一般地，若 $f(x) = e^{g(x)}$，则 $f'(x) = e^{g(x)} \cdot g'(x)$.

比如：$(e^{-x})' = e^{-x} \cdot (-x)' = -e^{-x}$　　　　$(e^{2x})' = e^{2x} \cdot (2x)' = 2e^{2x}$

$(e^{x^2})' = e^{x^2} \cdot (x^2)' = 2x e^{x^2}$　　　　$(e^{\frac{1}{x}})' = e^{\frac{1}{x}} \cdot \left(\dfrac{1}{x}\right)' = -\dfrac{1}{x^2} \cdot e^{\frac{1}{x}}$

d. $y = \sin x$,

$$y''=\cos x=\sin\left(x+\frac{\pi}{2}\right),$$

$$y'''=\cos\left(x+\frac{\pi}{2}\right)=\sin\left(x+\frac{\pi}{2}+\frac{\pi}{2}\right)=\sin\left(x+2\cdot\frac{\pi}{2}\right),$$

$$y'''=\cos\left(x+2\cdot\frac{\pi}{2}\right)=\sin\left(x+2\cdot\frac{\pi}{2}+\frac{\pi}{2}\right)=\sin\left(x+3\cdot\frac{\pi}{2}\right),$$

$$y^{(4)}=\cos\left(x+3\cdot\frac{\pi}{2}\right)=\sin\left(x+4\cdot\frac{\pi}{2}\right).$$

一般地，可得

$$y^{(n)}=\sin\left(x+n\cdot\frac{\pi}{2}\right),$$

即 $(\sin x)^{(n)}=\sin\left(x+n\cdot\frac{\pi}{2}\right)$.

用类似方法，可得 $(\cos x)^{(n)}=\cos\left(x+n\cdot\frac{\pi}{2}\right)$.

Example 4.11 抽象函数的导数.

Let f and g be differentiable functions such that

$f(1)=2,$ $f'(1)=3,$ $f'(2)=-4,$
$g(1)=2,$ $g'(1)=-3,$ $g'(2)=5.$

a. If $h(x)=f(g(x))$, then $h'(1)=?$

b. If f and g are twice differentiable and if $h(x)=f(g(x))$, then $h''(x)=?$

Solution

a. $h'(x)=f'(g(x))\cdot g'(x)$, then $h'(1)=f'(g(1))\cdot g'(1)=f'(2)\cdot g'(1)$
$\quad =-4\times(-3)=12.$

b. $h(x)=f(g(x))\Rightarrow h'(x)=f'(g(x))\cdot g'(x)$
$\quad \Rightarrow h''(x)=f''(g(x))\cdot [g'(x)]^2+f'(g(x))\cdot g''(x).$

4.5 Implicit Function Derivative 隐函数的导数

显函数：形如 $y=f(x)$ 的函数称为显函数，例如：$y=\sin x, y=\ln x+e^x$.

隐函数：由方程 $F(x,y)=0$ 所确定的函数称为隐函数.

例如，方程 $x+y^3-1=0$ 为隐函数，其对应的显函数为 $y=\sqrt[3]{1-x}$. 把一个隐函数化成显函数，叫作隐函数的显化. 隐函数的显化有时是很困难的，甚至是不可能的. 但在实际问题中，有时需要计算隐函数的导数，因此，我们希望有一种方法，不管隐函数能否显化，都能直接由方程算出它所确定的隐函数的导数来.

Example 4.12 隐函数求导.

求由方程 $e^y+xy-e=0$ 所确定的隐函数 y 的导数.

Solution

把方程两边的每一项对 x 求导数,得:$\dfrac{d(e^y)}{dx}+\dfrac{d(xy)}{dx}-0=0$

即 $e^y \cdot \dfrac{dy}{dx}+y+x \cdot \dfrac{dy}{dx}=0$

从而解出:$\dfrac{dy}{dx}=-\dfrac{y}{x+e^y}(x+e^y \neq 0)$.

Example 4.13

求椭圆 $\dfrac{x^2}{16}+\dfrac{y^2}{9}=1$ 在 $\left(2,\dfrac{3}{2}\sqrt{3}\right)$ 处的切线方程.

Solution

把椭圆方程的两边分别对 x 求导,得:$\dfrac{x}{8}+\dfrac{2}{9}y \cdot y'=0$.

从而 $y'=-\dfrac{9x}{16y}$.

当 $x=2$ 时,$y=\dfrac{3}{2}\sqrt{3}$,代入上式得到所求切线的斜率为

$$k=y'|_{x=2}=-\dfrac{\sqrt{3}}{4}.$$

所求的切线方程为

$$y-\dfrac{3}{2}\sqrt{3}=-\dfrac{\sqrt{3}}{4}(x-2), 即 \sqrt{3}x+4y-8\sqrt{3}=0.$$

Example 4.14 求反函数的导数.

利用隐函数求导法证明:$(\arcsin x)'=\dfrac{1}{\sqrt{1-x^2}}$.

Proof

$$y=\arcsin x \Rightarrow x=\sin y$$

方程 $x=\sin y$ 两边对 x 求导数,得:$1=\cos y \cdot \dfrac{dy}{dx}$

$$\dfrac{dy}{dx}=\dfrac{1}{\cos y}\left(\because -\dfrac{\pi}{2}<y<\dfrac{\pi}{2}, \therefore \cos y>0\right)$$

$$=\dfrac{1}{\sqrt{1-\sin^2 y}}$$

$$=\dfrac{1}{\sqrt{1-x^2}}(-1<x<1)$$

Example 4.15 求隐函数的二阶导数.

求由方程 $x-y+\dfrac{1}{2}\sin y=0$ 所确定的隐函数 y 的二阶导数.

Solution

方程两边对 x 求导,得: $1-\dfrac{dy}{dx}+\dfrac{1}{2}\cos y \cdot \dfrac{dy}{dx}=0$ ①

于是 $\dfrac{dy}{dx}=\dfrac{2}{2-\cos y}$ ②

②式两边再对 x 求导,得: $\dfrac{d^2y}{dx^2}=\dfrac{-2\sin y \cdot \dfrac{dy}{dx}}{(2-\cos y)^2}=\dfrac{-4\sin y}{(2-\cos y)^3}$.

或对①式两边对 x 求导,得:

$-\dfrac{d^2y}{dx^2}+\dfrac{1}{2}(-\sin y)\cdot\dfrac{dy}{dx}\cdot\dfrac{dy}{dx}+\dfrac{1}{2}\cos y\cdot\dfrac{d^2y}{dx^2}=0$

$(\cos y-2)\dfrac{d^2y}{dx^2}=\sin y\left(\dfrac{dy}{dx}\right)^2=\sin y\left(\dfrac{2}{2-\cos y}\right)^2$

$\therefore \dfrac{d^2y}{dx^2}=\dfrac{-4\sin y}{(2-\cos y)^3}$.

4.6 Logarithmic Differentiation(对数求导法)

对数求导法适用于求幂指函数 $y=[u(x)]^{v(x)}$ 的导数和多因子乘方的导数这两种类型. 这种方法是先在 $y=f(x)$ 的两边取对数,然后求出 y 的导数.

设 $y=f(x)$,两边取对数,得: $\ln y=\ln f(x)$,

两边对 x 求导,得: $\dfrac{1}{y}y'=[\ln f(x)]'$

$$y'=f(x)\cdot[\ln f(x)]'.$$

Example 4.16 求幂指函数的导数.

求 $y=x^{\sin x}(x>0)$ 的导数.

Solution

解法一:两边取对数,得: $\ln y=\sin x \cdot \ln x$,

上式两边对 x 求导,得: $\dfrac{1}{y}y'=\cos x\cdot\ln x+\sin x\cdot\dfrac{1}{x}$,

于是 $y'=y\left(\cos x\cdot\ln x+\sin x\cdot\dfrac{1}{x}\right)=x^{\sin x}\left(\cos x\cdot\ln x+\dfrac{\sin x}{x}\right)$.

解法二:这种幂指函数的导数也可按下面的方法求出:

$y=x^{\sin x}=e^{\sin x\cdot\ln x}$,

$y'=e^{\sin x\cdot\ln x}(\sin x\cdot\ln x)'=x^{\sin x}\left(\cos x\cdot\ln x+\dfrac{\sin x}{x}\right)$.

Example 4.17 多因子乘方的导数.

求函数 $y=\sqrt{\dfrac{(x-1)(x-2)}{(x-3)(x-4)}}\ (x>4)$ 的导数.

Solution

先在两边取对数,得:

$$\ln y = \frac{1}{2}[\ln(x-1)+\ln(x-2)-\ln(x-3)-\ln(x-4)],$$

上式两边对 x 求导,得:

$$\frac{1}{y}y' = \frac{1}{2}\left(\frac{1}{x-1}+\frac{1}{x-2}-\frac{1}{x-3}-\frac{1}{x-4}\right),$$

于是 $y' = \frac{y}{2}\left(\frac{1}{x-1}+\frac{1}{x-2}-\frac{1}{x-3}-\frac{1}{x-4}\right).$

$$y' = \frac{1}{2}\sqrt{\frac{(x-1)(x-2)}{(x-3)(x-4)}}\left(\frac{1}{x-1}+\frac{1}{x-2}-\frac{1}{x-3}-\frac{1}{x-4}\right)$$

4.7 Parametric Function Derivative(参数方程的导数)

If $x = f(t)$ and $y = g(t)$ are differentiable functions of t, then

$$\frac{dy}{dx} = \frac{\frac{dy}{dt}}{\frac{dx}{dt}} \quad \text{and} \quad \frac{d^2y}{dx^2} = \frac{d}{dx}\left(\frac{dy}{dx}\right) = \frac{\frac{d}{dt}\left(\frac{dy}{dx}\right)}{\frac{dx}{dt}}$$

Example 4.18 参数方程求切线.

求椭圆(ellipse) $\begin{cases} x = a\cos t \\ y = b\sin t \end{cases}$ 在 $t = \frac{\pi}{4}$ 处的切线方程.

Solution

$$\frac{dy}{dx} = \frac{(b\sin t)'}{(a\cos t)'} = \frac{b\cos t}{-a\sin t} = -\frac{b}{a}\cot t.$$

所求切线的斜率为 $\frac{dy}{dx}\Big|_{t=\frac{\pi}{4}} = -\frac{b}{a}.$

切点的坐标为: $x_0 = a\cos\frac{\pi}{4} = \frac{\sqrt{2}a}{2}, y_0 = b\sin\frac{\pi}{4} = \frac{\sqrt{2}b}{2}.$

切线方程为: $y - \frac{\sqrt{2}b}{2} = -\frac{b}{a}\left(x - \frac{\sqrt{2}a}{2}\right).$

即 $bx + ay - \sqrt{2}ab = 0.$

Example 4.19 参数方程的二阶导数.

计算由摆线(cycloid)的参数方程 $\begin{cases} x = a(t-\sin t) \\ y = a(1-\cos t) \end{cases}$ 所确定的函数 $y = f(x)$ 的二阶导数.

Solution

$$\frac{dy}{dx} = \frac{y'(t)}{x'(t)} = \frac{[a(1-\cos t)]'}{[a(t-\sin t)]'} = \frac{a\sin t}{a(1-\cos t)} = \frac{\sin t}{1-\cos t} = \cot\frac{t}{2}(t \neq 2k\pi, k \text{ 为整数}).$$

$$\frac{d^2y}{dx^2} = \frac{d}{dx}\left(\frac{dy}{dx}\right) = \frac{d}{dt}\left(\cot\frac{t}{2}\right) \cdot \frac{dt}{dx} = -\frac{1}{2\sin^2\frac{t}{2}} \cdot \frac{1}{a(1-\cos t)} = -\frac{1}{a(1-\cos t)^2}$$

($t \neq 2k\pi, k$ 为整数).

或 $\dfrac{dy}{dx} = \dfrac{y'(t)}{x'(t)} = \dfrac{[a(1-\cos t)]'}{[a(t-\sin t)]'} = \dfrac{a\sin t}{a(1-\cos t)} = \dfrac{\sin t}{1-\cos t}$,

$$\frac{d^2y}{dx^2} = \frac{d\left(\dfrac{\sin t}{1-\cos t}\right)}{dt} \cdot \frac{dt}{dx} = \frac{\dfrac{\cos t(1-\cos t) - \sin t \cdot \sin t}{(1-\cos t)^2}}{a(1-\cos t)} = \frac{\dfrac{\cos t - 1}{(1-\cos t)^2}}{a(1-\cos t)} = -\frac{1}{a(1-\cos t)^2}.$$

4.8　Polar Function Derivative(极坐标方程的导数)

极坐标方程对应的直角坐标方程为：$\begin{cases} x = r\cos\theta, \\ y = r\sin\theta. \end{cases}$

根据参数方程求导法,得：

$$\frac{dy}{dx} = \frac{\dfrac{dy}{d\theta}}{\dfrac{dx}{d\theta}} = \frac{\dfrac{dr}{d\theta}\sin\theta + r\cos\theta}{\dfrac{dr}{d\theta}\cos\theta - r\sin\theta}$$

Example 4.20　*极坐标方程求切线.*

求心形线(cardioid) $r = 2(1-\cos\theta)$ 在 $\theta = \dfrac{\pi}{2}$ 处的切线方程.

Solution

将极坐标方程化为参数方程,得：

$$\begin{cases} x = r\cos\theta = 2(1-\cos\theta)\cos\theta \\ y = r\sin\theta = 2(1-\cos\theta)\sin\theta \end{cases}, \text{即} \begin{cases} x = 2\cos\theta - 2\cos^2\theta \\ y = 2\sin\theta - \sin 2\theta \end{cases}.$$

于是 $\qquad \dfrac{dy}{dx} = \dfrac{\dfrac{dy}{d\theta}}{\dfrac{dx}{d\theta}} = \dfrac{\cos\theta - \cos 2\theta}{-\sin\theta + \sin 2\theta}$

$$\left.\frac{dy}{dx}\right|_{\theta=\frac{\pi}{2}} = -1$$

又当 $\theta = \dfrac{\pi}{2}$ 时, $r = 2, x = 0, y = 2$, 所以 $r = 2(1-\cos\theta)$ 在 $\theta = \dfrac{\pi}{2}$ 处的切线方程为

$$y - 2 = -x \Rightarrow y = 2 - x.$$

Practice Exercises(习题)

1. Let f and g be the functions that are differentiable everywhere. If g is the inverse function of f and if $g(-2) = 5$ and $f'(5) = -\dfrac{1}{2}$, then $g'(-2) =$

(A) 2　　　　(B) $\dfrac{1}{2}$　　　　(C) $\dfrac{1}{5}$　　　　(D) -2

2. 公式法求导练习题：

(1) $y = 5x^4 - 3x^2 + \dfrac{1}{x}$;

(2) $y = \sqrt{x} - \dfrac{1}{\sqrt{x}} + \dfrac{1}{x} - \sqrt{x^3}$;

(3) $y = (x+1)\sqrt{x}$;

(4) $y = (x^2 + \pi)\sqrt{x}$;

(5) $y = \dfrac{5x}{1+x^2}$;

(6) $y = \dfrac{3x^2}{1+x^3}$;

(7) $y = \dfrac{1-\ln x}{1+\ln x}$;

(8) $y = \dfrac{1-\ln x^2}{1+\ln x^2}$;

(9) $y = x^n \log_a x$;

(10) $y = e^x \log_3 x^2$;

(11) $y = \dfrac{1-x^3}{\sqrt[3]{x}}$;

(12) $y = \dfrac{x-x^3}{\sqrt[3]{x^2}}$;

(13) $y = 2^x \sin x$;

(14) $y = e^x \cos x$;

(15) $y = 3^{-x} \cdot x^2$;

(16) $y = 2^{-2x} \cdot x^3$;

(17) $y = \dfrac{x}{1-\cos x}$;

(18) $y = \dfrac{x}{1-\sec x}$;

(19) $y = \tan x - x \cot x$;

(20) $y = \cot x - x^2 \csc x$;

(21) $y = x \cdot \arctan x$;

(22) $y = x^2 \cdot \operatorname{arccot} x$;

(23) $y = e^x \arccos x$;

(24) $y = 5^x \arcsin x$;

(25) $y = \dfrac{\cos 2x}{\sin x + \cos x}$;

(26) $y = \dfrac{1+\sin 2x}{\sin x + \cos x}$;

(27) $y = x^2 \cdot \tan x \cdot \ln x$;

(28) $y = x^2 \cdot e^x \cdot \ln x$;

(29) $y = 2^{-x} + x^{-2} + \log_2 5$;

(30) $y = \log_3 x + x^{-3} + \pi^2$.

3. If $y = x^2 + 2$ and $u = 2x - 1$, then $\dfrac{dy}{du} =$

(A) $\dfrac{2x^2 - 2x + 4}{(2x-1)^2}$

(B) $6x^2 - 2x + 4$

(C) x

(D) $\dfrac{1}{x}$

4. If $y = \tan u$, $u = v - \dfrac{1}{v}$, and $v = \ln x$, what is the value of $\dfrac{dy}{dx}$ at $x = e$?

(A) 0　　　　(B) $\dfrac{1}{e}$　　　　(C) 1　　　　(D) $\dfrac{2}{e}$

5. If $y = \cos^2 3x$, then $\dfrac{dy}{dx} =$

(A) $-6\sin 3x \cos 3x$

(B) $-2\cos 3x$

(C) $2\cos 3x$

(D) $2\sin 3x \cos 3x$

6. If $f(x) = x \ln(x^2)$ then $f'(x) =$

(A) $\ln(x^2) + 1$　　(B) $\ln(x^2) + 2$　　(C) $\ln(x^2) + \dfrac{1}{x}$　　(D) $\dfrac{1}{x^2}$

7. $\dfrac{\mathrm{d}}{\mathrm{d}x}\ln\left|\cos\left(\dfrac{\pi}{x}\right)\right|$ is

   (A) $\dfrac{-\pi}{x^2\cos\left(\dfrac{\pi}{x}\right)}$ (B) $\dfrac{1}{\cos\left(\dfrac{\pi}{x}\right)}$ (C) $\dfrac{\pi}{x^2}\cdot\tan\left(\dfrac{\pi}{x}\right)$ (D) $\dfrac{-\pi}{x^2\tan\left(\dfrac{\pi}{x}\right)}$

8. If $f(x)=\mathrm{e}^{\tan^2 x}$, then $f'(x)=$

   (A) $\mathrm{e}^{\tan^2 x}$ (B) $\sec^2 x\cdot\mathrm{e}^{\tan^2 x}$

   (C) $2\tan x\cdot\sec^2 x\cdot\mathrm{e}^{\tan^2 x}$ (D) $\tan^2 x\cdot\mathrm{e}^{\tan^2 x-1}$

9. If $y=\arctan(\mathrm{e}^{2x})$, then $\dfrac{\mathrm{d}y}{\mathrm{d}x}=$

   (A) $\dfrac{2\mathrm{e}^{2x}}{\sqrt{1-\mathrm{e}^{4x}}}$ (B) $\dfrac{\mathrm{e}^{2x}}{1+\mathrm{e}^{4x}}$ (C) $\dfrac{1}{\sqrt{1-\mathrm{e}^{4x}}}$ (D) $\dfrac{2\mathrm{e}^{2x}}{1+\mathrm{e}^{4x}}$

10. If $h(x)=f^2(x)-g^2(x)$, $f'(x)=-g(x)$, and $g'(x)=f(x)$, then $h'(x)=$

    (A) $-4f(x)g(x)$ (B) $(-g(x))^2-(f(x))^2$

    (C) $-2(-g(x)+f(x))$ (D) 0

11. If f and g are twice differentiable functions such that $g(x)=\mathrm{e}^{f(x)}$ and $g''(x)=h(x)\mathrm{e}^{f(x)}$, then $h(x)=$

    (A) $f'(x)+f''(x)$ (B) $(f'(x))^2+f''(x)$

    (C) $(f'(x)+f''(x))^2$ (D) $2f'(x)+f''(x)$

12. If $3x^2+2xy+y^2=2$, then the value of $\dfrac{\mathrm{d}y}{\mathrm{d}x}$ at $x=1$ is

    (A) -2 (B) 0 (C) 2 (D) not defined

13. If $\sin x=\mathrm{e}^y$, $0<x<\pi$, what is $\dfrac{\mathrm{d}y}{\mathrm{d}x}$ in terms of x?

    (A) $-\tan x$ (B) $-\cot x$ (C) $\cot x$ (D) $\tan x$

14. If $xy^2+2xy=8$, then at the point $(1,2)$, y' is

    (A) $-\dfrac{5}{2}$ (B) $-\dfrac{4}{3}$ (C) $-\dfrac{1}{2}$ (D) 0

15. The slope of the line tangent to the graph of $\ln(xy)=x$ at the point where $x=1$ is

    (A) 0 (B) 1 (C) e (D) $1-\mathrm{e}$

16. The slope of the line tangent to the curve $y^2+(xy+1)^3=0$ at $(2,-1)$ is

    (A) $-\dfrac{3}{2}$ (B) $-\dfrac{4}{3}$ (C) $\dfrac{3}{2}$ (D) $\dfrac{3}{4}$

17. If $y=xy+x^2+1$, then when $x=-1$, $\dfrac{\mathrm{d}y}{\mathrm{d}x}$ is

    (A) $\dfrac{1}{2}$ (B) $-\dfrac{1}{2}$ (C) -1 (D) nonexistent

18. If $\dfrac{\mathrm{d}y}{\mathrm{d}x}=\sqrt{1-y^2}$, then $\dfrac{\mathrm{d}^2 y}{\mathrm{d}x^2}=$

    (A) $-2y$ (B) $\dfrac{-y}{\sqrt{1-y^2}}$ (C) $-y$ (D) y

19. 求下列函数的导数.

(1) $y = x^{\sqrt{x}}$;

(2) $y = \left(1 - \dfrac{1}{2x}\right)^x$;

(3) $y = \dfrac{\sqrt{x^2 + 4x}}{\sqrt[3]{x^3 + 2}}$;

(4) $y = \dfrac{x^2}{1-x}\sqrt[3]{\dfrac{3-x}{(3+x)^2}}$;

(5) $y = (x^2 + 1)^{(2-3x)}$;

(6) $y = x^{\ln x}$;

(7) $y = (x^2 + 1)^x$.

20. For $0 < x < \dfrac{\pi}{2}$, if $y = (\sin x)^x$, then $\dfrac{dy}{dx}$ is

(A) $x \ln(\sin x)$ (B) $(\sin x)^x \cot x$

(C) $(\sin x)^x (x \cos x + \sin x)$ (D) $(\sin x^2)(x \cot x + \ln(\sin x))$

21. The value of the derivative of $y = \dfrac{\sqrt[3]{x^2 + 8}}{\sqrt[4]{2x + 1}}$ at $x = 0$ is

(A) -1 (B) $-\dfrac{1}{2}$

(C) 0 (D) $\dfrac{1}{2}$

22. If $x = t^2 - 1$ and $y = 2e^t$, then $\dfrac{dy}{dx} =$

(A) $\dfrac{2e^t}{t}$ (B) $\dfrac{e^t}{t}$ (C) $\dfrac{e^{|t|}}{t^2}$ (D) $\dfrac{2e^t}{t-1}$

23. If $x = t^3 - t$ and $y = \sqrt{3t+1}$, then $\dfrac{dy}{dx}$ at $t = 1$ is

(A) $\dfrac{1}{8}$ (B) $\dfrac{3}{8}$ (C) $\dfrac{3}{4}$ (D) 8

24. If $x = t^2 + 1$ and $y = t^3$, then $\dfrac{d^2 y}{dx^2} =$

(A) $\dfrac{3}{2t}$ (B) $3t$ (C) $6t$ (D) $\dfrac{3}{4t}$

25. If $x = e^{2t}$ and $y = \sin(2t)$, then $\dfrac{dy}{dx} =$

(A) $\dfrac{e^{2t}}{\cos(2t)}$ (B) $\dfrac{\sin(2t)}{2e^{2t}}$ (C) $\dfrac{\cos(2t)}{e^{2t}}$ (D) $\dfrac{\cos(2t)}{2e^{2t}}$

26. For what values of t does the curve given by the parametric equations $x = t^3 - t^2 - 1$ and $y = t^4 + 2t^2 - 8t$ have a vertical tangent?

(A) 0 only   (B) 1 only   (C) 0 and $\dfrac{2}{3}$ only   (D) no value

27. In the xy-plane, the graph of the parametric equations $x=5t+2$ and $y=3t$, for $-3\leqslant x\leqslant 3$, is a line segment with slope

(A) $\dfrac{3}{5}$ (B) $\dfrac{5}{3}$ (C) 3   (D) 5

习题参考答案

1. D

2.

(1) $y'=20x^3-6x-\dfrac{1}{x^2}$;

(2) $y'=\dfrac{1}{2\sqrt{x}}+\dfrac{1}{2\sqrt{x^3}}-\dfrac{1}{x^2}-\dfrac{3}{2}\sqrt{x}$;

(3) $y'=\dfrac{3}{2}\sqrt{x}+\dfrac{1}{2\sqrt{x}}$;

(4) $y'=\dfrac{5}{2}\sqrt{x^3}+\dfrac{\pi}{2\sqrt{x}}$;

(5) $y'=\dfrac{5-5x^2}{(1+x^2)^2}$;

(6) $y'=\dfrac{6x-3x^4}{(1+x^3)^2}$;

(7) $y'=\dfrac{-2}{x(1+\ln x)^2}$;

(8) $y'=\dfrac{-4}{x(1+2\ln x)^2}$;

(9) $y'=nx^{n-1}\log_a x+\dfrac{x^{n-1}}{\ln a}$;

(10) $y'=e^x\log_3 x^2+e^x\dfrac{2}{x\ln 3}$;

(11) $y'=-\dfrac{1}{3\sqrt[3]{x^4}}-\dfrac{8}{3}\sqrt[3]{x^5}$;

(12) $y'=\dfrac{1}{3\sqrt[3]{x^2}}-\dfrac{7}{3}\sqrt[3]{x^4}$;

(13) $y'=2^x\ln 2\cdot \sin x+2^x\cos x$;

(14) $y'=e^x\cos x-e^x\sin x$;

(15) $y'=-3^{-x}\ln 3\cdot x^2+2x\cdot 3^{-x}$;

(16) $y'=2^{-2x}\cdot(-2\ln 2)\cdot x^3+2^{-2x}\cdot 3x^2$;

(17) $y'=\dfrac{1-\cos x-x\sin x}{(1-\cos x)^2}$;

(18) $y'=\dfrac{1-\sec x+x\cdot \sec x\cdot \tan x}{(1-\sec x)^2}$;

(19) $y=\sec^2 x-\cot x+x\cdot \csc^2 x$;

(20) $y=-\csc^2 x-2x\csc x+x^2\cdot \csc x\cdot \cot x$;

(21) $y=\arctan x+\dfrac{x}{1+x^2}$;

(22) $y'=2x\cdot \text{arccot}\, x-\dfrac{x^2}{1+x^2}$;

(23) $y'=e^x\arccos x-\dfrac{e^x}{\sqrt{1-x^2}}$;

(24) $y'=5^x\ln 5\cdot \arcsin x+5^x\cdot \dfrac{1}{\sqrt{1-x^2}}$;

(25) $y'=-\sin x-\cos x$;

(26) $y'=\cos x+\sin x$;

(27) $y'=2x\cdot \tan x\cdot \ln x+x^2\cdot \sec^2 x\cdot \ln x+x\tan x$;

(28) $y'=2x\cdot e^x\cdot \ln x+x^2\cdot e^x\cdot \ln x+xe^x$;

(29) $y=2^{-x}\ln\left(\dfrac{1}{2}\right)-2x^{-3}$;

(30) $y'=\dfrac{1}{x\ln 3}-3x^{-4}$.

3. C   4. D   5. A   6. B   7. C   8. C   9. D   10. A   11. B   12. D   13. C   14. B   15. A   16. D   17. B   18. C

19.

(1) $y'=x^{\sqrt{x}}\left(\dfrac{1}{\sqrt{x}}+\dfrac{\ln x}{2\sqrt{x}}\right)$;

(2) $y' = \left(1 - \dfrac{1}{2x}\right)^x \left[\ln\left(1 - \dfrac{1}{2x}\right) + \dfrac{1}{2x-1}\right]$;

(3) $y' = \dfrac{\sqrt{x^2+4x}}{\sqrt[3]{x^3+2}}\left(\dfrac{x+2}{x^2+4x} - \dfrac{x^2}{x^3+2}\right)$;

(4) $y' = \dfrac{x^2}{1-x}\sqrt[3]{\dfrac{3-x}{(3+x)^2}}\left[\dfrac{1}{x} + \dfrac{1}{1-x} + \dfrac{x-9}{3(9-x^2)}\right]$;

(5) $y' = (x^2+1)^{(2-3x)}\left[-3\ln(x^2+1) + (2-3x)\cdot\dfrac{2x}{x^2+1}\right]$;

(6) $y' = \dfrac{2x^{\ln x}\ln x}{x}$;

(7) $y' = (x^2+1)^x\left[\ln(x^2+1) + \dfrac{2x^2}{x^2+1}\right]$.

20.D 21.A 22.B 23.B 24.D 25.C 26.C 27.A

Chapter 5　Applications of Derivative(导数的应用)

5.1　Average and Instantaneous Rates of Change(平均变化率与瞬时变化率)

Average rate of change of $f(x)$ over the interval from a to $a+h$ is

$$\textbf{Difference quotient} = \frac{f(a+h)-f(a)}{h}.$$

函数 $y=f(x)$ 从 $x=a$ 到 $x=a+h$ 的平均变化率为：

$$差商 = \frac{f(a+h)-f(a)}{h}$$

The (instantaneous) rate of change of f at a is the derivative of f at a:

$$f'(a) = \lim_{h \to 0} \frac{f(a+h)-f(a)}{h}$$

函数 $y=f(x)$ 在 $x=a$ 处的瞬时变化率为 $f(x)$ 在 $x=a$ 处的导数(又称微商)：

$$f'(a) = \lim_{h \to 0} \frac{f(a+h)-f(a)}{h}$$

Example 5.1　*函数的平均变化率*.

Traffic flow is defined as the rate at which cars pass through an intersection, measured in cars per minute. The traffic flow at a particular intersection is modeled by the function F defined by $F(t) = 82 + 4\sin\left(\frac{t}{2}\right)$ for $0 \leqslant t \leqslant 30$, where $F(t)$ is measured in cars per minute and t is measured in minutes.

What is the average rate of change of the traffic flow over the time interval $10 \leqslant t \leqslant 15$? Indicate units of measure.

Solution

$\frac{F(15)-F(10)}{15-10} = 1.517$　or　1.518 cars/min^2

在 AP 考试中,对使用计算器求解得的结果取近似值的方法没有严格要求,既可以用四舍五入法(round to)，又可以用进一法(round up)，还可以用去尾法(round down).

Example 5.2 导函数的平均变化率.

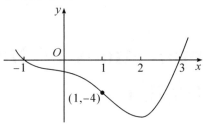

图 5.1 Graph of f'

Let f be twice-differentiable function defined on the interval $-1.2<x<3.2$ with $f(1)=2$. The graph of f', the derivative of f, is shown above. The graph of f' crosses the x-axis at $x=-1$ and $x=3$ and has a horizontal tangent at $x=2$. Let g be the function given by $g(x)=e^{f(x)}$. Find the average rate of change of g', the derivative of g, over the interval $[1,3]$.

Solution

$$\frac{g'(3)-g'(1)}{3-1}=\frac{e^{f(3)}f'(3)-e^{f(1)}f'(1)}{2}=\frac{e^{f(3)} \cdot 0-e^2 \cdot (-4)}{2}=2e^2$$

5.2 Tangents and Normals(切线和法线)

过曲线 $y=f(x)$ 的切点 (x_0,y_0) 的**切线方程为**：

$$y-y_0=f'(x_0)(x-x_0).$$

由于相互平行的两条直线的斜率相等，相互垂直的两条直线的斜率互为"负倒数"，因此**法线方程为**：

$$y-y_0=-\frac{1}{f'(x_0)}(x-x_0).$$

特殊地，

若 $y=f(x)$ 的切点 (x_0,y_0) 处有**水平切线(Horizontal tangent line)**，则：

$$\left.\frac{dy}{dx}\right|_{x=x_0}=0(即切线的斜率为0),切线方程为 y=y_0.$$

若 $y=f(x)$ 的切点 (x_0,y_0) 处有**垂直切线(Vertical tangent line)**，则：

$$\left.\frac{dy}{dx}\right|_{x=x_0}=\infty(\text{not differentiable})或\left.\frac{dx}{dy}\right|_{y=y_0}=0.$$

此时，**切线方程为** $x=x_0$.

Example 5.3 显函数求切线方程与法线方程.

Find the equations of the tangent and normal to the curve of $f(x)=x^3-3x^2$ at the point $(1,-2)$.

Solution

$\because f'(x)=3x^2-6x, \therefore f'(1)=3.$

Equation of the tangent line: $y+2=-3(x-1)$.

Equation of the normal line: $y+2=\dfrac{1}{3}(x-1)$.

Example 5.4 隐函数求切线方程.

Find the equations of the tangent to $x^2y-x=y^3-8$ at the point where $x=0$.

Solution

先求切点：当 $x=0$ 时，$y^3-8=0 \Rightarrow y=2$，切点为 $(0,2)$.

再求斜率：方程 $x^2y-x=y^3-8$ 两边对 x 求导数，得：
$$2xy+x^2y'-1=3y^2 \cdot y'$$

将 $x=0, y=2$ 代入上式，得：$-1=12y' \Rightarrow y'=-\dfrac{1}{12}$.

过切点 $(0,2)$ 的切线方程为：$y-2=-\dfrac{1}{12}x \Rightarrow y=-\dfrac{1}{12}x+2$.

过切点 $(0,2)$ 的法线方程为：$y-2=12x \Rightarrow y=12x+2$.

Example 5.5 求切点.

Find the coordinates of any point on the curve $y^2-4xy=x^2+5$ for which the tangent is horizontal.

Solution

对等式 $y^2-4xy=x^2+5$ 两边关于 x 求导，得：
$$2y \cdot y'-(4y+4xy')=2x \qquad ①$$

∵ 曲线 $y^2-4xy=x^2+5$ 有水平切线，∴ $y'=0$

将 $y'=0$ 代入①式得：$x=-2y$ ②

将②式代入 $y^2-4xy=x^2+5$，得：
$$y^2-4(-2y)y=4y^2+5 \Rightarrow 5y^2=5 \Rightarrow y=\pm 1$$

当 $y=1$ 时，$x=-2y=-2$

当 $y=-1$ 时，$x=-2y=2$

故曲线 $y^2-4xy=x^2+5$ 在点 $(2,-1)$ 和点 $(-2,1)$ 处有水平切线.

5.3 The Mean Value Theorem for Derivatives(微分中值定理)

5.3.1 Rolle's Theorem(罗尔定理)

Suppose that $f(x)$ is continuous at every point of the closed interval $[a,b]$ and differentiable at every point of its interior (a,b). If $f(a)=f(b)$, then there is at least one number c in (a,b) at which $f'(c)=0$.

罗尔定理可以简单地表达为：

$\left. \begin{array}{l} f(x)\text{在}[a,b]\text{连续} \\ f(x)\text{在}(a,b)\text{可导} \\ f(a)=f(b) \end{array} \right\} \Rightarrow$ 至少存在一个点 $c(a<c<b)$，使得 $f'(c)=0$（见图 5.2）

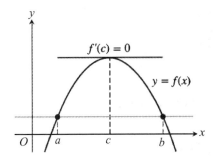 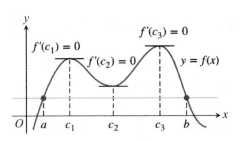

(a)在区间(a,b)只有1个满足罗尔定理的点.   (b)在区间(a,b)有3个满足罗尔定理的点.

图 5.2

罗尔定理的三个条件若有一条不满足,则定理就不成立(见图 5.3).

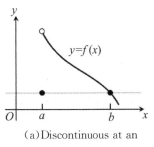 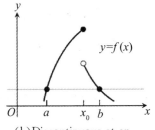 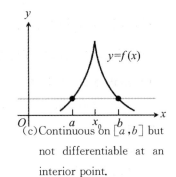

(a)Discontinuous at an endpoint of $[a,b]$.   (b)Discontinuous at an interior point of $[a,b]$.   (c)Continuous on $[a,b]$ but not differentiable at an interior point.

图 5.3

Example 5.6 微分中值定理(罗尔定理).

t(hours)	0	3	6	9	12	15	18	21	24
$R(t)$(gallons per hour)	9.6	10.4	10.8	11.2	11.4	11.3	10.7	10.2	9.6

The rate at which water flows out of a pipe, in gallons per hour, is given by a differentiable function R of time t. The table above shows the rate as measured every 3 hours for a 24-hour period. Is there some time t, $0<t<24$, such that $R'(t)=0$? Justify your answer.

Solution

Yes.

Since $R(0)=R(24)=9.6$, the Mean Value Theorem(MVT) guarantees that there is a t, $0<t<24$, such that $R'(t)=0$.

[**注**]在我国很多版本的高等数学教材中都介绍了罗尔定理,但是在国外专门针对AP考试的教材中没有介绍罗尔定理,因为罗尔定理是微分中值定理(the Mean Value Theorem)的特殊情况,所以在AP考试的自由问答的答题中都用微分中值定理加以作答.

5.3.2 The Mean Value Theorem for Derivatives(微分中值定理)

Assume that $f(x)$ is continuous on the closed interval $[a,b]$ and differentiable on (a,b). Then there exists at least one value c in (a,b) such that

$$f'(c) = \frac{f(b)-f(a)}{b-a}.$$

微分中值定理可以简单地表达为：

$\left.\begin{array}{l}f(x)在[a,b]连续\\f(x)在(a,b)可导\end{array}\right\} \Rightarrow$ 至少存在一个点 $c(a<c<b)$，使得 $f'(c) = \dfrac{f(b)-f(a)}{b-a}$

微分中值定理的几何意义(见图 5.4)：对于连续曲线 $y=f(x)$ 上的弧 $\overset{\frown}{AB}$，除端点外处处有不垂直于 x 轴的切线，则在这段弧上至少存在一点 C，使得过该点的切线与弦 AB 平行.

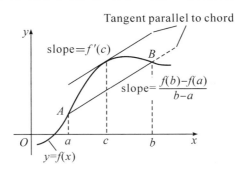

图 5.4

微分中值定理的物理意义：在时间段 (t_0, t_1) 内至少存在某一时刻 t_c，使得时刻 t_c 的瞬时速度(或瞬时变化率)等于从 t_0 到 t_1 的平均速度(又称平均变化率).

Example 5.7 求满足 MVT 的点的坐标.

(1) $f(x) = x^2 + 2x - 1$；

(2) $f(x) = \sqrt[3]{x^2}$, on $[0,1]$；

(3) $f(x) = \sqrt[3]{x}$, on $[-1,1]$；

(4) $f(x) = |x-1|$, on $[0,4]$；

(5) $f(x) = \arcsin x$, on $[-1,1]$；

(6) $f(x) = \ln(x-1)$, on $[2,4]$；

(7) $f(x) = \begin{cases} \cos x, & 0 \leqslant x < \dfrac{\pi}{2}, \\ \sin x, & \dfrac{\pi}{2} \leqslant x < \pi, \end{cases}$ on $[0,\pi]$；

(8) $f(x) = \begin{cases} \arcsin x, & -1 \leqslant x < 1, \\ \sin x, & 1 \leqslant x < 3, \end{cases}$ on $[-1,3]$.

(a) State whether or not the function satisfies the hypotheses of the Mean Value Theorem on the given interval, and (b) if it does, find each value of c in the interval (a,b) that satisfies $f'(c) = \dfrac{f(b)-f(a)}{b-a}$.

Solution

(1) (a) Yes.

(b) $f'(x) = (x^2 + 2x - 1)' = 2x + 2$,

$$2c+2=\frac{2-(-1)}{1-0}=3,$$
$$c=\frac{1}{2}.$$

(2)(a) Yes.

(b) $f'(x)=(x^{2/3})'=\frac{2}{3}x^{-1/3}$,
$$\frac{2}{3}c^{-1/3}=\frac{1-0}{1-0}=1,$$
$$c=\frac{8}{27}.$$

(3)(a) No. There is a vertical tangent at $x=0$.

(4)(a) No. There is a corner at $x=1$.

(5)(a) Yes.

(b) $f'(x)=(\arcsin x)'=\frac{1}{\sqrt{1-x^2}}$,
$$\frac{1}{\sqrt{1-c^2}}=\frac{\frac{\pi}{2}-(-\frac{\pi}{2})}{1-(-1)}=\frac{\pi}{2},$$
$$c=\sqrt{1-\frac{4}{\pi^2}}\approx 0.771.$$

(6)(a) Yes.

(b) $f'(x)=(\ln(x-1))'=\frac{1}{x-1}$,
$$\frac{1}{c-1}=\frac{\ln 3-\ln 1}{4-2},$$
$$c=\frac{4-2}{\ln 3-\ln 1}+1\approx 2.820.$$

(7)(a) No. The function is discontinuous at $x=\frac{\pi}{2}$.

(8)(a) No. The function is discontinuous at $x=1$.

Example 5.8 微分中值定理的应用.

t (minutes)	0	1	2	3	4	5	6
$C(t)$ (ounces)	0	5.3	8.8	11.2	12.8	13.8	14.5

Hot water is dripping through a coffeemaker, filling a large cup with coffee. The amount of coffee in the cup at time t, $0 \leqslant t \leqslant 6$, is given by a differentiable function C, where t is measured in minutes. Selected values of $C(t)$, measured in ounces, are given in the table above.

Is there a time t, $2 \leqslant t \leqslant 4$, at which $C'(t)=2$? Justify your answer.

Solution

$C(t)$ is differentiable $\Rightarrow C(t)$ is continuous (on the closed interval).

$$\frac{C(4)-C(2)}{4-2}=\frac{12.8-8.8}{2}=2.$$

Therefore, by the Mean Value Theorem, there is at least one time t, $2<t<4$, for which $C'(t)=2$.

Example 5.9 不满足微分中值定理的例子.

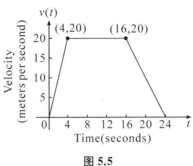

图 5.5

A car is traveling on a straight road. For $0 \leq t \leq 24$ seconds, the car's velocity $v(t)$, in meters per second, is modeled by the piecewise-linear function defined by the graph above.

Find the average rate of change of v over the interval $8 \leq t \leq 20$. Does the Mean Value Theorem guarantee a value of c, for $8<c<20$, such that $v'(c)$ is equal to this average rate of change? Why or why not?

Solution

The average rate of change of v over the interval $8 \leq t \leq 20$ is

$$\frac{v(20)-v(8)}{20-8}=-\frac{5}{6} \text{ m/sec}^2.$$

No, the Mean Value Theorem does not apply to v on the interval $[8,20]$, because v is not differentiable at $t=16$.

5.4 Related Rates(相关变化率)

设 $x=x(t)$ 及 $y=y(t)$ 都是可导函数,而变量 x 与 y 之间存在某种关系,从而变化率 $\dfrac{\mathrm{d}x}{\mathrm{d}t}$ 与 $\dfrac{\mathrm{d}y}{\mathrm{d}t}$ 之间也存在一定的关系.这两个相互依赖的变化率称为相关变化率.相关变化率问题就是研究这两个变化率之间的关系,以便利用一个变化率求出另一个变化率.

Example 5.10 相关变化率的几何应用.

a. Assume that the radius r of a sphere is a differentiable function of t and let V be the volume of the sphere. Find an equation that relates $\dfrac{\mathrm{d}V}{\mathrm{d}t}$ and $\dfrac{\mathrm{d}r}{\mathrm{d}t}$.

b. Assume that the radius r and height h of a cone are differentiable functions of t and let V be the volume of the cone. Find an equation that relates $\dfrac{dV}{dt}, \dfrac{dr}{dt}$ and $\dfrac{dh}{dt}$.

Solution

a. $V = \dfrac{4}{3}\pi r^3 \Rightarrow \dfrac{dV}{dt} = 4\pi r^2 \dfrac{dr}{dt}.$

b. $V = \dfrac{\pi}{3}r^2 h \Rightarrow \dfrac{dV}{dt} = \dfrac{\pi}{3}\left(r^2 \cdot \dfrac{dh}{dt} + 2r\dfrac{dr}{dt} \cdot h\right) = \dfrac{\pi}{3}\left(r^2 \dfrac{dh}{dt} + 2rh\dfrac{dr}{dt}\right).$

Example 5.11 相关变化率的应用.

A police cruiser, approaching a right-angled intersection from the north, is chasing a speeding car that has turned the corner and is now moving straight east(图 5.6). When the cruiser is 0.6 mi north of the intersection and the car is 0.8 mi to the east, the police determine with radar that the distance between them and the car is increasing at 20 mph. If the cruiser is moving at 60 mph at the instant of measurement, what is the speed of the car?

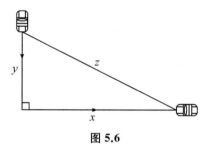

图 5.6

Solution

Let x be the distance of the speeding car from the intersection, let y be the distance of the police cruiser from the intersection, and let z be the distance between the car and the cruiser. Distances x and z are increasing, but distance y is decreasing; so $\dfrac{dy}{dt}$ is negative.

We know: $\dfrac{dz}{dt} = 20$ mph and $\dfrac{dy}{dt} = -60$ mph.

By the Pythagorean Theorem: $x^2 + y^2 = z^2$,

Differentiating implicitly with respect to t, we get

$$2x\dfrac{dx}{dt} + 2y\dfrac{dy}{dt} = 2z\dfrac{dz}{dt}, \text{ which reduces to } x\dfrac{dx}{dt} + y\dfrac{dy}{dt} = z\dfrac{dz}{dt}.$$

We now substitute the numerical values for x, y, $\dfrac{dz}{dt}$, $\dfrac{dy}{dt}$, and z (which equals $\sqrt{x^2+y^2}$):

$$(0.8)\dfrac{dx}{dt} + (0.6)(-60) = \sqrt{(0.8)^2 + (0.6)^2}\,(20)$$

$$\dfrac{dx}{dt} = 70$$

At the moment in question, the car's speed is 70 mph.

Example 5.12 相关变化率的几何应用.

Water runs into a conical tank at the rate of 9 ft/min. The tank stands point down and has a height of 10 ft and a base radius of 5 ft. How fast is the water level rising when the water is 6 ft deep?

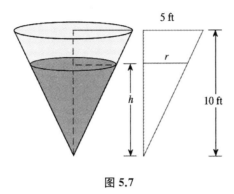

图 5.7

Solution

Let V be the volume, r the radius, and h the height of the cone of water(图 5.7).

$$\frac{r}{h}=\frac{5}{10}\Rightarrow r=\frac{h}{2}$$

$$V=\frac{1}{3}\pi r^2 h=\frac{1}{3}\pi\left(\frac{h}{2}\right)^2 h=\frac{1}{12}\pi h^3$$

Differentiate with respect to t: $\quad\dfrac{\mathrm{d}V}{\mathrm{d}t}=\dfrac{\pi}{4}h^2\dfrac{\mathrm{d}h}{\mathrm{d}t}$

$$\frac{\mathrm{d}V}{\mathrm{d}t}=9 \text{ ft}^3/\min$$

$$9=\frac{\pi}{4}(6)^2\frac{\mathrm{d}h}{\mathrm{d}t}\Rightarrow\frac{\mathrm{d}h}{\mathrm{d}t}=\frac{1}{\pi}\approx 0.32.$$

At the moment in question, the water level is rising at 0.32 ft/min.

5.5 L'Hôpital's Rule(洛必达法则)

在某个相同的极限过程中,$f(x)$,$g(x)$均以零为极限,但它们的商的极限值是否能确定呢?

考察：$\lim\limits_{x\to 0}\dfrac{1-\cos x}{x}=0, \lim\limits_{x\to 0}\dfrac{1-\cos x}{x^2}=\dfrac{1}{2}, \lim\limits_{x\to 0}\dfrac{\sin x}{x^2}=\infty$

虽然分子、分母在 $x\to 0$ 的过程中极限均为零,但商的极限各有不同,因此称之为未定式,并用"$\dfrac{0}{0}$"型来表示.

当分子、分母均在同一极限过程中以无穷大为极限,称之为"$\dfrac{\infty}{\infty}$"型未定式.

两个函数的积在同一过程中一个以零为极限,另一个以无穷大为极限,称为"$0 \cdot \infty$"型未定式.

两个函数的差，被减函数与减函数都以无穷大为极限，称为"$\infty-\infty$"型未定式.

还有幂指函数在同一过程中，底函数与指函数均以零为极限称为"0^0"型未定式；在同一极限过程中，底函数以 1 为极限，指数以无穷大为极限，称为"1^∞"型未定式，还有"∞^0"型未定式.

5.5.1 Theorem：L'Hôpital's Rule(洛必达法则)

设(1) $\lim\limits_{x\to a}f(x)=\lim\limits_{x\to a}g(x)=0$ 或 ∞；

(2) $f(x),g(x)$ 可导，且 $g'(x)\neq 0$；

(3) $\lim\limits_{x\to a}\dfrac{f'(x)}{g'(x)}$ 存在(或为 ∞)；

则 $\lim\limits_{x\to a}\dfrac{f(x)}{g(x)}=\lim\limits_{x\to a}\dfrac{f'(x)}{g'(x)}$.

[注]

(1)若 $\lim\limits_{x\to a}f(x)=\lim\limits_{x\to a}g(x)=0$，称 $\lim\limits_{x\to a}\dfrac{f(x)}{g(x)}$ 为 $\dfrac{0}{0}$ 型未定式 $\left(\dfrac{0}{0}\text{ indeterminate form}\right)$；

若 $\lim\limits_{x\to a}f(x)=\lim\limits_{x\to a}g(x)=\infty$，称 $\lim\limits_{x\to a}\dfrac{f(x)}{g(x)}$ 为 $\dfrac{\infty}{\infty}$ 型未定式 $\left(\dfrac{\infty}{\infty}\text{ indeterminate form}\right)$；

洛必达法则只适用于 $\dfrac{0}{0}$ 和 $\dfrac{\infty}{\infty}$ 两种未定式.

(2)定理适用于任意类型的自变量变化过程，即将 $x\to a$ 换为 $x\to\infty$ 或 $x\to a^-$，洛必达法则同样适用.

(3)只要满足洛必达法则的条件，洛必达法则可以反复使用.

Example 5.13 利用洛必达法则求极限.

求下列极限：

a. $\lim\limits_{x\to 0}\dfrac{\sin ax}{\sin bx}$；

b. $\lim\limits_{x\to 1}\dfrac{x^3-3x+2}{x^3-x^2-x+1}$；

c. $\lim\limits_{x\to 0}\dfrac{x-\sin x}{x^3}$；

d. $\lim\limits_{x\to +\infty}\dfrac{\dfrac{\pi}{2}-\arctan x}{\dfrac{1}{x}}$；

e. $\lim\limits_{x\to +\infty}\dfrac{\ln x}{x^\alpha}(\alpha>0)$；

f. $\lim\limits_{x\to +\infty}\dfrac{x^n}{e^{\lambda x}}(\lambda>0)$.

Solution

a. $\left(\dfrac{0}{0}\right)\lim\limits_{x\to 0}\dfrac{\sin ax}{\sin bx},b\neq 0$.

This is an indeterminate form of type $\dfrac{0}{0}$.

Hence, L'Hôpital's Rule applies.

$$\lim_{x\to 0}\dfrac{\sin ax}{\sin bx}=\lim_{x\to 0}\dfrac{(\sin ax)'}{(\sin bx)'}=\lim_{x\to 0}\dfrac{a\cos ax}{b\cos bx}=\dfrac{a}{b}, b\neq 0$$

b. $\left(\dfrac{0}{0}\right)\lim\limits_{x\to 1}\dfrac{x^3-3x+2}{x^3-x^2-x+1}=\lim\limits_{x\to 1}\dfrac{3x^2-3}{3x^2-2x-1}=\lim\limits_{x\to 1}\dfrac{6x}{6x-2}=\dfrac{6}{4}=\dfrac{3}{2}$.

注意：由于$\dfrac{6x}{6x-2}$不是$\dfrac{0}{0}$未定式，因此下面的计算是错误的！

$$\lim_{x\to 1}\dfrac{6x}{6x-2}=\lim_{x\to 1}\dfrac{(6x)'}{(6x-2)'}=\lim_{x\to 1}\dfrac{6}{6}=1.$$

c. $\left(\dfrac{0}{0}\right)\lim\limits_{x\to 0}\dfrac{x-\sin x}{x^3}=\lim\limits_{x\to 0}\dfrac{1-\cos x}{3x^2}=\lim\limits_{x\to 0}\dfrac{\sin x}{6x}=\lim\limits_{x\to 0}\dfrac{\cos x}{6}=\dfrac{1}{6}$.

d. $\left(\dfrac{0}{0}\right)\lim\limits_{x\to +\infty}\dfrac{\dfrac{\pi}{2}-\arctan x}{\dfrac{1}{x}}=\lim\limits_{x\to +\infty}\dfrac{-\dfrac{1}{1+x^2}}{-\dfrac{1}{x^2}}=\lim\limits_{x\to +\infty}\dfrac{x^2}{1+x^2}=\lim\limits_{x\to +\infty}\dfrac{2x}{2x}=1.$

e. $\left(\dfrac{\infty}{\infty}\right)\lim\limits_{x\to +\infty}\dfrac{\ln x}{x^\alpha}=\lim\limits_{x\to +\infty}\dfrac{\dfrac{1}{x}}{\alpha x^{\alpha-1}}=\lim\limits_{x\to +\infty}\dfrac{1}{\alpha x^\alpha}=0(\alpha>0).$

f. $\left(\dfrac{\infty}{\infty}\right)\lim\limits_{x\to +\infty}\dfrac{x^n}{e^{\lambda x}}=\lim\limits_{x\to +\infty}\dfrac{nx^{n-1}}{\lambda e^{\lambda x}}=\lim\limits_{x\to +\infty}\dfrac{n(n-1)x^{n-2}}{\lambda^2 e^{\lambda x}}=\cdots=\lim\limits_{x\to +\infty}\dfrac{n!}{\lambda^n e^{\lambda x}}=0(\lambda>0).$

5.5.2 利用其他未定式求极限

"$0\cdot\infty$""$\infty-\infty$""0^0""1^∞"和"∞^0"型未定式称为其他未定式，它们可以转化成$\dfrac{0}{0}$或$\dfrac{\infty}{\infty}$未定式计算.

Example 5.14 利用其他未定式求极限.

a. $\lim\limits_{x\to 0^+}x^\alpha\ln x(\alpha>0)$;

b. $\lim\limits_{x\to \frac{\pi}{2}}(\sec x-\tan x)$;

c. $\lim\limits_{x\to 0^+}x^x$;

d. $\lim\limits_{x\to 0^+}(\sin x)^x$;

e. $\lim\limits_{x\to 0^+}\left(\dfrac{1}{x}\right)^{\tan x}$;

f. $\lim\limits_{x\to 0}(1+2x)^{\csc x}$.

Solution

a. $(0\cdot\infty)\lim\limits_{x\to 0^+}x^\alpha\ln x=\lim\limits_{x\to 0^+}\dfrac{\ln x}{\dfrac{1}{x^\alpha}}=\lim\limits_{x\to 0^+}\dfrac{\ln x}{\dfrac{1}{x^\alpha}}=\lim\limits_{x\to 0^+}\dfrac{\dfrac{1}{x}}{-\alpha\dfrac{1}{x^{\alpha+1}}}=\lim\limits_{x\to 0^+}\dfrac{x^\alpha}{-\alpha}=0(\alpha>0).$

b. $(\infty-\infty)\lim\limits_{x\to\frac{\pi}{2}}(\sec x-\tan x)=\lim\limits_{x\to\frac{\pi}{2}}\dfrac{1-\sin x}{\cos x}=\lim\limits_{x\to\frac{\pi}{2}}\dfrac{-\cos x}{-\sin x}=0.$

c. $(0^0)\lim\limits_{x\to 0^+}x^x=\lim\limits_{x\to 0^+}\mathrm{e}^{x\ln x}=\mathrm{e}^{\lim\limits_{x\to 0^+}x\ln x}=\mathrm{e}^{\lim\limits_{x\to 0^+}\frac{\ln x}{\frac{1}{x}}}=\mathrm{e}^{\lim\limits_{x\to 0^+}\frac{\frac{1}{x}}{-\frac{1}{x^2}}}=\mathrm{e}^0=1.$

d. $(0^0)\lim\limits_{x\to 0^+}(\sin x)^x=\lim\limits_{x\to 0^+}\mathrm{e}^{x\ln\sin x}=\mathrm{e}^{\lim\limits_{x\to 0^+}x\ln\sin x}=\mathrm{e}^0=1.$

其中 $\lim\limits_{x\to 0^+}x\ln\sin x=\lim\limits_{x\to 0^+}\dfrac{\ln\sin x}{\frac{1}{x}}=\lim\limits_{x\to 0^+}\dfrac{\frac{\cos x}{\sin x}}{-\frac{1}{x^2}}=\lim\limits_{x\to 0^+}\dfrac{-x^2\cos x}{\sin x}=0.$

e. $(\infty^0)\lim\limits_{x\to 0^+}\left(\dfrac{1}{x}\right)^{\tan x}=\lim\limits_{x\to 0^+}\mathrm{e}^{-\tan x\ln x}=\mathrm{e}^{-\lim\limits_{x\to 0^+}\frac{\sin x}{x}\cdot\frac{1}{\cos x}x\ln x}=\mathrm{e}^0=1.$

f. $(1^\infty)\lim\limits_{x\to 0}(1+2x)^{\csc x}=\lim\limits_{x\to 0}\mathrm{e}^{\csc x\cdot\ln(1+2x)}=\mathrm{e}^{\lim\limits_{x\to 0}\frac{\ln(1+2x)}{\sin x}}=\mathrm{e}^{\lim\limits_{x\to 0}\frac{\frac{2}{1+2x}}{\cos x}}=\mathrm{e}^2,$

或

$\lim\limits_{x\to 0}(1+2x)^{\csc x}=\lim\limits_{x\to 0}(1+2x)^{\frac{1}{2x}\cdot\frac{2x}{\sin x}}=\mathrm{e}^2.$

[注]

(1)运用洛必达法则最好能与求极限的等价无穷小替换法结合使用,尽可能化简再计算.

$\lim\limits_{x\to 0}\dfrac{\tan x-x}{x^2\sin x}=\lim\limits_{x\to 0}\dfrac{\tan x-x}{x^3}\cdot\dfrac{x}{\sin x}=\lim\limits_{x\to 0}\dfrac{\tan x-x}{x^3}$

$=\lim\limits_{x\to 0}\dfrac{\sec^2 x-1}{3x^2}=\lim\limits_{x\to 0}\dfrac{2\sec^2 x\tan x}{6x}=\dfrac{1}{3}\lim\limits_{x\to 0}\dfrac{\tan x}{x}\sec^2 x$

$=\dfrac{1}{3}.$

(2)当洛必达法则条件不满足时,所求极限也可能存在.

例如,虽然 $\lim\limits_{x\to\infty}\dfrac{(x+\sin x)'}{(x)'}=\lim\limits_{x\to\infty}(1+\cos x)$ 不存在,但 $\lim\limits_{x\to\infty}\dfrac{x+\sin x}{x}=\lim\limits_{x\to\infty}\left(1+\dfrac{\sin x}{x}\right)=1.$

(3)指函数求极限的方法:

①重要极限法: $\lim\limits_{x\to\infty}\left(1+\dfrac{1}{x}\right)^x=\mathrm{e}.$

公式可以推广到:在极限 $\lim[1+\alpha(x)]^{\frac{1}{\alpha(x)}}$ 中,只要 $\lim\alpha(x)=0.$

②利用 $x=\mathrm{e}^{\ln x}$ 转化为指数函数,再求极限.

5.6 Monotony of Functions(函数的单调性)

5.6.1 Definitions of Monotonic Functions(单调函数的定义)

Let f be a function defined on an interval I and let x_1 and x_2 be any two points in I.

(1) If $f(x_1)<f(x_2)$ whenever $x_1<x_2$, then f is said to be **increasing**(单调递增) on I.

(2) If $f(x_2)>f(x_1)$ whenever $x_1<x_2$, then f is said to be **decreasing**(单调递减) on I.

A function that is increasing or decreasing on I is called **monotonic** on I.

5.6.2 First Derivative Test for Monotonic Functions(函数单调性的一阶导数判别法)

5.6.2.1 Theorem(定理)

Suppose that $f(x)$ is continuous on $[a,b]$ and differentiable on (a,b).

(1) If $f'(x)>0$ at each point $x\in(a,b)$, then $f(x)$ is increasing on $[a,b]$.
(2) If $f'(x)<0$ at each point $x\in(a,b)$, then $f(x)$ is decreasing on $[a,b]$.
(3) If $f'(x)=0$ at each point $x\in(a,b)$, then $f(x)$ is a constant on $[a,b]$.

5.6.2.2 Critical Point(驻点或临界点)

$f'(x)=0$ or $f'(x)$ undefined 的点称为函数的 **Critical Point(驻点或临界点)**.

Example 5.15

确定函数 $f(x)=2x^3-9x^2+12x-3$ 的单调区间.

Solution

这个函数的定义域为:$(-\infty,+\infty)$.

函数的导数为:$f'(x)=6x^2-18x+12=6(x-1)(x-2)$.

驻点有两个:$x_1=1, x_2=2$.

列表分析:

	$(-\infty,1)$	$(1,2)$	$(2,+\infty)$
$f'(x)$	+	−	+
$f(x)$	↗	↘	↗

函数 $f(x)$ 在区间 $(-\infty,1]$ 和 $[2,+\infty)$ 内单调增大,在区间 $[1,2]$ 上单调减小.

5.6.3 First Derivative Test for Local Extrema(函数极值的一阶导数判别法)

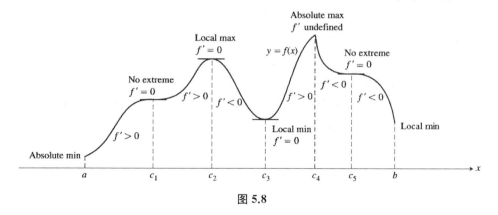

图 5.8

5.6.3.1 Local (or Relative) Extrema and Global (or Absolute) Extrema(极值与最值)

A function f has a **local(or relative) maximum**(极大值) value at an interior point c of its domain if $f(x)\leqslant f(c)$ for all x in some open interval containing c.

A function f has a **local(or relative) minimum**(极小值) value at an interior point c of its domain if $f(x)\geqslant f(c)$ for all x in some open interval containing c.

Let f be a function with domain D. Then f has an **global(or absolute) maximum(最大值)** value on D at a point
$$c \text{ if } f(x) \leqslant f(c) \text{ for all } x \text{ in } D.$$
and an **global(or absolute) minimum(最小值)** value on D at c if
$$f(x) \leqslant f(c) \text{ for all } x \text{ in } D$$

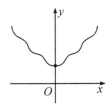

(a) f has an absolute minimum but no absolute maximum on $(-\infty, +\infty)$.

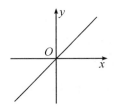

(b) f has no absolute extrema on $(-\infty, +\infty)$.

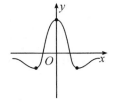

(c) f has an absolute maximum and minimum on $(-\infty, +\infty)$.

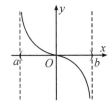

(d) f has no absolute extrema on (a, b).

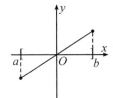

(e) f has an absolute maximum and minimum on $[a, b]$.

图 5.9

[注]闭区间连续的函数必有最值.

5.6.3.2 First Derivative Test for Local Extrema(极值的一阶导数判别法)

Suppose that c is a critical point of a continuous function f, and that f is differentiable at every point in some interval containing c except possibly at c itself.

Moving across c from left to right.

(1) if f' changes from **negative to positive** at c, then f has **a local minimum** at c;

(2) if f' changes from **positive to negative** at c, then f has **a local maximum** at c;

(3) if f' **does not change sign** at c (that is, is positive on both sides of c or negative on both sides), then f has **no local extrema** at c.

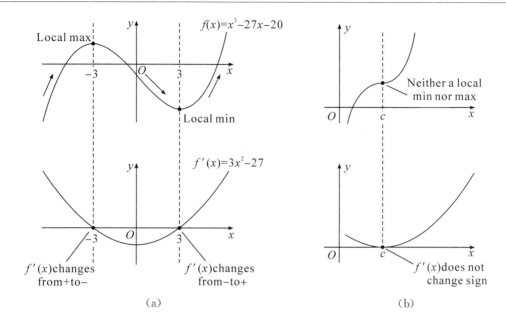

图 5.10 $f(x)-sf'(x)$ 图像的对应关系

Example 5.16 极值的一阶导数判别法.

Consider the function $f(x)=\dfrac{1}{x^2-kx}$, where k is a nonzero constant. The derivative of f is given by $f'(x)=\dfrac{k-2x}{(x^2-kx)^2}$.

a. Let $k=4$, so that $f(x)=\dfrac{1}{x^2-4x}$. Determine whether f has a relative minimum, a relative maximum, or neither at $x=2$. Justify your answer.

b. Find the value of k for which f has a critical point at $x=-5$.

Solution

a. $f'(x)=\dfrac{4-2x}{(x^2-4x)^2}$ $f'(2)=\dfrac{4-2\times 2}{(2^2-4\times 2)^2}=0$

$f'(x)$ changes sign from positive to negative at $x=2$.

Therefore, f has a relative maximum at $x=2$.

b. $f'(-5)=\dfrac{k-2\times(-5)}{[(-5)^2-k\cdot(-5)]^2}=0 \Rightarrow k=-10$.

[注]确定极值点和极值的步骤:

(1)求出导数 $f'(x)$.

(2)求出 $f(x)$ 的全部 Critical Points.

(3)考察 $f'(x)$ 的符号在每个 Critical Points 的左右两侧的情况,确定该点是否为极值点.

(4)求出函数的所有极值点和极值.

5.7 Concavity and the Point of Inflection(凹凸性与拐点)

5.7.1 Definition of Concavity(函数凹凸性的定义)

5.7.1.1 Definition of Concavity(函数凹凸性的定义)

Let $f(x)$ be a differentiable function on an open interval (a,b), then

f is **concave up**(凹的) on (a,b) if $f'(x)$ is increasing on (a,b).

f is **concave down**(凸的) on (a,b) if $f'(x)$ is decreasing on (a,b).

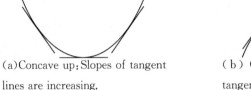
(a) Concave up: Slopes of tangent lines are increasing.

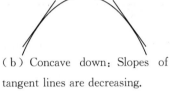
(b) Concave down: Slopes of tangent lines are decreasing.

图 5.11

5.7.1.2 Theorem: Test for Concavity(函数凹凸性的判断定理)

Assume that $f''(x)$ exists for all $x \in (a,b)$.

(1) If $f''(x) > 0$ for all $x \in (a,b)$, then f is **concave up** on (a,b).

(2) If $f''(x) < 0$ for all $x \in (a,b)$, then f is **concave down** on (a,b).

Example 5.17 判断函数的凹凸性.

Use the Concavity Test to determine the concavity of the given functions on the given intervals:

a. $y = x^2$ on $(3, 10)$; b. $y = 3 + \sin x$ on $(0, 2\pi)$.

Solution

a. Since $y'' = 2$ is always positive, the graph of $y = x^2$ is concave up on any interval. In particular, it is concave up on $(3, 10)$.

b. The graph of y is concave down on $(0, \pi)$, where $y'' = -\sin \pi$ is negative. It is concave up on $(\pi, 2\pi)$, where $y'' = -\sin x$ is positive.

Example 5.18 判断函数的凹凸性.

The derivative of a function f is given by $f'(x) = (x-3)e^x$ for $x > 0$, and $f(1) = 7$.

a. The function f has a critical point at $x = 3$. At this point, does f have a relative minimum, a relative maximum or neither? Justify your answer.

b. On what intervals, if any, is the graph of f both decreasing and concave up? Explain your reasons.

Solution

a. $f'(x) < 0$ for $0 < x < 3$ and $f'(x) > 0$ for $x > 3$.

Therefore, f has a relative minimum at $x = 3$.

b. $f''(x) = e^x + (x-3)e^x = (x-2)e^x$.

$f''(x) > 0$ for $x > 2$;

$f''(x) < 0$ for $0 < x < 2$.

Therefore, the graph of f is both decreasing and concave up on the interval $2 < x < 3$.

5.7.2　Inflection(拐点)

定义:拐点是函数凹凸区间的分界点.

[注]

(1)函数的二阶导数 $f''(x)$ 在拐点两侧异号.

(2)拐点两侧 $f'(x)$ 的单调性不同.

(3)拐点是 $f'(x)$ 的极值点.

(4)若 $x = x_0$ 为拐点,则 $f''(x_0) = 0$ or $f''(x_0)$ undefined.

Example 5.19

Find all points of inflection of the graph of $y = e^{-x^2}$.

Solution

First we find the second derivative, recalling the Chain and Product Rules:

$y = e^{-x^2}$;

$y' = e^{-x^2} \cdot (-2x)$;

$y'' = e^{-x^2} \cdot (-2x) \cdot (-2x) + e^{-x^2} \cdot (-2) = e^{-x^2}(4x^2 - 2)$.

The factor e^{-x^2} is always positive, while the factor $(4x^2 - 2)$ changes sign at $-\frac{\sqrt{2}}{2}$ and at $\frac{\sqrt{2}}{2}$. Since y'' must also change sign at these two numbers, the points of inflection are $\left(-\frac{\sqrt{2}}{2}, \frac{\sqrt{e}}{e}\right)$ and $\left(\frac{\sqrt{2}}{2}, \frac{\sqrt{e}}{e}\right)$.

5.7.3　Second Derivative Test for Local Extrema(极值的二阶导数判别法)

Suppose $f''(x)$ is continuous on an open interval that contain $x = c$.

(1) If $f'(c) = 0$ and $f''(c) < 0$, then $f(x)$ has a local maximum at $x = c$.

(2) If $f'(c) = 0$ and $f''(c) > 0$, then $f(x)$ has a local minimum at $x = c$.

(3) If $f'(c) = 0$ and $f''(c) = 0$, then the test fails. The function $f(x)$ may have a local maximum, a local minimum, or neither at $x = c$.

Example 5.20　*极值的二阶导数判别法.*

Analyze the critical points of $f(x) = (2x - x^2)e^x$.

Solution

First, solve $f'(x) = (2 - 2x)e^x + (2x - x^2)e^x = 0$.

The critical points are $c = \pm\sqrt{2}$.

Next, determine the sign of the second derivative at the critical points:

$$f''(x) = (2-x^2)e^x + e^x(-2x) = (2-2x-x^2)e^x$$
$$f''(-\sqrt{2}) = 2\sqrt{2}\,e^{-\sqrt{2}} > 0 \text{ (local min)}$$
$$f''(\sqrt{2}) = -2\sqrt{2}\,e^{\sqrt{2}} < 0 \text{ (local max)}$$

By the Second Derivative Test, $f(x)$ has a local min at $c = -\sqrt{2}$ and a local max at $c = \sqrt{2}$.

Example 5.21 二阶导数判别法失效的情形.

Analyze the critical points of $f(x) = x^5 - 5x^4$.

Solution

The first two derivatives are
$$f'(x) = 5x^4 - 20x^3 = 5x^3(x-4)$$
$$f''(x) = 20x^3 - 60x^2$$

The critical points are $c = 0, 4$, and the Second Derivative Test yields:
$$f''(0) = 0 \Rightarrow \text{Second Derivative Test fails}$$
$$f''(4) = 320 > 0 \Rightarrow f(4) \text{ is a local min}$$

The Second Derivative Test fails at $c = 0$, so we fall back on the First Derivative Test. Choosing test points to the left and right of $c = 0$, we find:

$f'(x)$ is positive on $(-\infty, 0)$

$f'(x)$ is negative on $(0, 4)$

Since $f'(x)$ changes from positive to negative at $c = 0$, $f(0)$ is a local max.

5.8 Curve Sketching(函数图形的描绘)

描绘函数图形的一般步骤:
(1)确定函数的定义域,并求函数的一阶和二阶导数.
(2)求出一阶、二阶导数为零的点和一阶、二阶导数不存在的点.
(3)列表分析,确定曲线的单调性和凹凸性.
(4)确定曲线的渐近性.
(5)确定并描出曲线上的极值点、拐点、与坐标轴的交点及其他关键点.
(6)用光滑的曲线连接这些点,画出函数的图形.

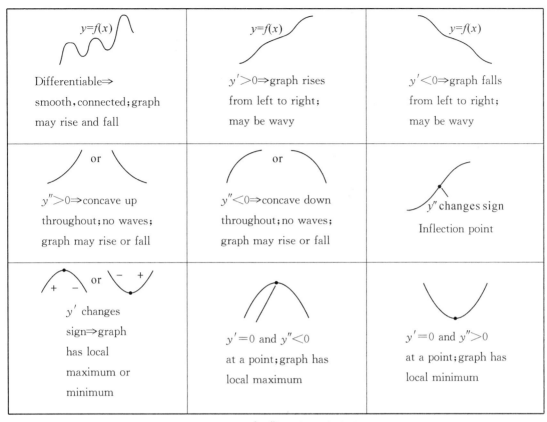

图 5.12　y,y',y'' 与函数图像的关系

Example 5.22

画出函数 $y=x^3-x^2-x+1$ 的图像.

Solution

(1)函数的定义域为 $(-\infty,+\infty)$.

(2)$f'(x)=3x^2-2x-1=(3x+1)(x-1)$, $f''(x)=6x-2=2(3x-1)$.

$f'(x)=0$ 的根为 $x=-1/3,1$; $f''(x)=0$ 的根为 $x=1/3$.

(3)列表分析:

x	$(-\infty,-1/3)$	$-1/3$	$(-1/3,1/3)$	$1/3$	$(1/3,1)$	1	$(1,+\infty)$
$f'(x)$	+	0	−	−	−	0	+
$f''(x)$	−	−	−	0	+	+	+
$f(x)$	⌢↗	极大值	⌢↘	拐点	⌣↘	极小值	⌣↗

(4)当 $x\to+\infty$ 时,$y\to+\infty$;当 $x\to-\infty$ 时,$y\to-\infty$.

(5)计算特殊点的函数值:$f(-1/3)=32/27,f(1/3)=16/27,f(1)=0,f(0)=1$; $f(-1)=0,f(3/2)=5/8$.

(6)描点连线,画出图像.

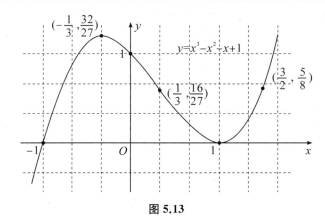

图 5.13

Example 5.23

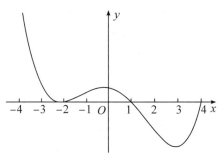

图 5.14 Graph of $f'(x)$

The graph of the derivative of a function f on the interval $[-4,4]$ is shown in 图 5.14. Answer the following questions about f, justifying each answer with information obtained from the graph of f'.

a. On what intervals is f increasing?

b. On what intervals is the graph of f concave up?

c. At which x-coordinates does f have local extrema?

d. What are the x-coordinates of all inflection points of the graph of f?

e. Sketch a possible graph of f on the interval.

Solution

Often, making a chart showing where f' is positive and negative and where f' is increasing and decreasing helps to understand the behavior of the function f (whose derivative is f').

Intervals	$-4 < x < -2$	$-2 < x \leq 0$	$0 < x \leq 1$	$1 < x \leq 3$	$3 < x \leq 4$
Sign of f'	positive	positive	positive	negative	negative
Graph of f'	decreasing	increasing	decreasing	decreasing	increasing

a. Since $f' > 0$ on the intervals $[-4, 2]$ and $(-2, 1)$, the function f must be increasing on the entire interval $[-4, 1]$ with a horizontal tangent at $x = -2$ (a "shelf point").

b. The graph of f is concave up on the intervals where f' is increasing. We see from the graph that f' is increasing on the intervals $(-2,0)$ and $(3,4)$.

c. By the First Derivative Test, there is a local maximum at $x=1$ because the sign of f' changes from positive to negative there. Note that there is no extremum at $x=-2$, since f' does not change sign. Because the function increases from the left endpoint and decreases to the right endpoint, there are local minima at the endpoints $x=-4$ and $x=4$.

d. The inflection points of the graph of f have the same x-coordinates as the turning points of the graph of f', namely $-2, 0,$ and 3.

e. A possible graph satisfying all the conditions is shown in figure 5.15.

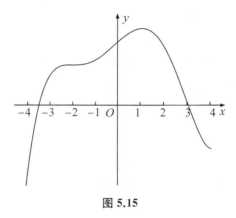

图 5.15

5.9 Absolute Minimum Value and Absolute Maximum Value (最大值与最小值)

5.9.1 极值与最值的关系

设函数 $f(x)$ 在闭区间 $[a,b]$ 上连续，则函数的最大值和最小值一定存在. 函数的最大值和最小值有可能在区间的端点取得，也可能出现在开区间 (a,b) 内. 在这种情况下，函数在闭区间 $[a,b]$ 上的最大值一定是函数的所有极大值和区间端点的函数值中的最大者. 同理，函数在闭区间 $[a,b]$ 上的最小值一定是函数的所有极小值和区间端点的函数值中的最小者.

5.9.2 求最值的步骤

(1) 求函数在定义域中的所有 critical points ($f'(x)=0$ or $f'(x)=0$ undefined).

(2) 比较 critical points 和区间端点的函数值，最大的为最大值，最小的为最小值.

Example 5.24 求最值.

Find the absolute minimum value of $h(x)=e^x-\ln x$ on the closed interval $\frac{1}{2} \leqslant x \leqslant 1$, and find the absolute maximum value of $h(x)$ on the closed interval $\frac{1}{2} \leqslant x \leqslant 1$. Show the analysis that leads to your answers.

Solution

$$h'(x) = e^x - \frac{1}{x} = 0 \Rightarrow x = 0.567143.$$

Absolute minimum value and absolute maximum value occur at the critical point or at the endpoints.

$h(0.567143) = 2.330$,

$h(0.5) = 2.3418$,

$h(1) = 2.718$,

The absolute minimum is 2.330.

The absolute maximum is 2.718.

5.9.3 Applied Optimization(最优化问题)

在工农业生产、工程技术和科学实验中,常常会遇到这样一类问题:在一定条件下,怎样使"产品最多""用料最省""成本最低""效率最高"等,这类问题在数学上可归结为求目标函数的最值问题,也称最优化问题.

Example 5.25 用料最省.

Design a cylindrical can of volume 900 cm³ so that it uses the least amount of metal. In other words, minimize the surface area of the can (including its top and bottom).

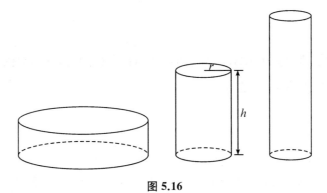

图 5.16

Solution

Let r be the radius and h the height. Let A be surface area of the can.

$$A = 2\pi r^2 + 2\pi rh = 2\pi r^2 + 2\pi r \cdot \left(\frac{V}{\pi r^2}\right)$$

$$\frac{dA}{dr} = 4\pi r - \frac{2V}{r^2} = \frac{4\pi r^3 - 2V}{r^2} = 0 \Rightarrow 4\pi r^3 = 2V = 2\pi r^2 h \Rightarrow h = 2r$$

Since $V = \pi r^2 h = 900$, so $\pi r^2 \cdot 2r = 900 \Rightarrow r = \left(\frac{450}{\pi}\right)^{1/3} \approx 5.23$ cm

$h = 2r \approx 10.46$ cm.

Notice that the optimal dimension satisfy $h = 2r$. In other words, the optimal can is as tall as its wide.

This function has no maximum value because it tends to infinity as $r \to \infty$.

Example 5.26 用时最短.

Your task is to build a road joining a ranch to a highway that enables drivers to reach the city in the shortest time. How should this be done if the speed limit is 60 km/h on the road and 110 km/h on the highway? The perpendicular distance from the ranch to the highway is 30 km, and the city is 50 km down the high way.

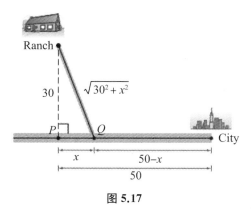

图 5.17

Solution

Let x be the distance from Q to the point P where the perpendicular joins the highway. Let $T(x)$ be the total number of hours for the trip.

$$T(x) = \frac{\sqrt{30^2 + x^2}}{60} + \frac{50 - x}{110}, x \in [0, 50]$$

$$T'(x) = \frac{x}{60\sqrt{30^2 + x^2}} - \frac{1}{110} = 0$$

$$\Rightarrow 110x = 60\sqrt{30^2 + x^2}$$
$$\Rightarrow 121x^2 = 36(30^2 + x^2)$$
$$\Rightarrow 85x^2 = 32400$$
$$\Rightarrow x = \sqrt{32400/85} \approx 19.52$$

To find the minimum value of $T(x)$, we compare the values of $T(x)$ at the critical point and the endpoints of $[0, 50]$:

$T(0) \approx 0.95$ h $\qquad T(19.52) \approx 0.87$ h $\qquad T(50) \approx 0.97$ h

We conclude that the travel time is minimized if the road joins the highway at a distance $x \approx 19.52$ km along the highway from P.

Example 5.27 最优定价.

All units in a 30-unit apartment building are rented out when the monthly rent is set at $r = \$1000$/month. A survey reveals that one unit becomes vacant with each $\$40$ increase in rent. Suppose that each occupied unit costs $\$120$/month in maintenance. Which rent r maximizes monthly profit?

Solution

Let $P(r)$ be the total monthly profit, $N(r)$ be the number of units occupied.

$$N(r) = 0 \text{ for } r = 40 \times 55 = 2200$$

$$N(r) = 30 - \frac{1}{40}(r - 1000) = 55 - \frac{1}{40}r, 1000 \leq r \leq 2200$$

$$P(r) = N(r)(r - 120) = \left(55 - \frac{1}{40}r\right)(r - 120) = -6600 + 58r - \frac{1}{40}r^2, 1000 \leq r \leq 2200$$

$$P'(r) = 0 \Rightarrow r = 1160$$

Compare values at the critical point and the endpoints:

$P(1000) = 26400$ $\qquad$ $P(1160) = 27040$ $\qquad$ $P(22000) = 0$

We conclude that the profit is maximized when the rent is set at $r = \$1160$. In this case, four units are left vacant.

5.10 Motion Problems(运动问题)

5.10.1 Move along a Line(直线运动)

令 $s = s(t)$ 表示质点运动的 **position function**(位置函数),则:

(1) displacement 位移:

从时刻 t_0 到时刻 t_1 的位移为:

$$\text{Displacement} = s(t_1) - s(t_0)$$

若位移大于零,表示 t_1 时刻质点的位置在 t_0 时刻的右边;反之,若位移小于零,表示 t_1 时刻质点的位置在 t_0 时刻的左边.

(2) velocity 速度 $v(t)$:

$$v(t) = \frac{ds}{dt} = \begin{cases} v > 0, \text{move to the right} \\ v < 0, \text{move to the left} \\ v = 0, \text{rest} \end{cases}$$

turning points 转向点: $v = 0$ 且在其两侧 v 异号时,此刻质点改变运动方向.

(3) acceleration 加速度 $a(t)$:

$$a(t) = \frac{dv}{dt} = \frac{d^2s}{dt^2} = \begin{cases} a > 0 \Rightarrow v \text{ is increasing} \\ a < 0 \Rightarrow v \text{ is decreasing} \end{cases}$$

(4) speed 速率 $|v(t)|$:

$$\text{speed} = |v(t)| \begin{cases} a \text{ 与 } v \text{ 同号} \Rightarrow \text{speed is increasing} \\ a \text{ 与 } v \text{ 异号} \Rightarrow \text{speed is decreasing} \end{cases}$$

Example 5.28

A particle moves along a line so that its position at any time $t \geq 0$ is given by the function, $s(t) = t^2 - 4t + 3$, where s is measured in meters and t is measured in seconds.

a. Find the displacement of the particle during the first 2 seconds.

b. Find the average velocity of the particle during the first 4 seconds.

c. Find the instantaneous velocity of the particle when $t=4$.

d. Find the acceleration of the particle when $t=4$.

e. At what values of t does the particle change directions?

Solution

a. The displacement is given by $s(2)-s(0)=(-1)-3=-4$. This value means that the particle is 4 units left of where it started.

b. The average velocity we seek is
$$\frac{s(4)-s(0)}{4-0}=\frac{3-3}{4}=0 \text{ m/sec.}$$

c. The velocity $v(t)$ at any time t is $v(t)=\dfrac{ds}{dt}=2t-4$. So $v(4)=4$ m/sec.

d. The acceleration $a(t)$ at any time t is $a(t)=\dfrac{dv}{dt}=2$ m/sec^2. So $a(4)=2$.

e. At $t=2, v(t)=0$, so the particle is at rest.

For $t>2, v(t)>0$, so the particle is moving to the right.

For $t<2, v(t)<0$, so the particle is moving to the left.

The particle changes direction at $t=2$ when $v=0$.

5.10.2　Move in a Plane(平面运动)

质点在平面运动时,规定逆时针方向(counter clockwise)为正方向.

5.10.2.1　Position Vector(位置向量)

$$\overrightarrow{R(t)}=\langle x(t),y(t)\rangle$$

5.10.2.2　Velocity Vector(速度向量)

$$\overrightarrow{v(t)}=\langle \frac{dx}{dt},\frac{dy}{dt}\rangle$$

The slope of $\overrightarrow{v(t)}=\dfrac{\frac{dy}{dt}}{\frac{dx}{dt}}$

$\dfrac{dx}{dt}>0$ 表示质点向右运动,$\dfrac{dx}{dt}<0$ 表示质点向左运动.

$\dfrac{dy}{dt}>0$ 表示质点向上运动,$\dfrac{dy}{dt}<0$ 表示质点向下运动.

5.10.2.3　Speed(速率)

速率是 magnitude of $\overrightarrow{v(t)}$(速度的模),是一个标量(只有大小,没有方向的量).

$$\text{speed}=|\overrightarrow{v(t)}|=\sqrt{\left(\frac{dx}{dt}\right)^2+\left(\frac{dy}{dt}\right)^2}$$

5.10.2.4 Acceleration Vector(加速度向量)

$$\vec{a(t)} = \langle \frac{d^2 x}{dt^2}, \frac{d^2 y}{dt^2} \rangle$$

magnitude of $\vec{a(t)}$: $|\vec{a(t)}| = \sqrt{\left(\frac{d^2 x}{dt^2}\right)^2 + \left(\frac{d^2 y}{dt^2}\right)^2}$

The slope of $\vec{a(t)} = \frac{d^2 y}{dt^2} / \frac{d^2 x}{dt^2}$

Example 5.29 参数方程的运动问题.

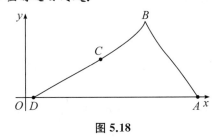

图 5.18

A particle starts at point A on the positive x-axis at time $t=0$ and travels along the curve from A to B to C to D, as shown in figure 5.18. The coordinates of the particle's position $(x(t), y(t))$ are differentiable functions of t, where $x'(t) = \frac{dx}{dt} = -9\cos\left(\frac{\pi t}{6}\right) \cdot \sin\left(\frac{\pi\sqrt{t+1}}{2}\right)$ and $y'(t) = \frac{dy}{dt}$ is not explicitly given. At time $t=9$, the particle reaches its final position at point D on the positive x-axis.

a. At point C, is $\frac{dy}{dt}$ positive? At point C, is $\frac{dx}{dt}$ positive? Give a reason for each answer.

b. The slope of the curve is undefined at point B. At what time t is particle at point B?

c. The line tangent to the curve at the point $(x(8), y(8))$ has equation $y = \frac{5}{9}x - 2$. Find the velocity vector and the speed of the particle at this point.

Solution

a. At point C, $\frac{dy}{dt}$ is not positive because $y(t)$ is decreasing along the arc BD as t increases.

At point C, $\frac{dx}{dt}$ is not positive because $x(t)$ is decreasing along the arc BD as t increases.

b. $\frac{dx}{dt} = 0$; $\cos\left(\frac{\pi t}{6}\right) = 0$ or $\sin\left(\frac{\pi\sqrt{t+1}}{2}\right) = 0$

$\frac{\pi t}{6} = \frac{\pi}{2}$ or $\frac{\pi\sqrt{t+1}}{2} = \pi$; $t=3$ for both.

Particle is at point B at $t=3$.

c. $x'(8) = -9\cos\left(\dfrac{4\pi}{3}\right)\sin\left(\dfrac{3\pi}{2}\right) = -\dfrac{9}{2}$

$\dfrac{y'(8)}{x'(8)} = \dfrac{\mathrm{d}y}{\mathrm{d}x} = \dfrac{5}{9}$

$y'(8) = \dfrac{5}{9}x'(8) = -\dfrac{5}{2}$

The velocity vector is $\langle -4.5, -2.5 \rangle$.

speed $= \sqrt{4.5^2 + 2.5^2} = 5.147$, or 5.148.

Example 5.30 极坐标方程的运动问题.

A particle moving with nonzero velocity along the polar curve given by $r = 3 + 2\cos\theta$ has position $(x(t), y(t))$ at time t, with $\theta = 0$ when $t = 0$. This particle moves along the curve so that $\dfrac{\mathrm{d}r}{\mathrm{d}t} = \dfrac{\mathrm{d}r}{\mathrm{d}\theta}$.

a. Find the value of $\dfrac{\mathrm{d}r}{\mathrm{d}t}$ at $\theta = \dfrac{\pi}{3}$ and interpret your answer in terms of the motion of the particle.

b. For the particle described in part a, $\dfrac{\mathrm{d}y}{\mathrm{d}t} = \dfrac{\mathrm{d}y}{\mathrm{d}\theta}$. Find the value of $\dfrac{\mathrm{d}y}{\mathrm{d}t}$ at $\theta = \dfrac{\pi}{3}$ and interpret your answer in terms of the motion of the particle.

Solution

a. $\left.\dfrac{\mathrm{d}r}{\mathrm{d}t}\right|_{\theta=\pi/3} = \left.\dfrac{\mathrm{d}r}{\mathrm{d}\theta}\right|_{\theta=\pi/3} = -1.732$

The particle is moving closer to the origin, since $\dfrac{\mathrm{d}r}{\mathrm{d}t} < 0$ and $r > 0$ when $\theta = \dfrac{\pi}{3}$.

b. $y = r\sin\theta = (3 + 2\cos\theta)\sin\theta$

$\left.\dfrac{\mathrm{d}y}{\mathrm{d}t}\right|_{\theta=\pi/3} = \left.\dfrac{\mathrm{d}y}{\mathrm{d}\theta}\right|_{\theta=\pi/3} = 0.5$

The particle is moving away from the x-axis, since $\dfrac{\mathrm{d}y}{\mathrm{d}t} > 0$ and $y > 0$ when $\theta = \dfrac{\pi}{3}$.

Practice Exercises(习题)

1. For what non-negative value of b is the line given by $y = -\dfrac{1}{3}x + b$ normal to the curve $y = x^3$?

   (A) 0 (B) 1 (C) $\dfrac{4}{3}$ (D) $\dfrac{10}{3}$

2. An equation for a tangent to the graph of $y = \arcsin\dfrac{x}{2}$ at the origin is

(A)$x-2y=0$　　(B)$x-y=0$　　(C)$x=0$　　(D)$y=0$

3. A curve in the plane is defined parametrically by the equations $x=t^3+t$ and $y=t^4+2t^2$. An equation of the line tangent to the curve at $t=1$ is

   (A)$y=2x$　　(B)$y=4x-5$　　(C)$y=8x+12$　　(D)$y=2x-1$

4. An equation of the line normal to the graph of $y=x^3+3x^2+7x-1$ at the point where $x=-1$ is

   (A)$4x+y=0$　　(B)$x+4y=-25$　　(C)$4x-y=2$　　(D)$x-4y=23$

5. Which of the following is an equation of the line tangent to the graph of $f(x)=x^4+2x^2$ at the point where $f'(x)=1$?

   (A)$y=8x-5$　　(B)$y=x+7$　　(C)$y=x+0.763$　　(D)$y=x-0.122$

6.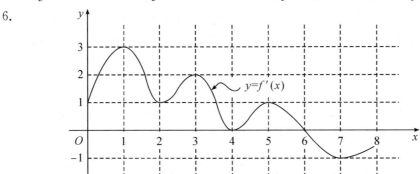

The function f is defined on the closed interval $[0,8]$. The graph of its derivative f' is shown above. The point $(3,5)$ is on the graph of $y=f(x)$. An equation of the line tangent to the graph of f at $(3,5)$ is

   (A)$y=2$　　　　　　　　　(B)$y=5$
   (C)$y-5=2(x-3)$　　　　　(D)$y+5=2(x+3)$

7. The Mean Value Theorem guarantees the existence of a special point on the graph of $y=\sqrt{x}$ between $(0,0)$ and $(4,2)$. What are the coordinate of this point?

   (A)$(2,1)$　　(B)$(2,\sqrt{2})$　　(C)$(1,1)$　　(D)$\left(\frac{1}{2},\frac{1}{\sqrt{2}}\right)$

8. Let f be the function given by $f(x)=x^3-3x^2$. What are all values of c that satisfy the conclusion of the Mean Value Theorem of differential calculus on the closed interval $[0,3]$?

   (A)0 only　　(B)2 only　　(C)0 and 3　　(D)2 and 3

9. If c is the number that satisfies the conclusion of the Mean Value Theorem for $f(x)=x^3-2x^2$ on the interval $[0,2]$, then $c=$

   (A)0　　(B)$\frac{1}{2}$　　(C)1　　(D)$\frac{4}{3}$

10. The volume of a cone of radius r and height h is given by $V=\frac{\pi}{3}r^2h$. If the radius and the height both increase at a constant rate of $\frac{1}{2}$ centimeter per second, at what

rate,in cubic centimeters per second,is the volume increasing when the height is 9 centimeters and the radius is 6 centimeters?

(A) $\dfrac{1}{2}\pi$ (B) 10π (C) 24π (D) 54π

11. The area of a circular region is increasing at a rate of 96π square meter per second. When the area of the region is 64π square meter, how fast, in meters per second, is the radius of the region increasing?

(A) 6   (B) 8   (C) 16   (D) $4\sqrt{3}$

12. A person 2 meters tall walks directly away from a streetlight that is 8 meters above the ground. If the person is walking at a constant rate and the person's shadow is lengthening at the rate of $\dfrac{4}{9}$ meters per second, at what rate, in meters per second, is the person walking?

(A) $\dfrac{4}{27}$ (B) $\dfrac{4}{9}$ (C) $\dfrac{3}{4}$ (D) $\dfrac{4}{3}$

13. If $f(x)=x+\dfrac{1}{x}$, then the set of values for which f increases is

(A) $(-\infty,-1]\cup[1,\infty)$ (B) $[-1,1]$
(C) $(0,\infty)$ (D) $(-\infty,\infty)$

14. Which of the following is true about the graph of $y=\ln|x^2-1|$ in the interval $(-1,1)$?

(A) It is increasing.
(B) It attains a relative minimum at $(0,0)$.
(C) It is concave down.
(D) It has an asymptote.

15. If $y=x^2 e^x$, then the graph of f is decreasing for all x such that
(A) $x<-2$ (B) $-2<x<0$ (C) $x>-2$ (D) $x<0$

16.

The graph of $y=h(x)$ is shown above. Which of the following could be the graph of $y'=h(x)$?

(A)   (B)   (C)   (D)

17.

The graph of f', the derivative of the function f, is shown above. Which of the following statements must be true?

I. f has a relative minimum at $x=-3$.

II. The graph of f has a point of inflection at $x=-2$.

III. The graph of f is concave down for $0<x<4$.

(A) I only   (B) II only   (C) I and II   (D) I and III only

18.

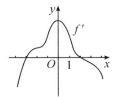

The graph of $y=f(x)$ on the closed interval $[2,7]$ is shown above. How many points of inflection does this graph have on this interval?

(A) one   (B) two   (C) three   (D) four

19.

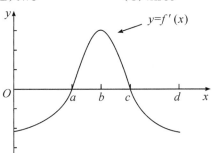

The graph of f', the derivative of a function f, is shown above. The domain of f is the open interval $0<x<d$. Which of the following statements is true?

(A) The graph of f has a point of inflection at $(b,f(b))$.

(B) f has a local minimum at $x=c$.

(C) f has a local maxmum at $x=b$.

(D) The graph of f is concave up on the open interval (c,d).

20. What is the area of the largest rectangle that can be inscribed in the ellipse $4x^2+9y^2=36$?

(A) $6\sqrt{2}$ (B) 12   (C) 24   (D) $24\sqrt{2}$

21. Consider all right circular cylinders for which the sum of the height and circumference is 30 centimeters. What is the radius of the one with maximum volume?

(A) 3 cm   (B) 10 cm   (C) 20 cm   (D) $\dfrac{10}{\pi}$ cm

22. A particle moves along the x-axis so that at any time t its position is given by $x(t)=te^{-2t}$. For what values of t is the particle at rest?

(A) 0 only (B) $\frac{1}{2}$ only (C) 1 only (D) 0 and $\frac{1}{2}$

23. A particle moves on the x-axis with velocity given by $v(t)=3t^4-11t^2+9t-2$ for $-3\leqslant x\leqslant 3$. How many times does the particle change directions as t increases from -3 to 3?

(A) zero (B) one (C) two (D) three

24.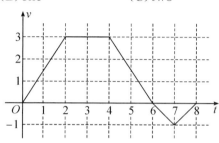

A bug begins to crawl up a vertical wire at time $t=0$. The velocity v of the bug at time t, $0\leqslant t\leqslant 8$, is given by the function whose graph is shown above. at what value of t does the bug change directions?

(A) 2 (B) 4 (C) 6 (D) 8

25. A particle moves on a plane curve so that at any time $t>0$ its x-coordinate is t^3-t and its y-coordinate is $(2t-1)^3$. The acceleration vector of the particle at $t=1$ is

(A) $\langle 2,3 \rangle$ (B) $\langle 2,6 \rangle$ (C) $\langle 6,12 \rangle$ (D) $\langle 6,24 \rangle$

26. The position of a particle moving in the xy-plane is given by the parametric equation $x(t)=t^3-3t^2$ and $y(t)=12t-3t^2$. At which of the following points (x, y) is the particle at rest?

(A) $(-4,12)$ (B) $(-3,6)$ (C) $(-2,9)$ (D) $(3,4)$

27. For time $t>0$, the position of a particle moving in the xy-plane is given by the vector $\left(\frac{1}{t}, e^{3t}\right)$. What is the velocity vector of the particle at time $t=2$?

(A) $\langle \frac{1}{4}, e^6 \rangle$ (B) $\langle \frac{1}{4}, 3e^6 \rangle$ (C) $\langle -\frac{1}{4}, e^6 \rangle$ (D) $\langle -\frac{1}{4}, 3e^6 \rangle$

习题参考答案

1.C 2.A 3.D 4.B 5.D 6.C 7.C 8.B 9.D 10.C 11.A 12.D 13.A 14.C
15.B 16.D 17.D 18.C 19.A 20.B 21.D 22.C 23.C 24.C 25.D 26.A 27.D

Chapter 6 Differential and Approximation (微分与近似计算)

6.1 Differentials(微分)

6.1.1 Definition of Differential(微分的定义)

Let $f(x)$ be a differentiable function, the differential dy is
$$dy = f'(x)dx.$$
$$\frac{dy}{dx} = f'(x).$$

通常把自变量 x 的增量 Δx 称为自变量的微分,记作 dx,即 $dx = \Delta x$.

函数的微分 dy 与自变量的微分 dx 之商等于该函数的导数,因此导数也叫作"微商".

Example 6.1 求函数的微分.

Finding the Differential dy.

a. Find dy if $y = x^5 + 37x$.

b. Find the value of dy when $x = 1$ and $dx = 0.2$.

Solution

a. $dy = (x^5 + 37x)'dx = (5x^4 + 37)dx$.

b. Substituting $x = 1$ and $dx = 0.2$ in the expression for dy, we have
$dy = (5 \times 1^4 + 37) \times 0.2 = 8.4$.

6.1.2 基本初等函数的微分公式

导数公式: 微分公式:

$(x^\mu)' = \mu\, x^{\mu-1}$ $d(x^\mu) = \mu\, x^{\mu-1}dx$

$(\sin x)' = \cos x$ $d(\sin x) = \cos x\, dx$

$(\cos x)' = -\sin x$ $d(\cos x) = -\sin x\, dx$

$(\tan x)' = \sec^2 x$ $d(\tan x) = \sec^2 x\, dx$

$(\cot x)' = -\csc^2 x$ $d(\cot x) = -\csc^2 x\, dx$

$(\sec x)' = \sec x \tan x$ $d(\sec x) = \sec x \tan x\, dx$

$(\csc x)' = -\csc x \cot x$　　　　　　　$d(\csc x) = -\csc x \cot x \, dx$

$(a^x)' = a^x \ln a$　　　　　　　　　　$d(a^x) = a^x \ln a \, dx$

$(e^x)' = e^x$　　　　　　　　　　　　$d(e^x) = e^x \, dx$

$(\log_a x)' = \dfrac{1}{x \ln a}$　　　　　　　　$d(\log_a x) = \dfrac{1}{x \ln a} dx$

$(\ln x)' = \dfrac{1}{x}$　　　　　　　　　$d(\ln x) = \dfrac{1}{x} dx$

$(\arcsin x)' = \dfrac{1}{\sqrt{1-x^2}}$　　　　　$d(\arcsin x) = \dfrac{1}{\sqrt{1-x^2}} dx$

$(\arccos x)' = -\dfrac{1}{\sqrt{1-x^2}}$　　　　$d(\arccos x) = -\dfrac{1}{\sqrt{1-x^2}} dx$

$(\arctan x)' = \dfrac{1}{1+x^2}$　　　　　　$d(\arctan x) = \dfrac{1}{1+x^2} dx$

$(\text{arccot } x)' = -\dfrac{1}{1+x^2}$　　　　　$d(\text{arccot } x) = -\dfrac{1}{1+x^2} dx$

6.1.3 函数的和、差、积、商的微分法则

求导法则:　　　　　　　　　　　　微分法则:

$(u \pm v)' = u' \pm v'$　　　　　　　　　$d(u \pm v) = du \pm dv$

$(Cu)' = Cu'$　　　　　　　　　　　$d(Cu) = C du$

$(u \cdot v)' = u'v + uv'$　　　　　　　　$d(u \cdot v) = v du + u dv$

$\left(\dfrac{u}{v}\right)' = \dfrac{u'v - uv'}{v^2} (v \ne 0)$　　　　　$d\left(\dfrac{u}{v}\right) = \dfrac{v du - u dv}{v^2} dx (v \ne 0)$

Example 6.2 求复合函数的微分.

Finding Differentials of Functions.

a. $y = \tan 2x$;

b. $y = \dfrac{x}{x+1}$;

c. $y = e^{1-3x} \cos x$.

Solution

a. $d(\tan 2x) = \sec^2(2x) d(2x) = 2\sec^2(2x) dx$.

b. $d\left(\dfrac{x}{x+1}\right) = \dfrac{(x+1)dx - x d(x+1)}{(x+1)^2} = \dfrac{x dx + dx - x dx}{(x+1)^2} = \dfrac{dx}{(x+1)^2}$.

c. $dy = d(e^{1-3x}) \cdot \cos x + e^{1-3x} d(\cos x)$

　　$= \cos x \, e^{1-3x} d(1-3x) + e^{1-3x}(-\sin x) dx$

　　$= -e^{1-3x}(3\cos x + \sin x) dx$.

6.1.4 微分的几何意义

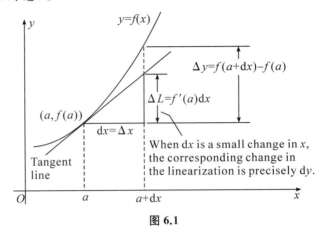

图 6.1

Geometrically, the differential dy is the change ΔL in the linearization of $f(x)$ when changes by an amount d$x=\Delta x$.

微分的几何意义:如图 6.1,当 Δy 是曲线 $y=f(x)$ 上某点的纵坐标的增量时,dy 就是曲线的切线上点的纵坐标的相应增量 ΔL.

当 $|\Delta x|$ 很小时,$|\Delta y-\mathrm{d}y|=|\Delta y-\Delta L|$ 比 $|\Delta x|$ 小得多,此时 $\Delta y\approx \mathrm{d}y$.同时,在点$(a,f(a))$附近,我们可以用切线段上的函数值来近似代替曲线段上对应的函数值.

6.2 Approximating a Derivative Value(导数的近似计算)

当 $\Delta x \to 0$ 时,$\Delta y \approx \mathrm{d}y$

从而: $\Delta x \to 0$ $\qquad \dfrac{\mathrm{d}y}{\mathrm{d}x} \approx \dfrac{\Delta y}{\Delta x}$ $\hfill (6-1)$

公式(6-1)是导数近似计算的基本公式,由公式(6-1)根据不同的已知条件可以得到求点导数 $f'(a)$ 的差商法和对称差商法公式.

6.2.1 Difference Quotient(差商法)

$$f'(a)\approx \dfrac{f(a+\Delta x)-f(a)}{\Delta x}(当\ \Delta x \to 0\ 时) \qquad (6-2)$$

公式(6-2)适用于 $f(a)$ 为已知的情形,$f'(a)$ 的近似值是 $x=a$ 与其最近的一点的函数值增量和自变量增量之商.

6.2.2 Symmetric Difference Quotient(对称差商法)

$$f'(a)\approx \dfrac{f(a+\Delta x)-f(a-\Delta x)}{2\Delta x}(当\ \Delta x \to 0\ 时) \qquad (6-3)$$

公式(6-3)是 $x=a-\Delta x$ 与 $x=a+\Delta x$ 这两个关于 $x=a$ 的对称点的函数值增量除以自变量增量,所以公式(6-3)称为对称差商法.

特别注意:若 $f(x)$ 在 $x=a$ 处不可导,则用差商法和对称差商法近似计算的结果就会产生错误,同学们可以试着用差商法和对称差商法近似计算 $y=|x|$ 在 $x=0$ 处的导数.

Example 6.3

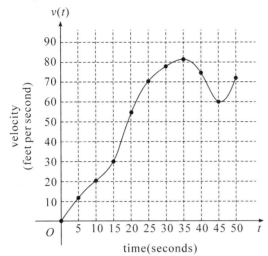

t (seconds)	$v(t)$ (feet per second)
0	0
5	12
10	20
15	30
20	55
25	70
30	78
35	81
40	75
45	60
50	72

The graph of the velocity $v(t)$, in ft/sec, of a car traveling on a straight road, for $0 \leqslant t \leqslant 50$, is shown above. A table of values for $v(t)$, at 5 second intervals of time t, is shown to the right of the graph.

a. During what intervals of time is the acceleration of the car positive? Give a reason for your answer.

b. Find the average acceleration of the car, in ft/sec², over the interval $0 \leqslant t \leqslant 50$.

c. Find one approximation for the acceleration of the car, in ft/sec², at $t=40$. Show the computations you used to arrive at your answer.

Solution

a. Acceleration is positive on $(0, 35)$ and $(45, 50)$ because the velocity $v(t)$ is increasing on $[0, 35]$ and $[45, 50]$.

b. Avg. Acc. $= \dfrac{v(50)-v(0)}{50-0} = \dfrac{72-0}{50} = \dfrac{72}{50}$ or 1.44 ft/sec².

c. Difference quotient; e.g.

$\dfrac{v(45)-v(40)}{45-40} = \dfrac{60-75}{5} = -3$ ft/sec²   or

$\dfrac{v(40)-v(35)}{40-35} = \dfrac{75-81}{5} = -\dfrac{6}{5}$ ft/sec²   or

$\dfrac{v(45)-v(35)}{45-35} = \dfrac{60-81}{10} = -\dfrac{21}{10}$ ft/sec²   or

slope of tangent line, e.g. through $(35, 90)$ and $(40, 75)$: $\dfrac{81-75}{35-40} = -3$ ft/sec².

6.3 Local Linear Approximation(局部线性近似)

6.3.1 局部线性近似公式

如果函数 $y=f(x)$ 在点 a 处的导数 $f'(a)\neq 0$，且 $|\Delta x|$ 很小时，我们有
$$\Delta y \approx dy$$
$$\Delta y = f(a+\Delta x) - f(a)$$
$$dy = f'(a)dx,$$
所以
$$f(a+\Delta x) \approx f(a) + f'(a)\Delta x.$$

If f is differentiable at $x=a$ and x is close to a, then
$f(x) \approx f(a) + f'(a)(x-a)$ —the local linear approximation of $f(x)$ at a

若 $x \approx a$，则 $f(x)$ 在 $x=a$ 处的线性近似为：
$$f(x) \approx f(a) + f'(a)(x-a)$$

6.3.2 Errors(误差)

$$\text{Error} = |\text{actual value} - \text{approximate value}|$$
$$\text{Percentage Error} = \left|\frac{\text{error}}{\text{actual value}}\right| \times 100\%$$

We see that the y value on the tangent line is an approximation for the actual or true value of f. Local linear approximation is therefore also called **tangent-line approximation**.

$f(x)$ is concave down at $x=a \Rightarrow$ approximates value $>$ actual value(over estimate).

$f(x)$ is concave up at $x=a \Rightarrow$ approximates value $<$ actual value(under estimate).

如图 6.2(a)，$f(x)$ 是 concave down 的，所以切线位于曲线 $y=f(x)$ 上方，曲线上 $y=f(x)$ 的值为真实值，切线上相同自变量对应的函数值为近似值，由于切线在上方，所以近似值大于真实值，此时线性近似会高估。在图 6.2(b) 中，$f(x)$ 是 concave up 的，所以切线位于曲线 $y=f(x)$ 下方，此时切线近似会低估真实值。

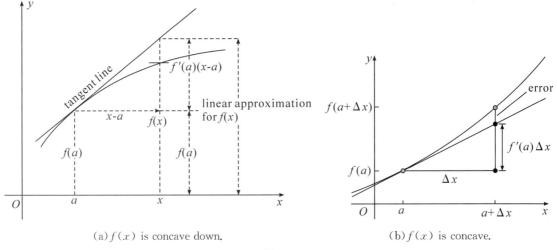

(a) $f(x)$ is concave down. (b) $f(x)$ is concave.

图 6.2

Example 6.4 线性近似.

Estimate $\tan\left(\dfrac{\pi}{4}+0.02\right)$ and compute the percentage error.

Solution

考虑 $f(x)=\tan x$,用过 $x=\dfrac{\pi}{4}$ 处的切线来近似计算.

因为 $f\left(\dfrac{\pi}{4}\right)=\tan\left(\dfrac{\pi}{4}\right)=1, f'\left(\dfrac{\pi}{4}\right)=\sec^2\left(\dfrac{\pi}{4}\right)=(\sqrt{2})^2=2$

所以切线近似公式为:
$$f(x)\approx f\left(\dfrac{\pi}{4}\right)+f'\left(\dfrac{\pi}{4}\right)\cdot\left(x-\dfrac{\pi}{4}\right)=1+2\cdot\left(x-\dfrac{\pi}{4}\right)$$

At $x=\dfrac{\pi}{4}+0.02$, the linearization yields the estimate:

$$\tan\left(\dfrac{\pi}{4}+0.02\right)\approx 1+2\times(0.02)=1.04$$

A calculator gives $\tan\left(\dfrac{\pi}{4}+0.02\right)\approx 1.0408$, so

Percentage Error $\approx \left|\dfrac{1.0408-1.04}{1.0408}\right|\times 100\approx 0.08\%$

Example 6.5

x	-1.5	-1.0	-0.5	0	0.5	1.0	1.5
$f(x)$	-1	-4	-6	-7	-6	-4	-1
$f'(x)$	-7	-5	-3	0	3	5	7

Let f be a function that is differentiable for all real numbers. The table above gives the values of f and its derivative f' for selected points x in the closed interval $-1.5\leqslant x\leqslant 1.5$. The second derivative of f has the property that $f''(x)>0$ for $-1.5\leqslant x\leqslant 1.5$.

Write an equation of the line tangent to the graph of f at the point where $x=1$. Use this line to approximate the value of $f(1.2)$. Is this approximation greater than or less than the actual value of $f(1.2)$? Give a reason for your answer.

Solution

$y=5(x-1)-4$

$f(1.2)\approx 5\times 0.2-4=-3$

The approximation is less than $f(1.2)$ because the graph of f is concave up on the interval $1<x<1.2$.

6.4 Newton's Method(牛顿法)

二次方程、三次方程、四次方程都有公式解,那么四次以上的多项式方程有公式解吗?

挪威数学家阿贝尔(Neils Henrik Abel)证明了对于次数大于四的多项式方程不可能有求解公式.

牛顿法是用曲线弧一端的切线来代替曲线弧,从而求出方程实根的近似值的一种迭代方法.

Procedure for Newton's Method(牛顿法的步骤)

Step 1:Choose initial guess x_0(close to the desired root if possible)(设定初始值为 x_0).

Step 2:Generate successive approximations $x_1, x_2, \cdots$(如图 6.3 所示),where

$$x_1 = x_0 - \frac{f(x_0)}{f'(x_0)}, x_2 = x_1 - \frac{f(x_1)}{f'(x_1)}, x_3 = x_2 - \frac{f(x_2)}{f'(x_2)}, \cdots$$

$$x_{n+1} = x_n - \frac{f(x_n)}{f'(x_n)} \text{[牛顿法求方程 } f(x)=0 \text{ 近似根的迭代公式]}$$

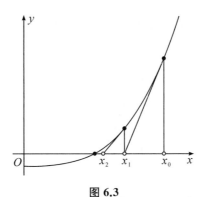

图 6.3

Example 6.6 牛顿法求方程的近似解.

Calculate the first three approximations x_1, x_2, x_3 to a root of $f(x) = x^2 - 5$ using the initial guess $x_0 = 2$.

Solution

We have $f'(x) = 2x$. Therefore,

$$x_{n+1} = x_n - \frac{f(x_n)}{f'(x_n)} = x_n - \frac{x_n^2 - 5}{2x_n}$$

We compute the successive approximations as follows:

$$x_1 = x_0 - \frac{f(x_0)}{f'(x_0)} = 2 - \frac{2^2 - 5}{2 \times 2} = 2.25$$

$$x_2 = x_1 - \frac{f(x_1)}{f'(x_1)} = 2.25 - \frac{2.25^2 - 5}{2 \times 2.25} \approx 2.23611$$

$$x_3 = x_2 - \frac{f(x_2)}{f'(x_2)} = 2.23611 - \frac{2.23611^2 - 5}{2 \times 2.23611} = 2.23606797789$$

This sequence provides successive approximations to a root of $x^2 - 5 = 0$, namely

$$\sqrt{5} = 2.23606797789\cdots$$

Observe that x_3 is accurate to within an error of less than 10^{-9}. This is impressive accuracy for just three iterations of Newton's Method.

Practice Exercises(习题)

1. The approximate value of $y=\sqrt{4+\sin x}$ at $x=0.12$, obtained from the tangent to the graph at $x=0$, is
   (A) 2.00 (B) 2.03 (C) 2.06 (D) 2.12

2. For small values of h, the function $\sqrt[4]{16+h}$ is best approximated by which of the following?
   (A) $2+\dfrac{h}{32}$ (B) $2-\dfrac{h}{32}$ (C) $\dfrac{h}{32}$ (D) $4+\dfrac{h}{32}$

3. Let f be a differentiable function such that $f(3)=2$ and $f'(3)=5$. If the tangent line to the graph of f at $x=3$, is used to find an approximation to a zero of f, that approximation is
   (A) 0.4 (B) 0.5 (C) 2.6 (D) 3.5

4. Use Newton's method to calculate the first three approximations x_1, x_2, x_3 to a root of $x^3+3x+1=0$ using the initial guess $x_0=-0.3$.

习题参考答案

1. B 2. A 3. C

4. Let $f(x)=x^3+3x+1$, then $f'(x)=3x^2+3$ and
$$x_{n+1}=x_n-\frac{f(x_n)}{f'(x_n)}=x_n-\frac{x_n^3+3x_n+1}{3x_n^2+3}$$

The graph of f from a calculator graph of $y=x^3+3x+1$ suggests that $x_0=-0.3$ is a good first approximation to the zero of f in the interval $[-1,0]$. Then,

$x_0=-0.3$

$x_1=-0.322324159$

$x_2=-0.3221853603$

$x_3=-0.3221853546$

The x_n for $n \geqslant 5$ all appear to equal x_4 on the calculator we used for our computations. We conclude that the solution to the equation $x^3+3x+1=0$ is about -0.3221853546.

Chapter 7　Antidifferentiation(不定积分)

7.1　Definition of Antidifferentiation(不定积分的定义)

7.1.1　Definition of Antiderivative(原函数的定义)

A function $F(x)$ is an **antiderivative** of a function $f(x)$ if $F'(x)=f(x)$ for all x in the domain of $f(x)$.

若 $F'(x)=f(x)$,则称 $F(x)$ 为 $f(x)$ 的一个原函数.函数 $f(x)$ 的原函数是不唯一的,若 $F(x)$ 和 $G(x)$ 均为 $f(x)$ 的原函数,则 $F(x)$ 和 $G(x)$ 仅相差一个常数,即

$$F'(x)=G'(x) \Leftrightarrow F(x)=G(x)+C(C \text{ is a constant})$$

证　设 $G(x)'F(x)$ 是 $f(x)$ 的任意两个原函数,即对任一 $x \in I$,有 $G'(x)=f(x)$.
则 $[G(x)-F(x)]'=G'(x)-F'(x)=f(x)-f(x)=0$.
所以 $G(x)-F(x)$ 为一常数.
令 $G(x)-F(x)=C$(常数),
则有 $G(x)=F(x)+C$.

Example 7.1

Which of the following are antiderivatives of $f(x)=\sin x \cdot \cos x$?

Ⅰ. $F(x)=\dfrac{\sin^2 x}{2}$

Ⅱ. $F(x)=\dfrac{\cos^2 x}{2}$

Ⅲ. $F(x)=\dfrac{-\cos(2x)}{4}$

(A) Ⅰ only　　　　(B) Ⅱ only　　　　(C) Ⅰ and Ⅱ　　　　(D) Ⅰ and Ⅲ

Solution

D

$\because \left(\dfrac{\sin^2 x}{2}\right)' = \dfrac{1}{2} \cdot 2 \cdot \sin x \cdot \cos x = \sin x \cdot \cos x$

$$\left(\frac{\cos^2 x}{2}\right)' = -\sin x \cdot \cos x$$

$$\left(\frac{-\cos(2x)}{4}\right)' = \frac{1}{4} \cdot \sin(2x) \cdot 2 = \frac{1}{2} \cdot 2 \cdot \sin x \cdot \cos x = \sin x \cdot \cos x$$

∴ Ⅰ and Ⅲ are antiderivatives of $f(x) = \sin x \cdot \cos x$.

7.1.2 Definition of Indefinite Integral(不定积分的定义)

$f(x)$ 的全体"原函数"称为 $f(x)$ 的 **indefinite integral or antidifferentiation**(不定积分)，记作：

$$\int f(x)\mathrm{d}x = F(x) + C \ (C \text{ is an arbitrary constant})$$

其中记号"$\int$"称为积分号，$f(x)$ 称为 **Integrand**(被积函数)，$f(x)\mathrm{d}x$ 称为被积表达式，x 称为积分变量. The constant C is called **the constant of integration**(积分常数).

$$\int f(x)\mathrm{d}x = F(x) + C \Leftrightarrow F'(x) = f(x)$$

$f(x)$ 称为 $F(x)$ 的导函数，$F(x)$ 称为 $f(x)$ 的一个"原函数".

7.1.3 Inverse Relationship between Integration and Differentiation(积分与微分的互逆关系)

从不定积分的定义即可知下述关系：

$$\frac{\mathrm{d}}{\mathrm{d}x}\left[\int f(x)\mathrm{d}x\right] = f(x) \qquad 或记作 \qquad \mathrm{d}\left[\int f(x)\mathrm{d}x\right] = f(x)\mathrm{d}x$$

又由于 $F(x)$ 是 $F'(x)$ 的原函数，所以

$$\int F'(x)\mathrm{d}x = F(x) + C \qquad 或记作 \qquad \int \mathrm{d}F(x) = F(x) + C$$

由此可见，微分运算(以记号 d 表示)与求不定积分的运算(以记号 $\int$ 表示)是互逆的. 当记号 $\int$ 与 d 连在一起时，可相互抵消，或者抵消后差一个常数.

7.1.4 基本积分表

(1) $\int k\,\mathrm{d}x = kx + C$ (k 是常数)

(2) $\int x^\mu \mathrm{d}x = \dfrac{1}{\mu+1}x^{\mu+1} + C$

常见的有：$\int \dfrac{1}{x^2}\mathrm{d}x = -\dfrac{1}{x} + C$； $\int \dfrac{1}{\sqrt{x}}\mathrm{d}x = 2\sqrt{x} + C$； $\int \sqrt{x}\,\mathrm{d}x = \dfrac{2}{3}x^{3/2} + C$

(3) $\int \dfrac{1}{x}\mathrm{d}x = \ln|x| + C$

(4) $\int \mathrm{e}^x \mathrm{d}x = \mathrm{e}^x + C$

(5) $\int a^x \mathrm{d}x = \dfrac{a^x}{\ln a} + C$

(6) $\int \cos x \, dx = \sin x + C$

(7) $\int \sin x \, dx = -\cos x + C$

(8) $\int \dfrac{1}{\cos^2 x} dx = \int \sec^2 x \, dx = \tan x + C$

(9) $\int \dfrac{1}{\sin^2 x} dx = \int \csc^2 x \, dx = -\cot x + C$

(10) $\int \dfrac{1}{1+x^2} dx = \arctan x + C$

(11) $\int \dfrac{1}{\sqrt{1-x^2}} dx = \arcsin x + C$

(12) $\int \sec x \tan x \, dx = \sec x + C$

(13) $\int \csc x \cot x \, dx = -\csc x + C$

(14) $\int \tan x \, dx = \ln|\sec x| + C$

(15) $\int \cot x \, dx = -\ln|\csc x| + C$

7.1.5 不定积分的性质

性质 1 函数的和的不定积分等各个函数的不定积分之和，即
$\int [f(x)+g(x)] dx = \int f(x) dx + \int g(x) dx.$

性质 2 求不定积分时，被积函数中不为零的常数因子可以提到积分号外面来，即
$\int k f(x) dx = k \int f(x) dx \, (k \text{ 是常数}, k \neq 0).$

Example 7.2 *直接积分法.*

求下列函数的不定积分：

a. $\int [(x+1)^2 - 2\cos x] dx$

b. $\int 2^x \cdot 5^x \cdot e^x \, dx$

c. $\int \left[\dfrac{1}{x\sqrt[3]{x}} - \sqrt{x\sqrt{x}} \right] dx$

d. $\int \dfrac{1+x+x^2}{x(1+x^2)} dx$

e. $\int \dfrac{dx}{\cos^2 x \sin^2 x}$

f. $\int \dfrac{1-\cos x}{1-\cos 2x} dx$

g. $\int \dfrac{x^4}{1+x^2} dx$

Solution

a.分析：要求将原题的被积函数进行变形，再利用上述三条性质，容易求出各个基本函数的积分.

$$\int [(x+1)^2 - 2\cos x] dx = \int (x^2 + 2x + 1 - 2\cos x) dx$$
$$= \int x^2 dx + \int 2x dx + \int 1 dx - 2\int \cos x dx$$
$$= \frac{1}{3}x^3 + x^2 + x - 2\sin x + C$$

[注] 在分项积分后，每个不定积分的结果本应都加上一个积分常数，但因为任意常数的和仍然是任意常数，所以只要在最后的结果中加上一个任意常数即可. 另外，$\int 1 dx$ 常记为 $\int dx$.

b. 因为 $2^x \cdot 5^x \cdot e^x = (10e)^x$，由公式 $\int a^x dx = \frac{a^x}{\ln a} + c$ 得：

$$\int 2^x \cdot 5^x \cdot e^x dx = \int (10e)^x dx = \frac{(10e)^x}{\ln(10e)} + c = \frac{10^x e^x}{\ln 10 + 1} + C$$

c. $\int \left[\frac{1}{x\sqrt[3]{x}} - \sqrt{x\sqrt{x}}\right] dx = \int x^{-\frac{4}{3}} dx - \int x^{\frac{3}{4}} dx = -3x^{-\frac{1}{3}} - \frac{4}{7}x^{\frac{7}{4}} + C = -\frac{3}{\sqrt[3]{x}} - \frac{4}{7}\sqrt[4]{x^7} + C$

d. $\int \frac{1+x+x^2}{x(1+x^2)} dx = \int \frac{(1+x^2)+x}{x(1+x^2)} dx = \int \frac{1}{x} dx + \int \frac{1}{1+x^2} dx = \ln|x| + \arctan x + C$

e. $\int \frac{dx}{\cos^2 x \sin^2 x}$

$= \int \frac{\cos^2 x + \sin^2 x}{\cos^2 x \sin^2 x} dx$

$= \int \left[\frac{1}{\sin^2 x} + \frac{1}{\cos^2 x}\right] dx$

$= \int \frac{1}{\sin^2 x} dx + \int \frac{1}{\cos^2 x} dx$

$= -\cot x + \tan x + C$

f. $\int \frac{1 - \cos x}{1 - \cos 2x} dx$

$= \int \frac{1 - \cos x}{2\sin^2 x} dx$

$= \frac{1}{2} \int \left[\frac{1}{\sin^2 x} - \frac{\cos x}{\sin^2 x}\right] dx$

$= \frac{1}{2} \int \frac{dx}{\sin^2 x} - \frac{1}{2} \int \cot x \csc x dx$

$= -\frac{1}{2} \cot x + \frac{1}{2} \csc x + C$

g. $\int \dfrac{x^4}{1+x^2}dx$

$= \int \dfrac{(x^4-1)+1}{1+x^2}dx = \int \left(x^2-1+\dfrac{1}{1+x^2}\right)dx$

$= \int x^2 dx - \int dx + \int \dfrac{1}{1+x^2}dx$

$= \dfrac{1}{3}x^3 - x + \arctan x + C$

7.2 Integral by Substitution(换元积分法)

7.2.1 U-Substitution(U 代换，又称第一类换元法)

$\int f[g(x)] \cdot g'(x)dx \xrightarrow{令 g(x)=u} \int f(u) \cdot u' dx$

$\xrightarrow{凑微分} \int f(u)du$

$\xrightarrow{求不定积分} F(u)+C$

$\xrightarrow{回代原变量} F[g(x)]+C$

[注]能够用 U 代换计算不定积分的被积函数由两个因式构成，其中一个因式为复合函数，另一个因式为其内层函数的导数，形如 $f[g(x)] \cdot g'(x)$. U 代换在国内的教材上称为凑微分法.

常见的 U 代换的类型

$\int f(ax+b)dx = \dfrac{1}{a} \int f(ax+b)d(ax+b) \quad (a \neq 0)$

$\int f(ax^n+b)x^{n-1}dx = \dfrac{1}{a^n} \int f(ax^n+b)d(ax^n+b)$

$\int f(\sqrt{x})\dfrac{1}{\sqrt{x}}dx = 2\int f(\sqrt{x})d(\sqrt{x})$

$\int f(e^x)e^x dx = \int f(e^x)d(e^x)$

$\int f(\ln x)\dfrac{1}{x}dx = \int f(\ln x)d(\ln x)$

$\int f(\cos x)\sin x\, dx = -\int f(\cos x)d(\cos x)$

$\int f(\sin x)\cos x\, dx = \int f(\sin x)d(\sin x)$

$\int f(\tan x)\sec^2 x\, dx = \int f(\tan x)d(\tan x)$

$\int f(\cot x)\csc^2 x\, dx = -\int f(\cot x)d(\cot x)$

$$\int f(\arcsin x)\frac{1}{\sqrt{1-x^2}}\mathrm{d}x = \int f(\arcsin x)\mathrm{d}(\arcsin x)$$

$$\int f(\arctan x)\frac{1}{1+x^2}\mathrm{d}x = \int f(\arctan x)\mathrm{d}(\arctan x)$$

Example 7.3 **U-Substitution**

a. $2\int \cos(2x+5)\mathrm{d}x$

b. $\int \dfrac{1}{3+2x}\mathrm{d}x$

c. $\int 2x\mathrm{e}^{x^2}\mathrm{d}x$

d. $\int x\sqrt{1-x^2}\mathrm{d}x$

e. $\int \dfrac{1}{a^2+x^2}\mathrm{d}x$

f. 当 $a>0$ 时, $\int \dfrac{1}{\sqrt{a^2-x^2}}\mathrm{d}x$

g. $\int \dfrac{1}{x^2-a^2}\mathrm{d}x$

h. $\int \dfrac{\mathrm{e}^{\sqrt[3]{x}}}{\sqrt{x}}\mathrm{d}x$

i. $\int \dfrac{\mathrm{e}^x}{2\mathrm{e}^x+1}\mathrm{d}x$

j. $\int \dfrac{\mathrm{d}x}{x(1+2\ln x)}$

k. $\int \tan x\,\mathrm{d}x$

l. $\int \sin^3 x\,\mathrm{d}x$

m. $\int \sin^2 x\,\cos^5 x\,\mathrm{d}x$

n. $\int \cos^2 x\,\mathrm{d}x$

o. $\int \cos^4 x\,\mathrm{d}x$

p. $\int \cos 3x \cos 2x\,\mathrm{d}x$

q. $\int \csc x\,\mathrm{d}x$

r. $\int \sec x\,\mathrm{d}x$

s. $\int \tan^5 x\,\sec^3 x\,\mathrm{d}x$

Solution

a. $2\int \cos(2x+5)\mathrm{d}x = \int \cos 2x \cdot (2x+5)'\mathrm{d}x = \int \cos(2x+5)\mathrm{d}(2x+5) \quad u=2x+5$

$= \int \cos u\,\mathrm{d}u = \sin u + C = \sin(2x+5) + C$

b. $\int \dfrac{1}{3+2x}\mathrm{d}x = \dfrac{1}{2}\int \dfrac{1}{3+2x}(3+2x)'\mathrm{d}x = \dfrac{1}{2}\int \dfrac{1}{3+2x}\mathrm{d}(3+2x) \quad u=3+2x$

$= \dfrac{1}{2}\int \dfrac{1}{u}\mathrm{d}u = \dfrac{1}{2}\ln|u| + C = \dfrac{1}{2}\ln|3+2x| + C$

c. $\int 2x\mathrm{e}^{x^2}\mathrm{d}x = \int \mathrm{e}^{x^2}(x^2)'\mathrm{d}x = \int \mathrm{e}^{x^2}\mathrm{d}(x^2) \quad u=x^2 = \int \mathrm{e}^u \mathrm{d}u = \mathrm{e}^u + C = \mathrm{e}^{x^2} + C$

d. $\int x\sqrt{1-x^2}\,\mathrm{d}x = \dfrac{1}{2}\int \sqrt{1-x^2}(x^2)'\mathrm{d}x = \dfrac{1}{2}\int \sqrt{1-x^2}\,\mathrm{d}x^2$

$= -\dfrac{1}{2}\int \sqrt{1-x^2}\,\mathrm{d}(1-x^2) \quad u=1-x^2 = -\dfrac{1}{2}\int u^{\frac{1}{2}}\mathrm{d}u = -\dfrac{1}{3}u^{\frac{3}{2}} + C$

$= -\dfrac{1}{3}(1-x^2)^{\frac{3}{2}} + C$

熟练之后，变量代换就不必再写出了.

e. $\int \dfrac{1}{a^2+x^2}\mathrm{d}x = \dfrac{1}{a^2}\int \dfrac{1}{1+\left(\dfrac{x}{a}\right)^2}\mathrm{d}x = \dfrac{1}{a}\int \dfrac{1}{1+\left(\dfrac{x}{a}\right)^2}\mathrm{d}\left(\dfrac{x}{a}\right) = \dfrac{1}{a}\arctan\dfrac{x}{a} + C$

即 $\int \dfrac{1}{a^2+x^2}\mathrm{d}x = \dfrac{1}{a}\arctan\dfrac{x}{a} + C$

f. 当 $a > 0$ 时，

$\int \dfrac{1}{\sqrt{a^2-x^2}}\mathrm{d}x = \dfrac{1}{a}\int \dfrac{1}{\sqrt{1-\left(\dfrac{x}{a}\right)^2}}\mathrm{d}x = \int \dfrac{1}{\sqrt{1-\left(\dfrac{x}{a}\right)^2}}\mathrm{d}\left(\dfrac{x}{a}\right) = \arcsin\dfrac{x}{a} + C$

即 $\int \dfrac{1}{\sqrt{a^2-x^2}}\mathrm{d}x = \arcsin\dfrac{x}{a} + C$

g. $\int \dfrac{1}{x^2-a^2}\mathrm{d}x = \dfrac{1}{2a}\int \left(\dfrac{1}{x-a} - \dfrac{1}{x+a}\right)\mathrm{d}x = \dfrac{1}{2a}\left(\int \dfrac{1}{x-a}\mathrm{d}x - \int \dfrac{1}{x+a}\mathrm{d}x\right)$

$= \dfrac{1}{2a}\left[\int \dfrac{1}{x-a}\mathrm{d}(x-a) - \int \dfrac{1}{x+a}\mathrm{d}(x+a)\right]$

$= \dfrac{1}{2a}[\ln|x-a| - \ln|x+a|] + C$

$= \dfrac{1}{2a}\ln\left|\dfrac{x-a}{x+a}\right| + C$

即 $\int \dfrac{1}{x^2-a^2}\mathrm{d}x = \dfrac{1}{2a}\ln\left|\dfrac{x-a}{x+a}\right| + C$

h. $\int \dfrac{\mathrm{e}^{\sqrt[3]{x}}}{\sqrt{x}}\mathrm{d}x = 2\int \mathrm{e}^{\sqrt[3]{x}}\mathrm{d}(\sqrt{x}) = \dfrac{2}{3}\int \mathrm{e}^{\sqrt[3]{x}}\mathrm{d}(3\sqrt{x}) = \dfrac{2}{3}\mathrm{e}^{\sqrt[3]{x}} + C$

i. $\int \dfrac{e^x}{2e^x+1}dx = \dfrac{1}{2}\int \dfrac{1}{2e^x+1}d(2e^x+1) = \dfrac{1}{2}\ln(2e^x+1)+C$

j. $\int \dfrac{dx}{x(1+2\ln x)} = \int \dfrac{d(\ln x)}{1+2\ln x} = \dfrac{1}{2}\int \dfrac{d(1+2\ln x)}{1+2\ln x} = \dfrac{1}{2}\ln|1+2\ln x|+C$

含三角函数的积分:

k. $\int \tan x\, dx = \int \dfrac{\sin x}{\cos x}dx = -\int \dfrac{1}{\cos x}d(\cos x) = -\ln|\cos x|+C$

即 $\int \tan x\, dx = -\ln|\cos x|+C$

类似地可得 $\int \cot x\, dx = \ln|\sin x|+C$

l. $\int \sin^3 x\, dx = \int \sin^2 x \cdot \sin x\, dx = -\int (1-\cos^2 x)\, d(\cos x)$

$= -\int d(\cos x) + \int \cos^2 x\, d(\cos x)$

$= -\cos x + \dfrac{1}{3}\cos^3 x + C$

m. $\int \sin^2 x \cos^5 x\, dx = \int \sin^2 x \cos^4 x\, d(\sin x) = \int \sin^2 x\, (1-\sin^2 x)^2\, d(\sin x)$

$= \int (\sin^2 x - 2\sin^4 x + \sin^6 x)\, d(\sin x)$

$= \dfrac{1}{3}\sin^3 x - \dfrac{2}{5}\sin^5 x + \dfrac{1}{7}\sin^7 x + C$

n. $\int \cos^2 x\, dx = \int \dfrac{1+\cos 2x}{2}dx = \dfrac{1}{2}\left(\int dx + \int \cos 2x\, dx\right)$

$= \dfrac{1}{2}\int dx + \dfrac{1}{4}\int \cos 2x\, d(2x) = \dfrac{1}{2}x + \dfrac{1}{4}\sin 2x + C$

o. $\int \cos^4 x\, dx = \int (\cos^2 x)^2\, dx = \int \left[\dfrac{1}{2}(1+\cos 2x)\right]^2 dx$

$= \dfrac{1}{4}\int (1+2\cos 2x+\cos^2 2x)\, dx$

$= \dfrac{1}{4}\int \left(\dfrac{3}{2}+2\cos 2x+\dfrac{1}{2}\cos 4x\right)dx$

$= \dfrac{1}{4}\left(\dfrac{3}{2}x+\sin 2x+\dfrac{1}{8}\sin 4x\right)+C$

$= \dfrac{3}{8}x+\dfrac{1}{4}\sin 2x+\dfrac{1}{32}\sin 4x+C$

p. $\int \cos 3x \cos 2x\, dx = \dfrac{1}{2}\int (\cos x + \cos 5x)\, dx = \dfrac{1}{2}\sin x + \dfrac{1}{10}\sin 5x + C$

q. $\int \csc x\, dx = \int \dfrac{1}{\sin x}dx = \int \dfrac{1}{2\sin\dfrac{x}{2}\cos\dfrac{x}{2}}dx = \int \dfrac{d\left(\dfrac{x}{2}\right)}{\tan\dfrac{x}{2}\cos^2\dfrac{x}{2}} = \int \dfrac{d\left(\tan\dfrac{x}{2}\right)}{\tan\dfrac{x}{2}} =$

$$\ln\left|\tan\frac{x}{2}\right|+C=\ln|\csc x-\cot x|+C$$

即 $\int \csc\,\mathrm{d}x = \ln|\csc x - \cot x| + C$

or

$$\int \csc\,\mathrm{d}x = \int \frac{\sin x}{\sin^2 x}\mathrm{d}x = \int \frac{-\mathrm{d}(\cos x)}{1-\cos^2 x} = \int \frac{\mathrm{d}(\cos x)}{\cos^2 x - 1}$$

$$=\frac{1}{2}\left[\int \frac{\mathrm{d}(\cos x-1)}{\cos x-1} - \int \frac{\mathrm{d}(\cos x+1)}{\cos x+1}\right]$$

$$=\frac{1}{2}\ln\left|\frac{\cos x-1}{\cos x+1}\right|+C$$

r. $\int \sec x\,\mathrm{d}x = \int \csc\left(x+\frac{\pi}{2}\right)\mathrm{d}x = \ln\left|\csc\left(x+\frac{\pi}{2}\right) - \cot\left(x+\frac{\pi}{2}\right)\right|+C = \ln|\sec x + \tan x|+C$

即 $\int \sec x\,\mathrm{d}x = \ln|\sec x + \tan x| + C$

or $\int \sec x\,\mathrm{d}x = \int \sec x \cdot \frac{\sec x + \tan x}{\sec x + \tan x}\mathrm{d}x = \int \frac{\sec^2 x + \sec x \cdot \tan x}{\sec x + \tan x}\mathrm{d}x$

$$=\int \frac{1}{\sec x + \tan x}\mathrm{d}(\sec x + \tan x) = \ln|\sec x + \tan x| + C$$

r. $\int \tan^5 x\, \sec^3 x\,\mathrm{d}x = \int \tan^4 x\, \sec^2 x\, \tan x\, \sec x\,\mathrm{d}x = \int (\sec^2 x - 1)^2\, \sec^2 x\,\mathrm{d}(\sec x)$

$$= \int (\sec^6 x - 2\sec^4 x + \sec^2 x)\,\mathrm{d}(\sec x)$$

$$= \frac{1}{7}\sec^7 x - \frac{2}{5}\sec^5 x + \frac{1}{3}\sec^3 x + C$$

[**注**]补充不定积分基本公式：

$$\int \frac{1}{a^2+x^2}\mathrm{d}x = \frac{1}{a}\arctan\frac{x}{a} + C$$

$$\int \frac{1}{\sqrt{a^2-x^2}}\mathrm{d}x = \arcsin\frac{x}{a} + C$$

$$\int \tan x\,\mathrm{d}x = -\ln|\cos x| + C$$

$$\int \cot x\,\mathrm{d}x = \ln|\sin x| + C$$

$$\int \csc x\,\mathrm{d}x = \ln|\csc x - \cot x| + C$$

$$\int \sec x\,\mathrm{d}x = \ln|\sec x + \tan x| + C$$

7.2.2 第二类换元法

定理(第二类换元法)

设 $x=x(t)$ 是单调可导的函数，并且 $x'(t)\neq 0$，若 $\int f[x(t)]x'(t)\mathrm{d}t = G(t) + C$

则 $\int f(x)\mathrm{d}x = G(t(x))+C$[其中 $t=t(x)$ 为 $x=x(t)$ 的反函数].

利用第二类换元法计算积分的思路可写成:

$$\int f(x)\mathrm{d}x \xrightarrow{x=x(t)} \int f(x(t))x'(t)\mathrm{d}t = G(t)+C \xrightarrow{t=t(x)} G(t(x))+C$$

第一、二类换元法的主要区别:第一类换元法从 $u=\varphi(x)$ 出发,且以 $u=\varphi(x)$ 结束,$\varphi(x)$ 可以没有反函数;而第二类换元法由 $x=x(t)$ 出发,到 $t=t(x)$ 结束,$x=x(t)$ 与 $t=t(x)$ 互为反函数.在第一类换元法中,u 是中间变量,x 是自变量,而在第二类换元法中,x 与 t 的中间变量与自变量的地位是变化的.

7.2.2.1 有理代换

将无理式通过适当的代换转化为有理式的方法称为有理代换.通过有理代换使根号消失,转化为有理函数的不定积分.

Example 7.4

a. $\int \dfrac{\sqrt{x-1}}{x}\mathrm{d}x$

b. $\int \dfrac{\mathrm{d}x}{1+\sqrt[3]{x+2}}$

c. $\int \dfrac{\mathrm{d}x}{(1+\sqrt[3]{x})\sqrt{x}}$

d. $\int \dfrac{1}{x}\sqrt{\dfrac{1+x}{x}}\mathrm{d}x$

Solution

a. Let $\sqrt{x-1}=u$, $x=u^2+1$, $\mathrm{d}x=2u\mathrm{d}u$, therefore

$$\int \dfrac{\sqrt{x-1}}{x}\mathrm{d}x = \int \dfrac{u}{u^2+1}\cdot 2u\mathrm{d}u = 2\int \dfrac{u^2}{u^2+1}\mathrm{d}u$$

$$= 2\int \left(1-\dfrac{1}{1+u^2}\right)\mathrm{d}u = 2(u-\arctan u)+C$$

$$= 2(\sqrt{x-1}-\arctan\sqrt{x-1})+C$$

b. Let $\sqrt[3]{x+2}=u$, so $x=u^3-2$, $\mathrm{d}x=3u^2\mathrm{d}u$,则

$$\int \dfrac{\mathrm{d}x}{1+\sqrt[3]{x+2}} = \int \dfrac{1}{1+u}\cdot 3u^2\mathrm{d}u = 3\int \dfrac{u^2-1+1}{1+u}\mathrm{d}u$$

$$= 3\int \left(u-1+\dfrac{1}{1+u}\right)\mathrm{d}u = 3\left(\dfrac{u^2}{2}-u+\ln|1+u|\right)+C$$

$$= \dfrac{3}{2}\sqrt[3]{(x+2)^2}-3\sqrt[3]{x+2}+\ln|1+\sqrt[3]{x+2}|+C$$

c. Let $x=t^6$, so $\mathrm{d}x=6t^5\mathrm{d}t$,从而

$$\int \dfrac{\mathrm{d}x}{(1+\sqrt[3]{x})\sqrt{x}} = \int \dfrac{6t^5}{(1+t^2)t^3}\mathrm{d}t = 6\int \dfrac{t^2}{1+t^2}\mathrm{d}t$$

$$= 6\int \left(1 - \frac{1}{1+t^2}\right) dt$$

$$= 6(t - \arctan t) + C$$

$$= 6(\sqrt[6]{x} - \arctan \sqrt[6]{x}) + C$$

d. Let $\sqrt{\dfrac{1+x}{x}} = t$, so $x = \dfrac{1}{t^2 - 1}$,

Therefore $\displaystyle\int \frac{1}{x}\sqrt{\frac{1+x}{x}} \, dx = \int (t^2 - 1) t \cdot \frac{-2t}{(t^2-1)^2} dt$

$$= -2\int \frac{t^2}{t^2-1} dt$$

$$= -2\int \left(1 + \frac{1}{t^2-1}\right) dt$$

$$= -2t - \ln\left|\frac{t-1}{t+1}\right| + C$$

$$= -2\sqrt{\frac{1+x}{x}} - \ln\frac{\sqrt{1+x}-\sqrt{x}}{\sqrt{1+x}+\sqrt{x}} + C$$

7.2.2.2 三角代换

Example 7.5

Evaluate $\displaystyle\int \sqrt{a^2 - x^2} \, dx \ (a > 0)$.

Solution

Let $x = a\sin t$, $-\dfrac{\pi}{2} < t < \dfrac{\pi}{2}$, so $\sqrt{a^2 - x^2} = \sqrt{a^2 - a^2\sin^2 t} = a\cos t$,

$dx = a\cos t \, dt$, therefore

$\displaystyle\int \sqrt{a^2 - x^2} \, dx = \int a\cos t \cdot a\cos t \, dt$

$$= a^2 \int \cos^2 t \, dt$$

$$= a^2 \left(\frac{1}{2}t + \frac{1}{4}\sin 2t\right) + C$$

Because $t = \arcsin \dfrac{x}{a}$, $\sin 2t = 2\sin t \cos t = 2 \cdot \dfrac{x}{a} \cdot \dfrac{\sqrt{a^2-x^2}}{a}$,

Therefore $\displaystyle\int \sqrt{a^2-x^2}\, dx = a^2\left(\frac{1}{2}t + \frac{1}{4}\sin 2t\right) + C = \frac{a^2}{2}\arcsin\frac{x}{a} + \frac{1}{2}x\sqrt{a^2-x^2} + C$.

[**注**]AP 考试中只要求掌握用三角代换求不定积分，此处求不定积分的三角代换法大家仅了解即可.

7.3 Integral by Parts(分部积分法)

7.3.1 Parts Formula(分部积分公式)

设函数 $u=u(x)$ 及 $v=v(x)$ 具有连续导数. 那么，两个函数乘积的导数公式为
$$(uv)'=u'v+uv',$$
移项得
$$uv'=(uv)'-u'v.$$
对这个等式两边求不定积分，得：
$$\int uv'\,dx=uv-\int u'v\,dx, 即 \int u\,dv=uv-\int v\,du.$$
上面两个公式是分部积分公式的两种形式.

分部积分法计算过程：$\int uv'\,dx=\int u\,dv=uv-\int v\,du=uv-\int u'v\,dx=\cdots$

Example 7.6 *直接用分部积分公式.*

a. $\int \ln x^3\,dx$

b. $\int \arccos x\,dx$

Solution

a. $\int \ln x^3\,dx=3\int \ln x\,dx=3x\ln x-3\int x\,d(\ln x)=3x\ln x-3\int x\cdot\frac{1}{x}\,dx=3x\ln x-3x+C$

b. $\int \arccos x\,dx = x\arccos x-\int x\,d(\arccos x)$
$= x\arccos x+\int x\,\frac{1}{\sqrt{1-x^2}}\,dx$
$= x\arccos x-\frac{1}{2}\int (1-x^2)^{-\frac{1}{2}}\,d(1-x^2)$
$= x\arccos x-\sqrt{1-x^2}+C$

Example 7.7 *先凑微分，再用分部积分公式.*

a. $\int x\cos x\,dx$

b. $\int x e^x\,dx$

c. $\int x^2 e^x\,dx$

d. $\int x\ln x\,dx$

e. $\int x\arctan x\,dx$

f. $\int xf(x)\,dx$

Solution

a. $\int x\cos x\,dx = \int x\,d(\sin x) = x\sin x - \int \sin x\,dx = x\sin x - \cos x + C$

b. $\int xe^x\,dx = \int x\,d(e^x) = xe^x - \int e^x\,dx = xe^x - e^x + C$

c. $\int x^2 e^x\,dx = \int x^2\,d(e^x) = x^2 e^x - \int e^x\,d(x^2)$
$= x^2 e^x - 2\int xe^x\,dx$
$= x^2 e^x - 2\int x\,de^x$
$= x^2 e^x - 2xe^x + 2\int e^x\,dx$
$= x^2 e^x - 2xe^x + 2e^x + C$
$= e^x(x^2 - 2x + 2) + C.$

d. $\int x\ln x\,dx = \frac{1}{2}\int \ln x\,d(x^2) = \frac{1}{2}x^2\ln x - \frac{1}{2}\int x^2\,d(\ln x)$
$= \frac{1}{2}x^2\ln x - \frac{1}{2}\int x^2\cdot\frac{1}{x}\,dx$
$= \frac{1}{2}x^2\ln x - \frac{1}{2}\int x\,dx = \frac{1}{2}x^2\ln x - \frac{1}{4}x^2 + C.$

e. $\int x\arctan x\,dx = \frac{1}{2}\int \arctan x\,d(x^2) = \frac{1}{2}x^2\arctan x - \frac{1}{2}\int x^2\cdot\frac{1}{1+x^2}\,dx$
$= \frac{1}{2}x^2\arctan x - \frac{1}{2}\int\left(1 - \frac{1}{1+x^2}\right)dx$
$= \frac{1}{2}x^2\arctan x - \frac{1}{2}x + \frac{1}{2}\arctan x + C.$

f. $\int xf(x)\,dx = \frac{1}{2}\int f(x)\,d(x^2) = \frac{x^2}{2}f(x) - \int \frac{x^2}{2}f'(x)\,dx.$

[注]先凑微分,再用分部积分公式的类型,其被积函数的表达式往往是两类函数的积的形式,按照"反、对、幂、三、指"的顺序用位于后面的函数去凑微分.

Example 7.8 出现循环的情形.

Find $\int e^x\cos x\,dx$.

Solution

Let $I = \int \cos x\,d(e^x) = \cos x\cdot e^x - \int e^x\,d(\cos x)$
$= \cos x\cdot e^x + \int e^x\cdot \sin x\,dx$
$= \cos x\cdot e^x + \int \sin x\,d(e^x)$
$= \cos x\cdot e^x + \sin x\cdot e^x - \int e^x\,d(\sin x)$

$$= \cos x \cdot e^x + \sin x \cdot e^x - \int e^x \cdot \cos x \, dx$$

$$\therefore I = \frac{e^x}{2}(\cos x + \sin x).$$

7.3.2 Tabular Integration by Parts(分部积分列表法)

分部积分列表法的步骤：

(1)一般按照"反、对、幂、三、指"的顺序，从左到右排列两类函数．

(2)排在第一列的函数逐次求导，直至导数为零为止；排在第二列的函数逐次积分．

(3)从第一行第一列向第二行第二列斜向画箭头，…，第 n 行第一列向第 $n+1$ 行第二列画箭头．

(4)在箭头上从上至下依次标上"＋""－"相间的符号．

(5)沿着箭头方向的两个函数之积的代数和(箭头上为＋则该项为正，箭头上为－则该项为负)便是积分结果．

Example 7.9 *分部积分列表法．*

a. $\int x e^x \, dx$

b. $\int x \cos x \, dx$

c. $\int x \sin(2x) \, dx$

d. $\int x e^{2x} \, dx$

e. $\int x^2 \sin x \, dx$

Solution

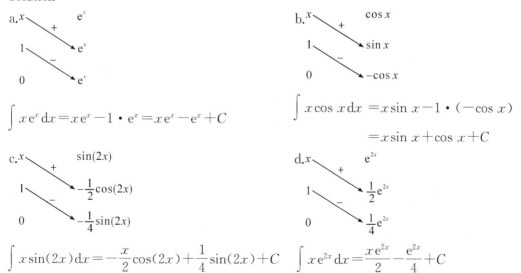

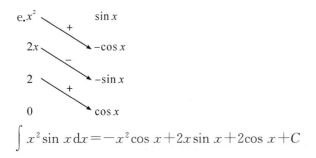

$$\int x^2 \sin x \, dx = -x^2 \cos x + 2x \sin x + 2\cos x + C$$

7.4 Indefinite Integral of Rational Functions(有理函数的不定积分)

有理函数(Rational Functions)是指由两个多项式(polynomials)的商所表示的函数,即具有如下形式的函数:

$$\frac{P(x)}{Q(x)} = \frac{a_0 x^n + a_1 x^{n-1} + \cdots + a_{n-1} x + a_n}{b_0 x^m + b_1 x^{m-1} + \cdots + b_{m-1} x + b_m},$$

其中 m 和 n 都是非负整数;$a_0, a_1, a_2, \cdots, a_n$ 及 $b_0, b_1, b_2, \cdots, b_m$ 都是实数,并且 $a_0 \neq 0, b_0 \neq 0$. 当 $n < m$ 时,称这样的有理函数是**真分式**(proper fraction);而当 $n \geq m$ 时,称这样的有理函数是**假分式**(improper fraction).

假分式总可以用多项式的除法化成一个多项式与一个真分式之和的形式. 例如:

$$\frac{x^3 + x + 1}{x^2 + 1} = \frac{x(x^2 + 1) + 1}{x^2 + 1} = x + \frac{1}{x^2 + 1}.$$

7.4.1 A Method of Partial Fractions (部分分式法)

理论上来说,任何一个真分式都可以表示成分母仅为一次式或二次式的真分式的线性表达式.

7.4.1.1 Linear Factor Rule(一次因式法则)

若真分式 $\frac{P(x)}{Q(x)}$ 的分母 $Q(x)$ 有因式 $(ax+b)^m$,则 $\frac{P(x)}{Q(x)}$ 拆分成部分分式的和的形式后必然包含以下 m 个**部分分式**(partial fractions)的和:

$$\frac{A_1}{ax+b} + \frac{A_2}{(ax+b)^2} + \cdots + \frac{A_m}{(ax+b)^m}$$

7.4.1.2 Quadratic Factor Rule(二次因式法则)

若真分式 $\frac{P(x)}{Q(x)}$ 的分母 $Q(x)$ 有因式 $(ax^2+bx+c)^m$,则 $\frac{P(x)}{Q(x)}$ 拆分成部分分式的和的形式后必然包含以下 m 个部分分式(partial fractions)的和:

$$\frac{A_1 x + B_1}{ax^2 + bx + c} + \frac{A_2 x + B_2}{(ax^2 + bx + c)^2} + \cdots + \frac{A_m x + B_m}{(ax^2 + bx + c)^m}$$

7.4.1.3 Integral by Partial Fractions(不定积分的部分分式法)

求真分式的不定积分时,如果分母可因式分解,则先因式分解,然后化成部分分式再积分,这种求不定积分的方法称为部分分式法.

Example 7.10 Integral by Partial Fractions.

a. Evaluate $\int \dfrac{x^2+2}{(x-1)(2x-8)(x+2)}\,\mathrm{d}x$.

b. Let $f(x)=\dfrac{1}{x^2-6x}$, find the partial fraction decomposition for the function f. Find $\int f(x)\,\mathrm{d}x$.

c. $\int \dfrac{x+4}{(x-1)(x^2+x+3)}\,\mathrm{d}x$.

Solution

Step 1: Find the partial fraction decomposition.

The decomposition has the form
$$\frac{x^2+2}{(x-1)(2x-8)(x+2)}=\frac{A}{x-1}+\frac{B}{2x-8}+\frac{C}{x+2}$$

Multiply by $(x-1)(2x-8)(x+2)$ to clear denominators:
$$x^2+2=A(2x-8)(x+2)+B(x-1)(x+2)+C(x-1)(2x-8)$$

We set $x=1$ to compute $A=-\dfrac{1}{6}$.

Set $x=4$ to compute $B=1$.

Set $x=-2$ to compute $C=\dfrac{1}{6}$.

The result is
$$\frac{x^2+2}{(x-1)(2x-8)(x+2)}=-\frac{\frac{1}{6}}{x-1}+\frac{1}{2x-8}+\frac{\frac{1}{6}}{x+2}.$$

Step 2: Carry out the integration.

$$\int \frac{x^2+2}{(x-1)(2x-8)(x+2)}\,\mathrm{d}x = -\frac{1}{6}\int\frac{1}{x-1}\,\mathrm{d}x+\int\frac{1}{2x-8}\,\mathrm{d}x+\frac{1}{6}\int\frac{1}{x+2}\,\mathrm{d}x$$

$$= -\frac{1}{6}\ln|x-1|+\frac{1}{2}\ln|x-4|+\frac{1}{6}\ln|x+2|+C.$$

b. $\dfrac{1}{x^2-6x}=\dfrac{1}{x(x-6)}=\dfrac{A}{x}+\dfrac{B}{x-6}\Rightarrow 1=A(x-6)+Bx$

$x=0\Rightarrow 1=A\cdot(-6)\Rightarrow A=-\dfrac{1}{6}$

$x=6\Rightarrow 1=B\cdot 6\Rightarrow B=\dfrac{1}{6}$

$\dfrac{1}{x(x-6)}=\dfrac{-1/6}{x}+\dfrac{1/6}{x-6}$

$\int f(x)\,\mathrm{d}x = \int\left(\dfrac{-1/6}{x}+\dfrac{1/6}{x-6}\right)\mathrm{d}x$

$\qquad = -\dfrac{1}{6}\ln|x|+\dfrac{1}{6}\ln|x-6|+C=\dfrac{1}{6}\ln\left|\dfrac{x-6}{x}\right|+C$

c. Let $\dfrac{x+4}{(x-1)(x^2+x+3)} = \dfrac{A}{x-1} + \dfrac{Bx+C}{x^2+x+3} \Rightarrow A=1, B=C=-1$

$$\int \dfrac{x+4}{(x-1)(x^2+x+3)} \mathrm{d}x = \int \dfrac{1}{x-1}\mathrm{d}x - \int \dfrac{x+1}{x^2+x+3}\mathrm{d}x$$

$\because \int \dfrac{x+1}{x^2+x+3}\mathrm{d}x = \dfrac{1}{2}\int \dfrac{2x+2}{x^2+x+3}\mathrm{d}x = \dfrac{1}{2}\int \dfrac{2x+1}{x^2+x+3}\mathrm{d}x + \dfrac{1}{2}\int \dfrac{\mathrm{d}x}{x^2+x+3}$

$$= \dfrac{1}{2}\ln(x^2+x+3) + \dfrac{1}{2}\int \dfrac{\mathrm{d}\left(x+\dfrac{1}{2}\right)}{\left(x+\dfrac{1}{2}\right)^2 + \dfrac{11}{4}}$$

$$= \dfrac{1}{2}\ln(x^2+x+3) + \dfrac{1}{2} \times \dfrac{2}{\sqrt{11}}\arctan \dfrac{2\left(x+\dfrac{1}{2}\right)}{\sqrt{11}} + C$$

$\therefore \int \dfrac{x+4}{(x+1)(x^2+x+3)}\mathrm{d}x = \ln|x-1| - \dfrac{1}{2}\ln(x^2+x+3) - \dfrac{\sqrt{11}}{11}\arctan \dfrac{\sqrt{11}(2x+1)}{11} + C$

7.4.2 有理函数积分的其他方法

求真分式的不定积分，如果分母不能进行因式分解，那么往往需要采用其他方法求积分．

Example 7.11

a. $\int \dfrac{x^2}{x^2+1}\mathrm{d}x$.

b. $\int \dfrac{x-4}{x^2}\mathrm{d}x$.

c. $\int \dfrac{x+3}{x^2+2x}\mathrm{d}x$.

Solution

a. $\int \dfrac{x^2}{x^2+1}\mathrm{d}x = \int \left(1 - \dfrac{1}{x^2+1}\right)\mathrm{d}x = x + \tan^{-1}x + C$.

b. $\int \dfrac{x-4}{x^2}\mathrm{d}x = \int \left(\dfrac{1}{x} - \dfrac{4}{x^2}\right)\mathrm{d}x = \ln|x| + \dfrac{4}{x} + C$.

c. $\int \dfrac{x+3}{x^2+2x}\mathrm{d}x = \dfrac{1}{2}\int \dfrac{(2x+2)}{x^2+2x}\mathrm{d}x + 2\int \dfrac{1}{x^2+2x}\mathrm{d}x$

$\qquad = \dfrac{1}{2}\ln|x^2+2x| + \int \dfrac{x+2-x}{x(x+2)}\mathrm{d}x$

$\qquad = \dfrac{1}{2}\ln|x^2+2x| + \int \left(\dfrac{1}{x} - \dfrac{1}{x+2}\right)\mathrm{d}x$

$\qquad = \dfrac{1}{2}\ln|x^2+2x| + \ln|x| - \ln|x+2| + C$.

[注]有理函数积分的形式可以分为以下几种类型：

(1) $\int \dfrac{1}{x-a}\mathrm{d}x = \ln|x-a| + C$.

(2) $\int \frac{1}{(x-a)^n}dx = -\frac{1}{n-1} \cdot \frac{1}{(x-a)^{n-1}} + C(n \neq 1)$.

(3) $\int \frac{1}{(x^2+px+q)^n}dx = \int \frac{1}{\left[\left(x+\frac{p}{2}\right)^2+\frac{4q-p^2}{4}\right]^n}dx \xrightarrow{\text{let } x+\frac{p}{2}=u, \frac{4q-p^2}{4}=a^2} \int \frac{1}{u^2+a^2}dx$.

(4) $\int \frac{x+\frac{p}{2}}{(x^2+px+q)^n}dx = \frac{1}{2}\int \frac{1}{(x^2+px+q)^n}d(x^2+px+q)$.

Practice Exercises(习题)

1. $\int (x^3-3x)dx =$

   (A) $3x^2-3+C$ (B) $4x^4-6x^2+C$ (C) $\frac{1}{4}x^4-3x^2+C$ (D) $\frac{1}{4}x^4-\frac{3}{2}x^2+C$

2. $\int \sec^2 x\, dx =$

   (A) $\tan x + C$ (B) $\csc^2 x + C$ (C) $\frac{\sec^3 x}{3}+C$ (D) $2\sec^2 x \tan x + C$

3. If $\frac{dy}{dx} = \tan x$, then $y=$

   (A) $\frac{1}{2}\tan^2 x + C$ (B) $\sec^2 x + C$ (C) $\ln|\sec x| + C$ (D) $\ln|\cos x| + C$

4. 求下列不定积分(直接积分法):

   (1) $\int (x+1)dx$;

   (2) $\int \left(3t^2+\frac{t}{2}\right)dt$;

   (3) $\int \left(\frac{1}{x}-\frac{5}{x^2+1}\right)dx$;

   (4) $\int \left(\frac{1}{x^2}-x^2-\frac{1}{3}\right)dx$;

   (5) $\int (e^x+4^x)dx$;

   (6) $\int (\sqrt{x}+\sqrt[3]{x})dx$;

   (7) $\int \left(\frac{2}{\sqrt{1-y^2}}-\frac{1}{y^{1/4}}\right)dx$;

   (8) $\int \frac{1}{\sqrt{9-x^2}}dx$.

5. 求下列不定积分(U 代换):

(1) $\int \dfrac{\mathrm{d}x}{x^2+2x+2}$；

(2) $\int \dfrac{1+\ln x}{(x\ln x)^2}\mathrm{d}x$；

(3) $\int \dfrac{1}{1+\mathrm{e}^{-x}}\mathrm{d}x$；

(4) $\int \dfrac{\mathrm{d}x}{x(1+2\ln x)}$；

(5) $\int \cos^3 x \mathrm{d}x$．

6．求下列不定积分（第二换元法）：

(1) $\int \dfrac{\mathrm{d}x}{1+\sqrt{x}}$；

(2) $\int \dfrac{\mathrm{d}x}{\sqrt{a^2+x^2}}(a>0)$．

7．求下列不定积分（分部积分法）：

(1) $\int x\sin x \mathrm{d}x$；

(2) $\int \arcsin x \mathrm{d}x$；

(3) $\int 4x^3 \arctan x \mathrm{d}x$；

(4) $\int \mathrm{e}^x \sin x \mathrm{d}x$；

(5) $\int \tan^2 x \sec x \mathrm{d}x$．

8．求下列不定积分（部分分式法）：

(1) $\int \dfrac{x+1}{x^2-3x+2}\mathrm{d}x$；

(2) $\int \dfrac{x-1}{2x^2+x-1}\mathrm{d}x$；

(3) $\int \dfrac{x^5+x^4-8}{x^3-x}\mathrm{d}x$．

习题参考答案

1．D　2．A　3．C

4．求下列不定积分（直接积分法）：

(1) $\int (x+1)\mathrm{d}x = \dfrac{x^2}{2}+x+C$；

(2) $\int \left(3t^2+\dfrac{t}{2}\right)\mathrm{d}t = t^3+\dfrac{t^2}{4}+C$；

(3) $\int \left(\dfrac{1}{x} - \dfrac{5}{x^2+1}\right) dx = \ln|x| - 5\tan^{-1}x + C$；

(4) $\int \left(\dfrac{1}{x^2} - x^2 - \dfrac{1}{3}\right) dx = -\dfrac{1}{x} - \dfrac{x^3}{3} - \dfrac{x}{3} + C$；

(5) $\int (e^x + 4^x) dx = e^x + \dfrac{4^x}{\ln 4} + C$；

(6) $\int (\sqrt{x} + \sqrt[3]{x}) dx = \int (x^{1/2} + x^{1/3}) dx = \dfrac{2}{3} x^{3/2} + \dfrac{3}{4} x^{4/3} + C$；

(7) $\int \left(\dfrac{2}{\sqrt{1-y^2}} - \dfrac{1}{y^{1/4}}\right) dx = 2\sin^{-1} y - \dfrac{4}{3} y^{3/4} + C$；

(8) $\int \dfrac{1}{\sqrt{9-x^2}} dx = \sin^{-1} \dfrac{x}{3} + C$.

5. 求下列不定积分(U 代换)：

(1) $\int \dfrac{dx}{x^2+2x+2} = \int \dfrac{d(x+1)}{(x+1)^2+1} \xlongequal{x+1=u} \int \dfrac{du}{u^2+1} = \arctan u + C = \arctan(x+1) + C$；

(2) 令 $u = x\ln x$，则 $du = d(x\ln x) = \left(x \cdot \dfrac{1}{x} + \ln x\right) dx = (1 + \ln x) dx$，从而

$\int \dfrac{1+\ln x}{(x\ln x)^2} dx = \int \dfrac{du}{u^2} = -\dfrac{1}{u} + C = -\dfrac{1}{x\ln x} + C$；

(3) $\int \dfrac{1}{1+e^{-x}} dx = \int \dfrac{e^x}{e^x+1} dx = \int \dfrac{d(e^x+1)}{e^x+1} = \ln(e^x+1) + C$；

(4) $\int \dfrac{dx}{x(1+2\ln x)} = \int \dfrac{d(\ln x)}{1+2\ln x} = \dfrac{1}{2} \int \dfrac{d(1+2\ln x)}{1+2\ln x} = \dfrac{1}{2} \ln|1+2\ln x| + C$；

(5) $\int \cos^3 x \, dx = \int \cos^2 x \cos x \, dx = \int (1-\sin^2 x) d(\sin x) = \sin x - \dfrac{1}{3} \sin^3 x + C$.

6. 求下列不定积分(第二换元法)：

(1) 令 $x = t^2 (t \geq 0)$，则 $dx = 2t \, dt$，从而

$$\int \dfrac{dx}{1+\sqrt{x}} \xlongequal{x=t^2} \int \dfrac{2t \, dt}{1+t}$$
$$= 2\int \dfrac{t+1-1}{1+t} dt = 2\int \left(1 - \dfrac{1}{1+t}\right) dt$$
$$= 2t - 2\ln|1+t| + C$$
$$= 2\sqrt{x} - 2\ln(1+\sqrt{x}) + C.$$

(当取 $t < 0$ 时，有同样的结果，此时 $t = -\sqrt{x}$)

(2) 令 $x = a\tan t (-\dfrac{\pi}{2} < t < \dfrac{\pi}{2})$，则 $dx = a \, t = \arctan \dfrac{x}{a}$.

$\int \dfrac{dx}{\sqrt{a^2+x^2}} = \int \dfrac{a\sec^2 t \, dt}{\sqrt{a^2+a^2\tan^2 t}} = \int \sec t \, dt = \ln|\sec t + \tan t| + C$，

而 $\tan t = \dfrac{x}{a}$，$\sec t = \sqrt{1+\tan^2 t} = \sqrt{1+\dfrac{x^2}{a^2}}$，所以

$$\int \frac{\mathrm{d}x}{\sqrt{a^2+x^2}} = \ln\left|\sqrt{1+\frac{x^2}{a^2}}+\frac{x}{a}\right|+C_1 = \ln(\sqrt{a^2+x^2}+x)+C.$$

7. 求下列不定积分(分部积分法):

(1) 设 $u=x, v'=\sin x$, 则 $u'=1, v=-\cos x$, 从而

$$\int x\sin x = \int uv'\mathrm{d}x = uv - \int v\mathrm{d}u = x(-\cos x) - \int(-\cos x)\mathrm{d}x$$

$$= -x\cos x + \int \cos x\,\mathrm{d}x = -x\cos x + \sin x + C.$$

(2) 设 $u=\arcsin x, \mathrm{d}v=\mathrm{d}x$, 则 $\mathrm{d}u = \frac{1}{\sqrt{1-x^2}}\mathrm{d}x, v=x$, 于是

$$\int \arcsin x\,\mathrm{d}x = x\arcsin x - \int \frac{x}{\sqrt{1-x^2}}\mathrm{d}x$$

$$= x\arcsin x + \frac{1}{2}\int \frac{\mathrm{d}(1-x^2)}{\sqrt{1-x^2}} = x\arcsin x + \sqrt{1-x^2} + C.$$

(3) $\int 4x^3 \arctan x\,\mathrm{d}x = \int \arctan x\,\mathrm{d}(x^4) = \arctan x \cdot x^4 - \int x^4 \cdot \frac{1}{1+x^2}\mathrm{d}x$

$$= x^4 \arctan x - \int \frac{x^4-1+1}{1+x^2}\mathrm{d}x$$

$$= x^4 \arctan x - \frac{1}{3}x^3 + x - \arctan x + C.$$

(4) $\int e^x \sin x\,\mathrm{d}x = \int e^x \mathrm{d}(-\cos x) = -e^x \cos x - \int (-\cos x)e^x\,\mathrm{d}x$

$$= -e^x \cos x + \int e^x \cos x\,\mathrm{d}x,$$

而 $\int e^x \cos x\,\mathrm{d}x = \int e^x \mathrm{d}(\sin x) = e^x \sin x - \int \sin x\,e^x\,\mathrm{d}x$, 所以

$$\int e^x \cos x\,\mathrm{d}x = -e^x \cos x + e^x \sin x - \int \sin x\,e^x\,\mathrm{d}x.$$

$$2\int e^x \sin x\,\mathrm{d}x = e^x(\sin x - \cos x).$$

$$\int e^x \sin x\,\mathrm{d}x = \frac{1}{2}e^x(\sin x - \cos x) + C.$$

(5) $\int \tan^2 x \sec x\,\mathrm{d}x = \int \tan x\,\mathrm{d}(\sec x) = \tan x \sec x - \int \sec x\,\mathrm{d}(\tan x)$

$$= \tan x \sec x - \int \sec x \sec^2 x\,\mathrm{d}x$$

$$= \tan x \sec x - \int \sec x(\tan^2 x + 1)\,\mathrm{d}x$$

$$= \tan x \sec x - \int \tan^2 x \sec x\,\mathrm{d}x - \int \sec x\,\mathrm{d}x$$

$$= \tan x \sec x - \int \tan^2 x \sec x\,\mathrm{d}x - \ln|\sec x + \tan x|,$$

所以 $\int \tan^2 x \sec x\,\mathrm{d}x = \frac{1}{2}(\tan x \sec x - \ln|\sec x + \tan x|) + C.$

8.求下列不定积分(部分分式法):

(1) $\int \dfrac{x+1}{x^2-3x+2}dx = \int \left(\dfrac{3}{x-2}-\dfrac{2}{x-1}\right)dx = 3\ln|x-2|-2\ln|x-1|+C$
$= \ln\left|\dfrac{(x-2)^3}{(x-1)^2}\right|+C.$

(2) $\int \dfrac{x-1}{2x^2+x-1}dx = \int \left(-\dfrac{1}{6}\cdot\dfrac{1}{2x-1}+\dfrac{2}{3}\cdot\dfrac{1}{x+1}\right)dx$
$= -\dfrac{1}{6}\ln|2x-1|+\dfrac{2}{3}\ln|x+1|+C.$

(3) $\int \dfrac{x^5+x^4-8}{x^3-x}dx = \int \left(x^2+x+1+\dfrac{8}{x}-\dfrac{4}{1+x}-\dfrac{3}{x-1}\right)dx$
$= \dfrac{1}{3}x^3+\dfrac{1}{2}x^2+x+8\ln|x|-4\ln|1+x|-3\ln|x-1|+C.$

Chapter 8 Definite Integrals(定积分)

8.1 Riemann Sums and Definite Integrals(黎曼和与定积分)

8.1.1 Sigma Notation($\sum$ 符号)

Sigma notation enables us to write a sum with many terms in the compact form:
$$\sum_{k=1}^{n} a_k = a_1 + a_2 + a_3 + \cdots + a_n$$

The Greek letter $\sum$ (capital sigma, corresponding to our letter S), stands for "sum". The **index of summation** k tells us where the sum begins (at the number below the symbol) and where it ends (at the number above). Any letter can be used to denote the index, but the letters i, j, and k are customary.

$$\underset{\substack{\text{The summation symbol}\\\text{(Greek letter sigma)}}}{} \overset{\overset{\text{The index } k \text{ ends at } k=n.}{n}}{\underset{\underset{\text{The index } k \text{ starts at } k=1.}{k=1}}{\sum}} a_k \;\; \text{—} \;\; a_k \text{ is a formula for the } k\text{th term.}$$

例如：$\sum\limits_{k=0}^{3} 2k = 2\times 0 + 2\times 1 + 2\times 2 + 2\times 3$

$\sum\limits_{k=1}^{4}(-1)^k k = -1+2-3+4$

$\sum\limits_{i=5}^{7} i^2 = 5^2 + 6^2 + 7^2$

Algebra Rules for Finite Sums(有限和的代数运算法则)：

(1) **Sum Rule**: $\sum\limits_{i=1}^{n}(a_i + b_i) = \sum\limits_{i=1}^{n} a_i + \sum\limits_{i=1}^{n} b_i$.

(2) **Difference Rule**: $\sum\limits_{i=1}^{n}(a_i - b_i) = \sum\limits_{i=1}^{n} a_i - \sum\limits_{i=1}^{n} b_i$.

(3) Constant Multiple Rule: $\sum_{i=1}^{n} Ca_i = C \cdot \sum_{i=1}^{n} a_i$ (any number C).

(4) Constant Value Rule: $\sum_{i=1}^{n} C = n \cdot C$ (C is any constant value).

Summation Formulas(求和公式):

If n is a positive integer, then:

(1) $\sum_{i=1}^{n} a = an$.

(2) $\sum_{i=1}^{n} i = \dfrac{n(n+1)}{2}$.

(3) $\sum_{i=1}^{n} i^2 = \dfrac{n(n+1)(2n+1)}{6}$.

(4) $\sum_{i=1}^{n} i^3 = \dfrac{n^2(n+1)^2}{4}$.

(5) $\sum_{i=1}^{n} i^4 = \dfrac{n(n+1)(6n^3 + 9n^2 + n - 1)}{30}$.

Example 8.1

Evaluate $\sum_{i=1}^{n} \dfrac{i(i+1)}{n}$.

Solution

Rewrite $\sum_{i=1}^{n} \dfrac{i(i+1)}{n}$ as $\dfrac{1}{n} \sum_{i=1}^{n} (i^2 + i)$.

$$\dfrac{1}{n} \sum_{i=1}^{n} (i^2 + i) = \dfrac{1}{n} \left(\sum_{i=1}^{n} i^2 + \sum_{i=1}^{n} i \right) = \dfrac{1}{n} \left[\dfrac{n(n+1)(2n+1)}{6} + \dfrac{n(n+1)}{2} \right]$$
$$= \dfrac{(n+1)(2n+4)}{6} = \dfrac{(n+1)(n+2)}{3}.$$

8.1.2 Riemann Sums and Definite Integrals(黎曼和与定积分)

8.1.2.1 曲边梯形的面积

(1)曲边梯形.

设函数 $y = f(x)$ 在区间 $[a,b]$ 上非负、连续,由直线 $x = a$, $x = b$, $y = 0$ 及曲线 $y = f(x)$ 所围成的图形(如图 8.1)称为曲边梯形,其中曲线弧称为曲边.

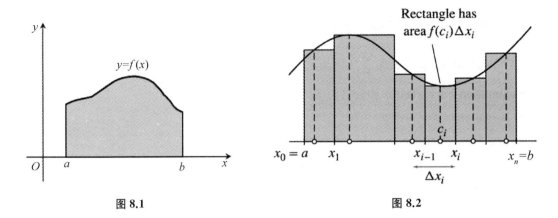

图 8.1　　　　　　　　　　　图 8.2

(2) 曲边梯形的面积的近似值.

将曲边梯形分割成一些小的曲边梯形(如图 8.2),每个小曲边梯形都用一个等宽的小矩形代替,每个小曲边梯形的面积都近似地等于小矩形的面积,则所有小矩形面积的和就是曲边梯形面积的近似值. 具体方法如下:在区间$[a,b]$中任意插入若干个分点:

$$a=x_0<x_1<x_2<\cdots<x_{n-1}<x_n=b,$$

把$[a,b]$分成n个小区间:

$$[x_0,x_1],[x_1,x_2],[x_2,x_3],\cdots,[x_{n-1},x_n]$$

它们的长度依次为$\Delta x_1=x_1-x_0,\Delta x_2=x_2-x_1,\cdots,\Delta x_n=x_n-x_{n-1}$.

经过每一个分点作平行于y轴的直线段,把曲边梯形分成n个小曲边梯形. 在每个小区间$[x_{i-1},x_i]$上任取一点c_i,以$[x_{i-1},x_i]$为底、$f(c_i)$为高的小矩形近似替代第i个小曲边梯形$(i=1,2,\cdots,n)$,把这样得到的n个小矩阵形面积之和作为所求曲边梯形面积A的近似值,即

$$A\approx f(c_1)\Delta x_1+f(c_2)\Delta x_2+\cdots+f(c_n)\Delta x_n$$

$$=\sum_{i=1}^{n}f(c_i)\Delta x_i \ (\textbf{The Riemann Sum 黎曼和}).$$

(3) 曲边梯形的面积的精确值.

显然,分点越多、每个小曲边梯形越窄,所求得的曲边梯形面积A的近似值就越接近于曲边梯形面积A的精确值. 因此,要求曲边梯形面积A的精确值,只需无限地增加分点,使每个小曲边梯形的宽度趋于近零.

记$\lambda=\max\{\Delta x_1,\Delta x_2,\cdots,\Delta x_n\}$,当$\lambda\to 0$时,每个小曲边梯形的宽度都趋于零,所以曲边梯形面积的精确值为:

$$A=\lim_{\lambda\to 0}\sum_{i=1}^{n}f(c_i)\Delta x_i.$$

8.1.2.2　Definition of Definite Integrals(定积分的定义)

Let $f(x)$ be defined on $[a,b]$ and x_i be points on $[a,b]$ such that $x_0=a, x_n=b$, and $a<x_1<x_2<x_3<\cdots<x_{n-1}<b$(如图 8.3).

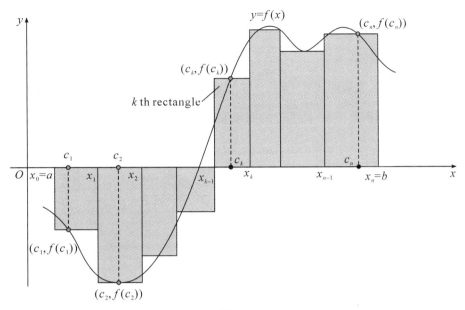

图 8.3

The set
$$P = \{x_0, x_1, x_2, \cdots, x_{n-1}, x_n\}$$
is called **a partition** of $[a, b]$.

The partition P divides $[a, b]$ into n closed subintervals:
$$[x_0, x_1], [x_1, x_2], [x_2, x_3], \cdots, [x_{n-1}, x_n],$$
The first of these subintervals is $[x_0, x_1]$, the second is $[x_1, x_2]$ and the kth **subinterval of P** is $[x_{k-1}, x_k]$, for k an integer between 1 and n.

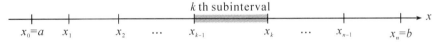

The width of the first subinterval $[x_0, x_1]$ is denoted Δx_1, the width of the second $[x_1, x_2]$ is denoted Δx_2, and the width of the kth subinterval is $\Delta x_k = x_k - x_{k-1}$.

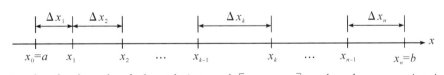

Let Δx_i be the length of the ith interval $[x_{i-1}, x_i]$ and c_i be any point in the ith interval.

$$\textbf{The Riemann Sum} = \sum_{i=1}^{n} f(c_i) \Delta x_i$$

If $\max \Delta x_i$ is the length of the largest subinterval in the partition and $\lim\limits_{\max \Delta x_i \to 0} \sum_{i=1}^{n} f(c_i) \Delta x_i$ exists, then the limit is denoted by:

$$\lim_{\max \Delta x_i \to 0} \sum_{i=1}^{n} f(c_i) \Delta x_i = \int_a^b f(x) \, dx$$

$\int_a^b f(x)\,dx$ is the definite integral of $f(x)$ from a to b.

8.1.2.3 定积分的几何意义

定积分 $\int_a^b f(x)\,dx$ 表示曲线 $y=f(x)$ 与 x 轴及 $x=a$，$x=b$ 所围成的曲边梯形面积的代数和，x 轴上方的面积取正，x 轴下方的面积取负（如图 8.4）.

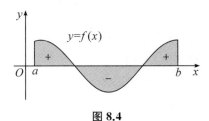

图 8.4

Example 8.2

a. If n is a positive integer, then $\lim\limits_{n\to\infty}\dfrac{1}{n}\left[\left(\dfrac{1}{n}\right)^2+\left(\dfrac{2}{n}\right)^2+\cdots+\left(\dfrac{3n}{n}\right)^2\right]$ can be expressed as

(A) $\int_0^1 \dfrac{1}{x^2}\,dx$ (B) $3\int_0^1 \left(\dfrac{1}{x}\right)^2 dx$ (C) $\int_0^1 \left(\dfrac{1}{x}\right)^2 dx$ (D) $\int_0^3 x^2\,dx$

b. Evaluate $\int_{-2}^{2} \sqrt{4-x^2}\,dx$.

Solution

a. **D**

The expression is a Riemann sum with $\Delta x = \dfrac{1}{n}$ and $f(x)=x^2$（每个小区间的间隔为 $\dfrac{1}{n}$）.

The evaluation points are: $\dfrac{1}{n}, \dfrac{2}{n}, \dfrac{3}{n}, \cdots, \dfrac{3n}{n}$（取每个小区间右端点的函数值为高）.

Thus the right Riemann sum is for $x=0$ to $x=3$. The limit is equal to $\int_0^3 x^2\,dx$.

b. 由定积分的几何意义：

$\because \int_{-2}^{2}\sqrt{4-x^2}\,dx$ 表示以原点为圆心，以 2 为半径的圆与 x 轴所围成的上半圆的面积，

$\therefore \int_{-2}^{2}\sqrt{4-x^2}\,dx = \dfrac{\pi\cdot 2^2}{2}=2\pi$.

一般地，有 $\int_{-R}^{R}\sqrt{R^2-x^2}\,dx = \dfrac{\pi\cdot R^2}{2}$.

8.2 Approximation of Definite Integral(定积分的近似计算)

8.2.1 Method of Approximation of Definite Integral(定积分近似计算的方法)

根据定积分的几何意义可知,函数 $f(x)$ 在某一区间的定积分等于曲边梯形面积的代数和,在近似计算曲边梯形面积的时候可以用矩形面积去近似代替(如图 8.5),也可以用梯形面积去近似代替(如图 8.6),由此产生了定积分近似计算的矩形法和梯形法.

Let $f(x)$ be defined on $[a,b]$ and x_i be points on $[a,b]$ such that $x_0=a$, $x_n=b$, and $a<x_1<x_2<x_3<\cdots<x_{n-1}<b$, Δx_i be the length of the i th interval $[x_{i-1},x_i]$.

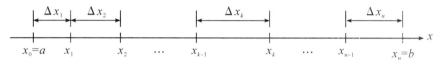

8.2.1.1 矩形法

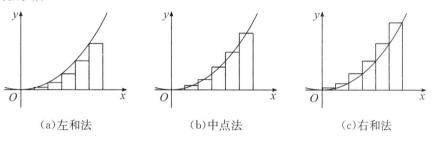

(a)左和法　　　　　　(b)中点法　　　　　　(c)右和法

图 8.5

(1) Left-Riemann Sum Rule(左和法).

如图 8.5(a)所示,计算小矩形面积的时候,选择每一个小区间的左端点的函数值作为小矩形的高,以每个小区间的长度作为宽,然后把所有小矩形的面积相加得到的结果叫作左和法.

左和法的计算公式为:

$$\int_a^b f(x)\mathrm{d}x \approx L(n) = f(x_0)\Delta x_1 + f(x_1)\Delta x_2 + \cdots + f(x_{n-1})\Delta x_n.$$

(2) Midpoint Sum Rule(中点法).

如图 8.5(b)所示,计算小矩形面积的时候选择每一个小区间的中点的函数值作为小矩形的高,以每个小区间的长度作为宽,然后把所有小矩形的面积相加的方法叫作中点法.

中点法的计算公式为:

$$\int_a^b f(x)\mathrm{d}x \approx M(n) = f(\bar{x}_1)\Delta x_1 + f(\bar{x}_2)\Delta x_2 + \cdots + f(\bar{x}_n)\Delta x_n.$$

Where $\bar{x}_i = \frac{1}{2}(x_{i-1}+x_i) = $ midpoint of $[x_{i-1},x_i]$.

(3) Right-Riemann Sum Rule(右和法).

如图 8.5(c)所示,计算小矩形面积的时候,选择每一个小区间的右端点的函数值作为小矩形的高,以每个小区间的长度作为宽,然后把所有小矩形的面积相加的方法叫作右和法.

右和法的计算公式为：
$$\int_a^b f(x)\mathrm{d}x \approx R(n) = f(x_1)\Delta x_1 + f(x_2)\Delta x_2 + \cdots + f(x_n)\Delta x_n.$$

8.2.1.2　Trapezoid Sum Rule(梯形法)

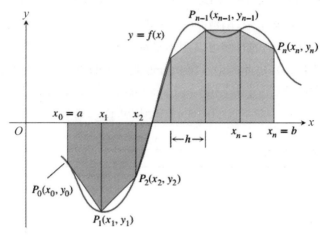

图 8.6

如图 8.6 所示，选择每一个小区间的两个端点的函数值作为梯形的上底和下底，以每个小区间的长度作为高，然后把所有小梯形的面积相加的方法叫作梯形法．

梯形法的计算公式为：
$$\int_a^b f(x)\mathrm{d}x \approx T(n) = \frac{f(x_0)+f(x_1)}{2}\Delta x_1 + \frac{f(x_1)+f(x_2)}{2}\Delta x_2 + \cdots + \frac{f(x_{n-1})+f(x_n)}{2}\Delta x_n.$$

若为 n 等分时，小区间长度均为 $h = \dfrac{b-a}{n}$，此时上面的公式可以简化为：
$$\int_a^b f(x)\mathrm{d}x \approx T(n) = \frac{h}{2}(y_0 + 2y_1 + 2y_2 + \cdots + 2y_{n-1} + y_n).$$

Example 8.3　定积分的近似计算．

a.

x	2	3	5	8	13
$f(x)$	6	-2	-1	3	9

The function f is continuous on the closed interval $[2,13]$ and has values as shown in the table above. Using the intervals $[2,3]$, $[3,5]$, $[5,8]$ and $[8,13]$, what is the approximation of $\int_2^{13} f(x)\mathrm{d}x$ obtained from a left Riemann sum?

b. Use a midpoint Riemann sum with three subdivisions of equal length to find the approximate value of $\int_0^6 x^2\mathrm{d}x$.

c. Use the Trapezoidal Rule with four subintervals with equal length to estimate $\int_1^2 x^2\mathrm{d}x$.

Solution

a. $\int_2^{13} f(x)\mathrm{d}x \approx 6\times 1 + (-2)\times 2 + (-1)\times 3 + 3\times 5 = 14$.

b. $\Delta x = \dfrac{6-0}{3} = 2, f(x) = x^2$.

midpoints are $x = 1, 3$ and 5.

$\int_0^6 x^2 \mathrm{d}x \approx f(1)\Delta x + f(3)\Delta x + f(5)\Delta x = 1\times 2 + 9\times 2 + 25\times 2 = 70$.

c. $\int_1^2 x^2 \mathrm{d}x \approx \dfrac{1}{4}\times \dfrac{1}{2}\left[f(1) + 2f\left(\dfrac{5}{4}\right) + 2f\left(\dfrac{6}{4}\right) + 2f\left(\dfrac{7}{4}\right) + f(2)\right]$

$= \dfrac{1}{8}\times\left(1 + 2\times\dfrac{25}{16} + 2\times\dfrac{36}{16} + 2\times\dfrac{49}{16} + 4\right) = \dfrac{75}{32}$.

8.2.2 Comparing Approximating Sums(近似和的比较)

8.2.2.1 利用单调性比较 $L(n)$ 与 $R(n)$

$$f(x) \text{ is increasing} \Rightarrow L(n) < \int_a^b f(x)\mathrm{d}a < R(n)$$

$$f(x) \text{ is decreasing} \Rightarrow L(n) < \int_a^b f(x)\mathrm{d}a < R(n)$$

8.2.2.2 利用凹凸性比较 $M(n)$ 与 $T(n)$

$$f(x) \text{ is concave up} \Rightarrow M(n) < \int_a^b f(x)\mathrm{d}x < T(n)$$

$$f(x) \text{ is concave down} \Rightarrow M(n) < \int_a^b f(x)\mathrm{d}x < T(n)$$

Example 8.4 近似和的比较.

t(minutes)	0	2	5	7	11	12
r/t(feet per minute)	5.7	4.0	2.0	1.2	0.6	0.5

The volume of a spherical hot air balloon expands as the air inside the balloon is heated. The radius of the balloon, in feet, is modeled by a twice-differentiable function r of time t, where t is measured in minutes. For $0 < t < 12$, the graph of r is concave down. The table above gives selected values of the rate of change, r/t, of the radius of the balloon over the time interval $0 \leqslant t \leqslant 12$. The radius of the balloon is 30 feet when $t = 5$. (Note: The volume of a sphere of radius r is given by $V = \dfrac{4}{3}\pi r^3$.)

a. Use a right Riemann sum with the five subintervals indicated by the data in the table to approximate $\int_0^{12} \dfrac{r}{t} \mathrm{d}t$.

b. Is your approximation in part (a) greater than or less than $\int_0^{12} \dfrac{r}{t} \mathrm{d}t$? Given a reason for your answer.

Solution

a. $\int_0^{12} \dfrac{r}{t} dt \approx 2 \times 4.0 + 3 \times 2.0 + 2 \times 1.2 + 4 \times 0.6 + 1 \times 0.5 = 19.3$ ft.

b. Since r is concave down, $\dfrac{r}{t}$ is decreasing on $0 < t < 12$. Therefore, this approximation, 19.3 ft, is less than $\int_0^{12} r'(t) dt$.

8.3 Properties of Definite Integrals(定积分的性质)

两点规定:

(1)当 $a = b$ 时,$\int_a^b f(x) dx = 0.$

(2)当 $a > b$ 时,$\int_a^b f(x) dx = -\int_b^a f(x) dx.$

性质 1 函数的和(差)的定积分等于它们的定积分的和(差),即
$$\int_a^b [f(x) \pm g(x)] dx = \int_a^b f(x) dx \pm \int_a^b g(x) dx.$$

证明 $\int_a^b [f(x) \pm g(x)] dx = \lim_{\lambda \to 0} \sum_{i=1}^n [f(c_i) \pm g(c_i)] \Delta x_i$
$$= \lim_{\lambda \to 0} \sum_{i=1}^n f(c_i) \Delta x_i \pm \lim_{\lambda \to 0} \sum_{i=1}^n g(c_i) \Delta x_i$$
$$= \int_a^b f(x) dx \pm \int_a^b g(x) dx.$$

性质 2 被积函数的常数因子可以提到积分号外面,即
$$\int_a^b kf(x) dx = k \int_a^b f(x) dx.$$

证明 $\int_a^b kf(x) dx = \lim_{\lambda \to 0} \sum_{i=1}^n kf(c_i) \Delta x_i = k \lim_{\lambda \to 0} \sum_{i=1}^n f(c_i) \Delta x_i = k \int_a^b f(x) dx.$

性质 3 如果将积分区间分成两部分,则在整个区间上的定积分等于这两部分区间上的定积分之和,即
$$\int_a^b f(x) dx = \int_a^c f(x) dx + \int_c^b f(x) dx.$$

这个性质表明,定积分对于积分区间具有可加性.

值得注意的是,不论 a, b, c 的相对位置如何,总有等式 $\int_a^b f(x) dx = \int_a^c f(x) dx + \int_c^b f(x) dx$ 成立.

例如,当 $a < b < c$ 时,有:
$$\int_a^b f(x) dx = \int_a^c f(x) dx - \int_b^c f(x) dx = \int_a^c f(x) dx + \int_c^b f(x) dx.$$

性质 4 如果在区间 $[a\ b]$ 上 $f(x) \equiv 1$,则

$$\int_a^b 1\mathrm{d}x = \int_a^b \mathrm{d}x = b-a.$$

性质 5 如果在区间 $[a,b]$ 上 $f(x) \geqslant 0$,则
$$\int_a^b f(x)\mathrm{d}x \geqslant 0 (a<b).$$

推论 1 如果在区间 $[a,b]$ 上 $f(x) \leqslant g(x)$,则
$$\int_a^b f(x)\mathrm{d}x \leqslant \int_a^b g(x)\mathrm{d}x (a<b).$$

证明 因为 $g(x)-f(x) \geqslant 0$,从而 $\int_a^b g(x)\mathrm{d}x - \int_a^b f(x)\mathrm{d}x = \int_a^b [g(x)-f(x)]\mathrm{d}x \geqslant 0$,

所以 $\int_a^b f(x)\mathrm{d}x \leqslant \int_a^b g(x)\mathrm{d}x$.

推论 2 $\left|\int_a^b f(x)\mathrm{d}x\right| \leqslant \int_a^b |f(x)|\mathrm{d}x (a<b).$

证明 因为 $-|f(x)| \leqslant f(x) \leqslant |f(x)|$,

所以 $-\int_a^b |f(x)|\mathrm{d}x \leqslant \int_a^b f(x)\mathrm{d}x \leqslant \int_a^b |f(x)|\mathrm{d}x$,

即 $\left|\int_a^b f(x)\mathrm{d}x\right| \leqslant \int_a^b |f(x)|\mathrm{d}x$.

性质 6 设 M 和 m 分别是函数 $f(x)$ 在区间 $[a,b]$ 上的最大值和最小值,则
$$m(b-a) \leqslant \int_a^b f(x)\mathrm{d}x \leqslant M(b-a) (a<b).$$

证明 因为 $m \leqslant f(x) \leqslant M$,所以 $\int_a^b m\mathrm{d}x \leqslant \int_a^b f(x)\mathrm{d}x \leqslant \int_a^b M\mathrm{d}x$,

从而 $m(b-a) \leqslant \int_a^b f(x)\mathrm{d}x \leqslant M(b-a)$.

性质 7 **The Mean Value Theorem for Integrals(积分中值定理)**.

If $f(x)$ is continuous on $[a,b]$, then at some point c in $[a,b]$,
$$\int_a^b f(x)\mathrm{d}x = f(c)(b-a).$$

这个公式叫作积分中值公式.

证明 由性质 6 得:$m(b-a) \leqslant \int_a^b f(x)\mathrm{d}x \leqslant M(b-a)$,

各项除以 $b-a$,得:$m \leqslant \dfrac{1}{b-a}\int_a^b f(x)\mathrm{d}x \leqslant M$.

再由连续函数的介值定理,在 $[a,b]$ 上至少存在一点 c,使
$$f(c) = \frac{1}{b-a}\int_a^b f(x)\mathrm{d}x,$$

于是两端乘以 $(b-a)$ 得中值公式:
$$\int_a^b f(x)\mathrm{d}x = f(c)(b-a).$$

[注]

(1) 积分中值公式的几何解释:

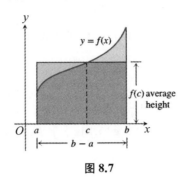

图 8.7

The value $f(c)$ in the Mean Value Theorem is, in a sense, the average (or mean) height of f on $[a,b]$. When the area of the rectangle is the area under the graph of f from a to b (如图 8.7),

$$\int_a^b f(x)dx = f(c)(b-a).$$

(2) Average Value of a Continuous Function(连续函数的平均值):

If f is integrable on $[a,b]$, its average (mean) value on $[a,b]$ is

The average value of $f(x)$ on $[a,b] = \dfrac{1}{b-a}\int_a^b f(x)dx$.

8.4 Fundamental Theorem of Calculus(微积分基本定理)

8.4.1 First Fundamental Theorem of Calculus(微积分基本定理,FTC)

If $f(x)$ is continuous on $[a,b]$ and $F(x)$ is an antiderivative of $f(x)$ on $[a,b]$, then

$$\int_a^b f(x)dx = F(x)\Big|_a^b = F(b) - F(a).$$

Example 8.5 微积分基本定理.

a. Evaluate $\int_0^2 (4x^3 + x - 1)dx$.

b. Evaluate $\int_{-\pi}^{\pi} \sin x \, dx$.

c. If $\int_{-2}^k (4x+1)dx = 30, k > 0$, find k.

d. Evaluate $\int_0^4 \dfrac{1}{x-1}dx$.

Solution

a. $\int_0^2 (4x^3 + x - 1)dx = \left(x^4 + \dfrac{x^2}{2} - x\right)\Big|_0^2 = \left(2^4 + \dfrac{2^2}{2} - 2\right) - 0 = 16.$

b. $\int_{-\pi}^{\pi} \sin x \, dx = -\cos x \Big|_{-\pi}^{\pi} = (-\cos \pi) - [-\cos(-\pi)] = -(-1) - [-(-1)]$

$$=1-1=0.$$

c. $\int_{-2}^{k}(4x+1)\mathrm{d}x=(2x^2+x)\,|_{-2}^{k}=2k^2+k-6.$

Let $2k^2+k-6=30 \Rightarrow 2k^2+k-36=0 \Rightarrow (2k+9)(k-4)=0 \Rightarrow k=-\dfrac{9}{2}$ (eliminate) or $k=4$.

d. $\int_{0}^{4}\dfrac{1}{x-1}\mathrm{d}x$ cannot evaluate using the First Fundamental Theorem of Calculus since $f(x)=\dfrac{1}{x-1}$ is discontinuous at $x=1$.

Example 8.6 连续函数的平均值.

Find the average value of $f(x)=4-x^2$ on $[0,3]$. Does f actually take on this value at some point in the given interval?

Solution

The average value of $f(x)$ on $[a,b]=\dfrac{1}{b-a}\int_{a}^{b}f(x)\mathrm{d}x=\dfrac{1}{3-0}\int_{0}^{3}(4-x^2)\mathrm{d}x=1.$

The average value of $f(x)=4-x^2$ on $[0,3]$ is 1. The function assumes this value when $4-x^2=1$ or $x=\pm\sqrt{3}$. Since $x=\sqrt{3}$ lies in the interval $[0,3]$, the function take its average value in the given interval.

8.4.2 Second Fundamental Theorem of Calculus(微积分第二基本定理)

8.4.2.1 Accumulation Function(累积函数,又称变上限积分)

We name $F(x)=\int_{a}^{x}f(t)\mathrm{d}t$ accumulation functions because they are accumulating area under a curve based on the value of x.

8.4.2.2 Second Fundamental Theorem of Calculus

If $f(x)$ is continuous on $[a,b]$ and $F(x)=\int_{a}^{x}f(t)\mathrm{d}t$, then

$$\dfrac{\mathrm{d}}{\mathrm{d}x}\int_{a}^{x}f(t)\mathrm{d}t=f(x) \quad \text{or} \quad \left(\int_{a}^{x}f(t)\mathrm{d}t\right)'=f(x)$$

at every point x in $[a,b]$.

一般地,变上限积分的导数为:

$$\dfrac{\mathrm{d}}{\mathrm{d}x}\int_{a(x)}^{b(x)}f(t)\mathrm{d}t=f[b(x)]\cdot b'(x)-f[a(x)]\cdot a'(x)$$

or $\left(\int_{a(x)}^{b(x)}f(t)\mathrm{d}t\right)'=f[b(x)]\cdot b'(x)-f[a(x)]\cdot a'(x).$

Example 8.7 变上限积分的导数.

Using the Second Fundamental Theorem of Calculus to find the following values.

a. $\dfrac{\mathrm{d}}{\mathrm{d}x}\int_{0}^{x}\dfrac{1}{1+t^2}\mathrm{d}t.$

b. $\dfrac{d}{dx}\displaystyle\int_{-\pi}^{x}\cos t\,dt$.

c. $\dfrac{d}{dx}\displaystyle\int_{1}^{x^2}\cos t\,dt$.

d. $\dfrac{d}{dx}\displaystyle\int_{x}^{5}3t\sin t\,dt$.

e. $\dfrac{d}{dx}\displaystyle\int_{2x}^{x^2}\dfrac{1}{2+e^t}\,dt$.

f. $\displaystyle\lim_{x\to 0}\dfrac{\displaystyle\int_{\cos x}^{1}e^{-t^2}\,dt}{\sin x\tan x}$.

Solution

a. $\dfrac{d}{dx}\displaystyle\int_{0}^{x}\dfrac{1}{1+t^2}\,dt=\dfrac{1}{1+x^2}$.

b. $\dfrac{d}{dx}\displaystyle\int_{-\pi}^{x}\cos t\,dt=\cos x$.

c. $\dfrac{d}{dx}\displaystyle\int_{1}^{x^2}\cos t\,dt=\cos(x^2)\cdot 2x=2x\cos x^2$.

d. $\dfrac{d}{dx}\displaystyle\int_{x}^{5}3t\sin t\,dt=-3x\sin x$.

e. $\dfrac{d}{dx}\displaystyle\int_{2x}^{x^2}\dfrac{1}{2+e^t}\,dt=\dfrac{1}{2+e^{x^2}}\cdot 2x-\dfrac{1}{2+e^{2x}}\cdot 2=\dfrac{2x}{2+e^{x^2}}-\dfrac{2}{2+e^{2x}}$.

f. $\displaystyle\lim_{x\to 0}\dfrac{\int_{\cos x}^{1}e^{-t^2}\,dt}{\sin x\tan x}=\lim_{x\to 0}\dfrac{\int_{\cos x}^{1}e^{-t^2}\,dt}{x^2}=\lim_{x\to 0}\dfrac{-e^{-\cos^2 x}(-\sin x)}{2x}=\lim_{x\to 0}\dfrac{e^{-\cos^2 x}\sin x}{2x}=\dfrac{1}{2e}$.

8.4.3 Net Change as the Integral of a Rate(净增量即变化率的定积分)

The net change in $f(t)$ over an interval $[t_1,t_2]$ is given by the integral

$$\underbrace{\int_{t_1}^{t_2}f'(t)\,dt}_{\text{Integral of the rate of change}}=\underbrace{f(t_2)-f(t_1)}_{\text{Net change over}[t_1,t_2]}$$

Example 8.8

A cylindrical can of radius 10 millimeters is used to measure rainfall in Stormville. The can is initially empty, and rain enters the can during a 60-day period. The height of water in the can is modeled by the function S, where $S(t)$ is measured in millimeters an t is measured in days for $0\leqslant t\leqslant 60$. The rate at which the height of the water is rising in the can is given by $S'(t)=2\sin(0.03t)+1.5$.

a. According to the model, what is the height of the water in the can at the end of the 60-day period?

b. According to the model, what is the average rate of change in the height of water in the can over the 60-day period? Show the computations that lead to your answer. Indicate units of measure.

c. Assuming no evaporation occurs, at what rate is the volume of water in the can changing at time $t=7$?

Solution

a. $S(60) = \int_0^{60} S'(t) \, dt = 171.813$ mm.

b. $\dfrac{S(60) - S(0)}{60} = 2.863$ or 2.864 mm/day.

c. $V(t) = 100\pi S(t)$,

$V'(7) = 100\pi S'(7) = 602.218$.

The volume of water in the can is increasing at a rate of 602.218 mm^3/day.

Example 8.9

For time $t \geq 0$ hours, let $r(t) = 120(1 - e^{-10t^2})$ represent the speed, in kilometers per hour, at which a car travels along a straight road. The number of liters of gasoline used by the car to travel x kilometers is modeled by $g(x) = 0.05x(1 - e^{-x/2})$.

a. How many kilometers does the car travel during the first 2 hours?

b. Find the rate of change with respect to time of the number of liters of gasoline used by the car when $t = 2$ hours. Indicate units of measure.

c. How many liters of gasoline have been used by the car when it reaches a speed of 80 kilometers per hour?

Solution

a. $\int_0^2 r(t) \, dt = 206.370$ kilometers.

b. $\dfrac{dg}{dt} = \dfrac{dg}{dx} \cdot \dfrac{dx}{dt}, \dfrac{dx}{dt} = r(t)$.

$\left. \dfrac{dg}{dt} \right|_{t=2} = \left. \dfrac{dg}{dx} \right|_{x=206.370} \cdot r(2) = 0.050 \times 120 = 6$ liters/hour.

c. Let T be the time at which the car's speed reaches 80 kilometers per hour.
Then, $r(T) = 80$ or $T = 0.331453$ hours.

At time T, the car has gone

$x(T) = \int_0^T r(t) \, dt = 10.794097$ kilometers

and has consumed

$g(x(T)) = 0.537$ liters of gasoline.

[注]

(1) 令 $v(t)$ 表示质点做**直线运动**在 t 时刻的瞬时速度，则：

从 $t=a$ 到 $t=b$ 时刻质点运动的**距离**(**distance**)：distance $= \int_a^b |v(t)| \, dt$；

从 $t=a$ 到 $t=b$ 时刻质点运动的**位移**(**displacement**)：displacement $= \int_a^b v(t) \, dt$.

(2) 令 $\overrightarrow{v(t)} = \langle \dfrac{\mathrm{d}x}{\mathrm{d}t}, \dfrac{\mathrm{d}y}{\mathrm{d}t} \rangle$ 表示质点做**平面运动**时的速度向量,则:

从 $t=a$ 到 $t=b$ 时刻,质点运动的距离(**distance**):

$$\text{distance} = \int_a^b \left| \overrightarrow{v(t)} \right| \mathrm{d}t = \int_a^b \sqrt{\left(\dfrac{\mathrm{d}x}{\mathrm{d}t}\right)^2 + \left(\dfrac{\mathrm{d}y}{\mathrm{d}t}\right)^2} \, \mathrm{d}t$$

向量 $\overrightarrow{v(t)} = \langle \dfrac{\mathrm{d}x}{\mathrm{d}t}, \dfrac{\mathrm{d}y}{\mathrm{d}t} \rangle$ 的斜率也就是运动曲线上某点的斜率:

slope of the curve: $\dfrac{\mathrm{d}y}{\mathrm{d}x} = \dfrac{\dfrac{\mathrm{d}y}{\mathrm{d}t}}{\dfrac{\mathrm{d}x}{\mathrm{d}t}}$.

8.5 Operations on Definite Integrals(定积分的计算)

8.5.1 Integration by Substitution(换元法)

8.5.1.1 U-Substitution(U 代换,又称第一换元法或凑微分法)

Example 8.10 定积分的第一换元法.

Evaluate the Integrations:

a. $\int_0^8 \dfrac{\mathrm{d}x}{\sqrt{1+x}}$.

b. $\int_0^1 (x+1)\mathrm{e}^{x^2+2x} \mathrm{d}x$.

c. $\int_1^2 \dfrac{x+1}{x^2+2x} \mathrm{d}x$.

d. $\int_2^3 \dfrac{3}{(x-1)(x+2)} \mathrm{d}x$.

Solution

a. $\int_0^8 \dfrac{\mathrm{d}x}{\sqrt{1+x}} = \int_0^8 \dfrac{1}{\sqrt{1+x}} \mathrm{d}(1+x) = 2\sqrt{1+x} \Big|_0^8 = 2 \times (3-1) = 4.$

b. $\int_0^1 (x+1)\mathrm{e}^{x^2+2x} \mathrm{d}x = \dfrac{1}{2}\int_0^1 (2x+2)\mathrm{e}^{x^2+2x} \mathrm{d}x = \dfrac{1}{2}\int_0^1 \mathrm{e}^{x^2+2x} \mathrm{d}(x^2+2x) = \dfrac{1}{2}\mathrm{e}^{x^2+2x} \Big|_0^1$

$= \dfrac{1}{2}(\mathrm{e}^3 - \mathrm{e}^0) = \dfrac{\mathrm{e}^3 - 1}{2}.$

c. $\int_1^2 \dfrac{x+1}{x^2+2x} \mathrm{d}x = \dfrac{1}{2}\int_1^2 \dfrac{1}{x^2+2x} \mathrm{d}(x^2+2x) = \dfrac{1}{2}\ln|x^2+2x| \Big|_1^2 = \dfrac{1}{2}(\ln 8 - \ln 3).$

d. $\int_2^3 \dfrac{3}{(x-1)(x+2)} \mathrm{d}x = \int_2^3 (\dfrac{1}{x-1} - \dfrac{1}{x+2}) \mathrm{d}x = \int_2^3 \dfrac{1}{x-1} \mathrm{d}(x-1) - \int_2^3 \dfrac{1}{x+2} \mathrm{d}(x+2)$

$= (\ln|x-1| - \ln|x+2|) \Big|_2^3$

$= \ln 2 - \ln 5 - \ln 1 + \ln 4 = \ln \dfrac{8}{5}.$

8.5.1.2 Integration by Substitution(第二换元法或换元积分法)

假设函数 $f(x)$ 在区间 $[a,b]$ 上连续,函数 $x=\varphi(t)$ 满足条件:

(1) $\varphi(\alpha)=a$, $\varphi(\beta)=b$;

(2) $\varphi(t)$ 在 $[\alpha,\beta]$(或 $[\beta,\alpha]$)上具有连续导数,且其值域不越出 $[a,b]$,则有

$$\int_a^b f(x)\mathrm{d}x = \int_\alpha^\beta f[\varphi(t)]\varphi'(t)\mathrm{d}t.$$

这个公式叫作定积分的换元公式.

Example 8.11 定积分的第二换元法.

a. $\int_0^a \sqrt{a^2-x^2}\,\mathrm{d}x\,(a>0)$(三角代换).

b. $\int_0^4 \dfrac{x+2}{\sqrt{2x+1}}\mathrm{d}x$(有理代换).

c. 证明:若 $f(x)$ 在 $[-a,a]$ 上连续,则

$$\int_{-a}^a f(x)\mathrm{d}x = \begin{cases} 2\int_0^a f(x)\mathrm{d}x, & f(x) \text{ 为偶函数} \\ 0, & f(x) \text{ 为奇函数} \end{cases} \quad (\text{如图 8.8})$$

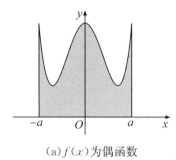

(a) $f(x)$ 为偶函数   (b) $f(x)$ 为奇函数

图 8.8

Solution

a. $\int_0^a \sqrt{a^2-x^2}\,\mathrm{d}x \xrightarrow{\text{令 } x=a\sin t} \int_0^{\frac{\pi}{2}} a\cos t \cdot a\cos t\,\mathrm{d}t$

$= a^2 \int_0^{\frac{\pi}{2}} \cos^2 t\,\mathrm{d}t = \dfrac{a^2}{2}\int_0^{\frac{\pi}{2}}(1+\cos 2t)\mathrm{d}t$

$= \dfrac{a^2}{2}\left[t+\dfrac{1}{2}\sin 2t\right]_0^{\frac{\pi}{2}} = \dfrac{1}{4}\pi a^2.$

提示: $\sqrt{a^2-x^2}=\sqrt{a^2-a^2\sin^2 t}=a\cos t$, $\mathrm{d}x=a\cos t$, 当 $x=0$ 时 $t=0$, 当 $x=a$ 时 $t=\dfrac{\pi}{2}$, 在用第二换元法计算定积分时,换元和换积分上、下限同时进行.

当被积函数是 $\sqrt{a^2+x^2}$, $\sqrt{a^2-x^2}$ 和 $\sqrt{x^2-a^2}$ 或可以转化为这三种形式时,可以参考图 8.9 中所示的辅助三角形进行三角代换.

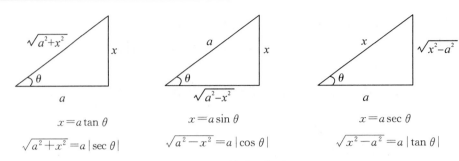

图 8.9 辅助三角形

b. $\int_0^4 \dfrac{x+2}{\sqrt{2x+1}}\,dx \xrightarrow{\text{令}\sqrt{2x+1}=t} \int_1^3 \dfrac{\frac{t^2-1}{2}+2}{t}\cdot t\,dt = \dfrac{1}{2}\int_1^3 (t^2+3)\,dt$

$= \dfrac{1}{2}\left[\dfrac{1}{3}t^3+3t\right]\Big|_1^3 = \dfrac{1}{2}\left[\left(\dfrac{27}{3}+9\right)-\left(\dfrac{1}{3}+3\right)\right] = \dfrac{22}{3}.$

提示: $x=\dfrac{t^2-1}{2}, dx=t\,dt$；当 $x=0$ 时 $t=1$，当 $x=4$ 时 $t=3$.

c. **证明** 因为 $\int_{-a}^a f(x)\,dx = \int_{-a}^0 f(x)\,dx + \int_0^a f(x)\,dx$，

而 $\int_{-a}^0 f(x)\,dx \xrightarrow{\text{令}x=-t} -\int_a^0 f(-t)\,dt = \int_0^a f(-t)\,dt = \int_0^a f(-x)\,dx$，

所以 $\int_{-a}^a f(x)\,dx = \int_0^a f(-x)\,dx + \int_0^a f(x)\,dx.$

若 $f(x)$ 为偶函数，则 $f(-x)=f(x)$，从而

$\int_{-a}^a f(x)\,dx = \int_0^a [f(-x)+f(x)]\,dx = \int_{-a}^a 2f(x)\,dx = 2\int_0^a f(x)\,dx.$

若 $f(x)$ 为奇函数，则 $f(-x)+f(x)=0$，从而

$\int_{-a}^a f(x)\,dx = \int_0^a [f(-x)+f(x)]\,dx = 0.$

8.5.2 Integration by Parts(分部积分法)

设 $u(x), v(x)$ 在 $[a,b]$ 上具有连续导数，则

$\int_a^b u(x)v'(x)\,dx = [u(x)v(x)]\Big|_a^b - \int_a^b u'(x)v(x)\,dx$（定积分的分部积分公式）.

Example 8.12 定积分的分部积分法.

a. $\int_0^{\frac{\pi}{2}} x\sin x\,dx.$

b. $\int_0^{\frac{1}{2}} \arcsin x\,dx.$

c. $\int_0^1 e^{\sqrt{x}}\,dx.$

Solution

a. $\int_0^{\frac{\pi}{2}} x\sin x\,dx = -\int_0^{\frac{\pi}{2}} x\,d\cos x = -x\cos x\Big|_0^{\frac{\pi}{2}} + \int_0^{\frac{\pi}{2}} \cos x\,dx$

$$= \sin x \big|_0^{\frac{\pi}{2}} = 1.$$

b. $\displaystyle\int_0^{\frac{1}{2}} \arcsin x \, dx = x \arcsin x \big|_0^{\frac{1}{2}} - \int_0^{\frac{1}{2}} x \cdot \frac{dx}{\sqrt{1-x^2}}$

$$= \frac{1}{2} \cdot \frac{\pi}{6} + \sqrt{1-x^2} \big|_0^{\frac{1}{2}}$$

$$= \frac{\pi}{12} + \frac{\sqrt{3}}{2} - 1.$$

c. 令 $\sqrt{x} = t$,则 $x = t^2$, $dx = 2t \, dt$,

且当 $x=0$ 时, $t=0$;当 $x=1$ 时, $t=1$,

故 $\displaystyle\int_0^1 e^{\sqrt{x}} dx = \int_0^1 e^t \cdot 2t \, dt = 2 \int_0^1 t e^t \, dt$.

再由分部积分公式,有

$\displaystyle\int_0^1 t e^t \, dt = \int_0^1 t \, de^t = t e^t \big|_0^1 - \int_0^1 e^t \, dt = e - e^t \big|_0^1 = 1$,

所以 $\displaystyle\int_0^1 e^{\sqrt{x}} dx = 2$.

Example 8.13 参数方程的定积分.

a. Evaluate $\displaystyle\int_{-2}^{2} y \, dx$, where $x = 2\sin\theta$ and $y = 2\cos\theta$.

b. Evaluate $2\pi \displaystyle\int_0^1 xy \, dx$, where $x = \ln t$ and $y = t^3$.

Solution

a. Note that $dx = 2\cos\theta \, d\theta$, that $\theta = -\frac{\pi}{2}$ when $x = -2$, and that $\theta = \frac{\pi}{2}$ when $x = 2$.

Then $\displaystyle\int_{-2}^{2} y \, dx = \int_{-\pi/2}^{\pi/2} 2\cos\theta(2\cos\theta) d\theta = 4 \int_{-\pi/2}^{\pi/2} \frac{1 + \cos 2\theta}{2} d\theta$

$$= 2\left(\theta + \frac{\sin 2\theta}{2}\right) \Big|_{-\pi/2}^{\pi/2} = 2\pi.$$

b. $dx = \dfrac{1}{t} dt$. For $x = 0$, we solve $\ln t = 0$ to get $t = 1$.

For $x = 1$, we solve $\ln t = 1$ to get $t = e$.

Then $2\pi \displaystyle\int_0^1 xy \, dx = 2\pi \int_1^e (\ln t)(t^3)\left(\frac{1}{t} dt\right) = 2\pi \int_1^e t^2 \ln t \, dt$

$$= \frac{2\pi}{3} \int_1^e \ln t \, d(t^3) = \frac{2\pi}{3} \left[\ln t \cdot t^3 \Big|_1^e - \int_1^e t^3 \, d(\ln t)\right]$$

$$= \frac{2\pi}{3}\left[e^3 - \int_1^e t^2 \, dt\right] = \frac{2\pi}{3}\left[e^3 - \frac{e^3}{3} + \frac{1}{3}\right]$$

$$= \frac{4\pi}{9} e^3 + \frac{2\pi}{9}.$$

8.6 Improper Integral(反常积分)

回忆:First Fundamental Theorem of Calculus (FTC)

If $f(x)$ is continuous on $[a,b]$ and $F(x)$ is an antiderivative of $f(x)$ on $[a,b]$, then
$$\int_a^b f(x)\mathrm{d}x = F(x)\Big|_a^b = F(b) - F(a).$$

由于 $f(x)$ 在闭区间 $[a,b]$ 连续,则 $f(x)$ 在闭区间 $[a,b]$ 必有最值,所以 FTC 要求 $f(x)$ 在闭区间 $[a,b]$ 为有界函数,且积分区间的上下限为常数.将这两个限制条件推广后的定积分称为反常积分,又称为广义积分.

反常积分有两类类型:

第一类反常积分(Type Ⅰ Improper Integrals):积分限至少有一个为**无穷大**的定积分称为无穷限反常积分(Infinite Limits of Integration).

第二类反常积分(Type Ⅱ Improper Integrals):被积函数在积分区间为**无界函数**的定积分称为反常积分(Integrands with Infinite Discontinuities).

8.6.1 Infinite Limits of Integration(无穷限反常积分)

Integrals with infinite limits of integration are improper integrals of Type I.

(1) If $f(x)$ is continuous on $[a,\infty)$, then
$$\int_a^\infty f(x)\mathrm{d}x = \lim_{b\to\infty} \int_a^b f(x)\mathrm{d}x.$$

(2) If $f(x)$ is continuous on $(-\infty,b]$, then
$$\int_{-\infty}^b f(x)\mathrm{d}x = \lim_{a\to-\infty} \int_a^b f(x)\mathrm{d}x.$$

(3) If $f(x)$ is continuous on $(-\infty,\infty)$, then
$$\int_{-\infty}^\infty f(x)\mathrm{d}x = \int_{-\infty}^c f(x)\mathrm{d}x + \int_c^\infty f(x)\mathrm{d}x = \lim_{a\to-\infty}\int_a^c f(x)\mathrm{d}x + \lim_{b\to\infty}\int_c^b f(x)\mathrm{d}x$$

where c is any real number.

在这三种情况下,若极限存在,我们称为反常积分**收敛(converges)**;若极限不存在,则称为反常积分**发散(diverges)**.

Example 8.14 *无穷限反常积分.*

Evaluating the Improper Integrals:

a. $\int_1^\infty \dfrac{\ln x}{x^2}\mathrm{d}x.$

b. $\int_{-\infty}^\infty \dfrac{\mathrm{d}x}{1+x^2}.$

Solution

a. We find the area under the curve from $x=1$ to $x=b$ and examine the limit as $b\to\infty$. If the limit is finite, we take it to be the area under the curve. The area (如图 8.10) from 1 to b is

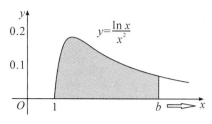

图 8.10

$$\int_1^b \frac{\ln x}{x^2}\mathrm{d}x = \ln x(-\frac{1}{x})\Big|_1^b - \int_1^b (-\frac{1}{x}) \cdot \frac{1}{x}\mathrm{d}x = -\frac{\ln b}{b} - \frac{1}{x}\Big|_1^b = -\frac{\ln b}{b} - \frac{1}{b} + 1.$$

The limit of the area as $b \to \infty$ is

$$\int_1^\infty \frac{\ln x}{x^2}\mathrm{d}x = \lim_{b \to \infty}\int_1^b \frac{\ln x}{x^2}\mathrm{d}x = \lim_{b \to \infty}(-\frac{\ln b}{b} - \frac{1}{b} + 1)$$

$$= -\lim_{b \to \infty}\frac{\ln b}{b} - 0 + 1 = -\lim_{b \to \infty}\frac{1/b}{1} + 1 = 0 + 1 = 1.$$

Thus, the improper integral converges and the area has finite value 1.

b. $\int_{-\infty}^\infty \frac{\mathrm{d}x}{1+x^2} = \int_{-\infty}^0 \frac{\mathrm{d}x}{1+x^2} + \int_0^\infty \frac{\mathrm{d}x}{1+x^2}.$

$\because \int_{-\infty}^0 \frac{\mathrm{d}x}{1+x^2} = \lim_{a \to -\infty}\int_a^0 \frac{\mathrm{d}x}{1+x^2} = \lim_{a \to -\infty}\tan^{-1}x \Big|_a^0 = \lim_{a \to -\infty}(\tan^{-1}0 - \tan^{-1}a)$

$$= 0 - (-\frac{\pi}{2}) = \frac{\pi}{2}.$$

$\int_0^\infty \frac{\mathrm{d}x}{1+x^2} = \lim_{b \to \infty}\int_0^b \frac{\mathrm{d}x}{1+x^2} = \lim_{b \to \infty}\tan^{-1}x \Big|_0^b = \lim_{b \to \infty}(\tan^{-1}b - \tan^{-1}0) = \frac{\pi}{2} - 0 = \frac{\pi}{2}.$

$\therefore \int_{-\infty}^\infty \frac{\mathrm{d}x}{1+x^2} = \int_{-\infty}^0 \frac{\mathrm{d}x}{1+x^2} + \int_0^\infty \frac{\mathrm{d}x}{1+x^2} = \frac{\pi}{2} + \frac{\pi}{2} = \pi.$

Since $\frac{1}{1+x^2} > 0$, the improper integral can be interpreted as the (finite) area beneath the curve and above the x-axis(图 8.11).

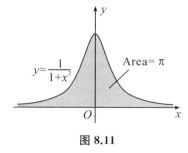

图 8.11

Theorem 1 The p-Integral over $[1, \infty)$.

$$\int_1^\infty \frac{1}{x^p}\mathrm{d}x = \begin{cases} \dfrac{1}{p-1}, & \text{if } p > 1 \\ \text{diverges}, & \text{if } p \leqslant 1 \end{cases} \quad (\text{如图 } 8.13(a))$$

Proof

If $p \neq 1$,

$$\int_1^b \frac{1}{x^p} dx = \frac{x^{-p+1}}{-p+1}\Big|_1^b - \int_1^b = \frac{1}{1-p}(b^{-p+1}-1) = \frac{1}{1-p}\left(\frac{1}{b^{p-1}}-1\right).$$

Thus, $\int_1^\infty \frac{1}{x^p} dx = \lim_{b \to \infty} \frac{1}{1-p}\left(\frac{1}{b^{p-1}}-1\right) = \begin{cases} \dfrac{1}{p-1}, & p > 1, \\ \infty, & p < 1, \end{cases}$

because $\lim_{b \to \infty} \frac{1}{b^{p-1}} = \begin{cases} 0, & p > 1, \\ \infty, & p < 1. \end{cases}$

Therefore, the integral converges to the value $1/(p-1)$ if $p > 1$ and it diverges if $p < 1$.

If $p = 1$,

$$\int_1^\infty \frac{1}{x} dx = \lim_{b \to \infty} \int_1^b \frac{1}{x} dx = \lim_{b \to \infty} \ln x \Big|_1^b - \int_1^b = \lim_{b \to \infty}(\ln b - \ln 1) = \infty \text{ and the integral diverges.}$$

8.6.2 Integrands with Infinite Discontinuities(无界函数的反常积分，又称瑕积分)

Integrals of functions that become infinite at a point within the interval of integration are improper integrals of Type Ⅱ.

(1) If $f(x)$ is continuous on $(a, b]$ and is discontinuous at a, then

$$\int_a^b f(x) dx = \lim_{c \to a^+} \int_c^b f(x) dx.$$

(2) If $f(x)$ is continuous on $[a, b)$ and is discontinuous at b, then

$$\int_a^b f(x) dx = \lim_{c \to b^-} \int_a^c f(x) dx.$$

(3) If $f(x)$ is discontinuous at c, where $a < c < b$, and continuous on $[a, c) \cup (c, b]$, then

$$\int_a^b f(x) dx = \int_a^c f(x) dx + \int_c^b f(x) dx.$$

In each case, if the limit is finite we say the improper integral **converges** and that the limit is the **value** of the improper integral. If the limit does not exist, the integral **diverges**.

Example 8.15 瑕积分.

Evaluating the Improper Integrals:

a. $\int_0^{1/2} \frac{1}{x} dx$.

b. $\int_0^1 \frac{dx}{\sqrt{1-x^2}}$.

Solution

a. $\because \int_0^{1/2} \frac{1}{x} dx = \lim_{a \to 0^+} \int_a^{1/2} \frac{1}{x} dx = \lim_{a \to 0^+} \ln x \Big|_a^{1/2} = \lim_{a \to 0^+}\left(\ln \frac{1}{2} - \ln a\right)$

$= \ln \frac{1}{2} - \lim_{a \to 0^+} \ln a = \infty,$

$\therefore \int_0^{1/2} \dfrac{1}{x} \mathrm{d}x$ diverges.

b. This integral is improper with an infinite discontinuity at $x=1$.

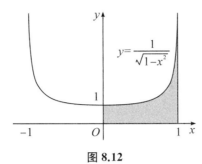

图 8.12

$\therefore \int_0^1 \dfrac{\mathrm{d}x}{\sqrt{1-x^2}} = \lim\limits_{b \to 1^-} \int_0^b \dfrac{\mathrm{d}x}{\sqrt{1-x^2}} = \lim\limits_{b \to 1^-}(\sin^{-1}b - \sin^{-1}0) = \sin^{-1}1 - \sin^{-1}0 = \dfrac{\pi}{2} - 0 = \dfrac{\pi}{2}$,

$\therefore \int_0^1 \dfrac{\mathrm{d}x}{\sqrt{1-x^2}}$ converges to $\dfrac{\pi}{2}$ (如图 8.12). The infinite shaded region has area $\dfrac{\pi}{2}$.

Theorem 2 The p-Integral over $[0,1]$.

$$\int_0^1 \dfrac{1}{x^p}\mathrm{d}x = \begin{cases} \dfrac{1}{1-p}, p<1 \\ \text{diverges if}, p \geqslant 1 \end{cases} \quad \text{[如图 8.13(b)]}.$$

[注]

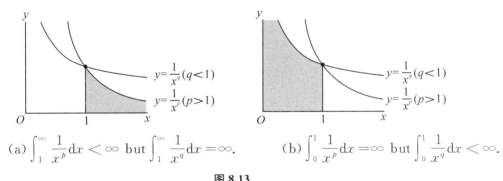

(a) $\int_1^\infty \dfrac{1}{x^p}\mathrm{d}x < \infty$ but $\int_1^\infty \dfrac{1}{x^q}\mathrm{d}x = \infty$. \quad (b) $\int_0^1 \dfrac{1}{x^p}\mathrm{d}x = \infty$ but $\int_0^1 \dfrac{1}{x^q}\mathrm{d}x < \infty$.

图 8.13

Theorem 3 Comparison Test for Improper Integrals(反常积分的比较判别法).

Assume that $f(x) \geqslant g(x) \geqslant 0$ for $x \geqslant a$.

(1) If $\int_a^\infty f(x)\mathrm{d}x$ converges, then $\int_a^\infty g(x)\mathrm{d}x$ also converges.

(2) If $\int_a^\infty g(x)\mathrm{d}x$ diverges, then $\int_a^\infty f(x)\mathrm{d}x$ also diverges.

The Comparison Test is also valid for improper integrals with infinite discontinuities at the endpoints.

Example 8.16 Using the Comparison Test.

1. $\int_1^\infty \frac{\sin^2 x}{x^2} dx$ converges because $0 \leq \frac{\sin^2 x}{x^2} \leq \frac{1}{x^2}$ on $[1,\infty)$ and $\int_1^\infty \frac{1}{x^2} dx$ converges.

2. $\int_1^\infty \frac{1}{\sqrt{x^2-0.1}} dx$ diverges because $\frac{1}{\sqrt{x^2-0.1}} \geq \frac{1}{x}$ on $[1,\infty)$ and $\int_1^\infty \frac{1}{x} dx$ diverges.

Practice Exercises(习题)

1. If three equal subdivision of $[-4,2]$ are used, what is the trapezoidal approximation of $\int_{-4}^{2} \frac{e^{-x}}{2} dx$?

   (A) $e^4 + e^2 + e^0$
   (B) $e^4 + 2e^2 + 2e^0 + e^{-2}$
   (C) $\frac{1}{2}(e^4 + e^2 + e^0 + e^{-2})$
   (D) $\frac{1}{2}(e^4 + 2e^2 + 2e^0 + e^{-2})$

2.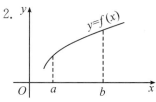

   If f is the continuous, strictly increasing function on the interval $a \leq x \leq b$ as shown above, Which of the following must be true?

   I. $\int_a^b f(x) dx < f(b)(b-a)$

   II. $\int_a^b f(x) dx > f(a)(b-a)$

   III. $\int_a^b f(x) dx = f(c)(b-a)$ for some number c such that $a < c < b$.

   (A) I only
   (B) II only
   (C) I and II
   (D) I, II and III

3. If $\int_a^b f(x) dx = 5$ and $\int_a^b g(x) dx = -1$, which of the following must be true?

   I. $f(x) > g(x)$ for $a \leq x \leq b$

   II. $\int_a^b (f(x) + g(x)) dx = 4$

   III. $\int_a^b (f(x) + g(x)) dx = -5$

   (A) I only
   (B) II only
   (C) I and II
   (D) I, II and III

4. If $f(x) = g(x) + 7$ for $3 \leq x \leq 5$, then $\int_3^5 [f(x) + g(x)] dx =$

   (A) $\int_3^5 g(x) dx + 7$
   (B) $2\int_3^5 g(x) dx + 7$

(C) $2\int_3^5 g(x)dx + 14$ (D) $2\int_3^5 g(x)dx + 28$

5. If p is a polynomial of degree n, $n>0$, what is the degree of the polynomial $Q(x) = \int_0^x p(t)dt$?

(A) 1 (B) n (C) $n+1$ (D) $n-1$

6. If $F(x) = \int_1^{x^2} \sqrt{1+t^3}\, dt$, then $F'(x) = $

(A) $2x\sqrt{1+x^6}$ (B) $2x\sqrt{1+x^3}$
(C) $\sqrt{1+x^6}$ (D) $\sqrt{1+x^3}$

7.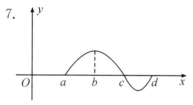

The graph of f is shown in the figure above. If $g(x) = \int_a^x f(t)dt$, for what value of x does $g(x)$ have a maximum?

(A) a (B) b (C) c (D) d

8. If $0 \leqslant x \leqslant 4$, of the following, which is the greatest value of x such that $\int_0^x (t^2 - 2t)dt \geqslant \int_2^x t\, dt$?

(A) 1.35 (B) 1.38 (C) 1.41 (D) 1.48

9. If the substitution $\sqrt{x} = \sin y$ is made in the integrand of $\int_0^{1/2} \frac{\sqrt{x}}{\sqrt{1-x}}dx$, the resulting integral is

(A) $\int_0^{1/2} \sin^2 y\, dy$ (B) $\int_0^{\pi/4} \sin^2 y\, dy$
(C) $2\int_0^{\pi/4} \sin^2 y\, dy$ (D) $2\int_0^{\pi/6} \sin^2 y\, dy$

10. If $\int_1^2 f(x-C)dx = 5$ where C is a constant, then $\int_{1-C}^{2-C} f(x)dx = $

(A) 5 (B) $5-C$
(C) $C-5$ (D) $5+C$

11. If the substitution $u = \frac{x}{2}$ is made, the integral $\int_2^4 \frac{1-(x/2)^2}{x}dx = $

(A) $\int_2^4 \frac{1-u^2}{u}du$ (B) $\int_1^2 \frac{1-u^2}{2u}du$
(C) $\int_1^2 \frac{1-u^2}{u}du$ (D) $\int_1^2 \frac{1-u^2}{4u}du$

12. If $\int_1^4 f(x)dx = 6$, what is the value of $\int_1^4 f(5-x)dx$?

(A) 6   (B) 3   (C) 0   (D) −1

13. $\int_0^1 (4-x^2)^{-\frac{3}{2}} dx =$

(A) $\dfrac{2-\sqrt{3}}{3}$ (B) $\dfrac{2\sqrt{3}-3}{4}$ (C) $\dfrac{\sqrt{3}}{12}$ (D) $\dfrac{\sqrt{3}}{3}$

14. $\int_0^1 \dfrac{5x+8}{x^2+3x+2} dx$ is

(A) ln8   (B) $\ln\dfrac{27}{2}$ (C) ln18   (D) ln288

x	2	4
$f(x)$	7	13
$g(x)$	2	9
$g'(x)$	1	7
$g''(x)$	5	8

15. The table above gives selected values of twice-differentiable functions f and g, as well as the first two derivative of g. If $f'(x) = 3$ for all values of x, what is the value of $\int_2^4 f(x)g''(x)dx$?

(A) 63   (B) 69   (C) 78   (D) 84

16.

t (hours)	0	1	3	6	8
$R(t)$ (liters per hour)	1340	1190	950	740	700

Water is pumped into a tank at a rate modeled by $R(t)$ liters per hour, where R is differentiable and decreasing on $0 \leqslant t \leqslant 8$. Selected values of $R(t)$ are shown in the table above. At time $t=0$, there are 50,000 liters of water in the tank.

Use a left Riemann sum with the four subintervals indicated by the table to estimate the total amount of water removed from the tank during the 8 hours. Is this an overestimate or an underestimate of the total amount of water removed? Give a reason for your answer.

17. Find $\dfrac{dy}{dx}$:

(1) $y = \int_0^x \sqrt{1+t^2}\, dt$;

(2) $y = \int_{\sin x}^{\cos x} e^{t^2}\, dt$;

(3) $y = \int_0^{x^2} x^2 \sin t^2\, dt$.

18. Use the comparison test to determine whether the improper integral converges or diverges.

(1) $\int_1^\infty \dfrac{\ln x}{x}\,\mathrm{d}x$;

(2) $\int_1^\infty \dfrac{\mathrm{e}^{-x}}{\sqrt{x}}\,\mathrm{d}x$.

19. Evaluate the improper integral or state that it diverges.

(1) $\int_1^\infty \dfrac{1}{x^{3/2}}\,\mathrm{d}x$;

(2) $\int_{-\infty}^{-1} \dfrac{3}{3x-x^2}\,\mathrm{d}x$;

(3) $\int_{-\infty}^\infty \dfrac{1}{\mathrm{e}^x+\mathrm{e}^{-x}}\,\mathrm{d}x$;

(4) $\int_0^1 \ln x\,\mathrm{d}x$;

(5) $\int_{-1}^1 \dfrac{1}{x^{2/3}}\,\mathrm{d}x$.

20. Evaluate the integrals：

(1) $\int_{\pi/3}^\pi \sin\left(x+\dfrac{\pi}{3}\right)\mathrm{d}x$;

(2) $\int_0^{\pi/2} \sin x\cos^3 x\,\mathrm{d}x$;

(3) $\int_0^5 \dfrac{x^3}{x^2+1}\,\mathrm{d}x$;

(4) $\int_0^1 t\mathrm{e}^{-\frac{t^2}{2}}\,\mathrm{d}t$;

(5) $\int_1^{\mathrm{e}^2} \dfrac{\mathrm{d}x}{x\sqrt{1+\ln x}}$;

(6) $\int_0^{\sqrt{2}} \sqrt{2-x^2}\,\mathrm{d}x$.

21. Evaluate the integrals：

(1) $\int_0^1 x\mathrm{e}^{-x}\,\mathrm{d}x$;

(2) $\int_1^{\mathrm{e}} x\ln x\,\mathrm{d}x$;

(3) $\int_0^1 x\arctan x\,\mathrm{d}x$;

(4) $\int_0^5 \dfrac{x^3}{x^2+1}\,\mathrm{d}x$.

习题参考答案

1.D 2.D 3.B 4.C 5.C 6.A 7.C 8.B 9.C 10.A 11.C 12.A 13.C 14.C

15. A

16. The total amount of water removed is given by $\int_0^8 R(t)\,dt$.

$$\int_0^\infty R(t)\,dt \approx 1 \cdot R(0) + 2 \cdot R(1) + 3 \cdot R(3) + 2 \cdot R(6)$$
$$= 1 \times 1340 + 2 \times 1190 + 3 \times 950 + 2 \times 740$$
$$= 8050.$$

This is an overestimate since R is a decreasing function.

17. (1) $\dfrac{d}{dx}\int_0^x \sqrt{1+t^2}\,dt = \sqrt{1+x^2}$.

   (2) $\dfrac{d}{dx}\int_{\sin x}^{\cos x} e^{t^2}\,dt = e^{\cos^2 x} \cdot (-\sin x) - e^{\sin^2 x}\cos x$.

   (3) $\dfrac{d}{dx}\int_0^{x^2} x^2 \sin t^2\,dt = \dfrac{d}{dx}(x^2 \int_0^{x^2} \sin t^2\,dt) = x^2 (\int_0^{x^2} \sin t^2\,dt)' + (x^2)' \int_0^{x^2} \sin t^2\,dt$
   $= x^2 \sin x^4 \cdot 2x + 2x \int_0^{x^2} \sin t^2\,dt$
   $= 2x^3 \sin x^4 + 2x \int_0^{x^2} \sin t^2\,dt$.

18. (1) Because $\int_1^\infty \dfrac{\ln x}{x}\,dx = \int_1^e \dfrac{\ln x}{x}\,dx + \int_e^\infty \dfrac{\ln x}{x}\,dx > \int_1^e \dfrac{\ln x}{x}\,dx + \int_e^\infty \dfrac{1}{x}\,dx = \infty$,

   Therefore $\int_1^\infty \dfrac{\ln x}{x}\,dx$ diverges.

   (2) Because $0 \leqslant \int_1^\infty \dfrac{e^{-x}}{\sqrt{x}}\,dx \leqslant \int_1^\infty e^{-x}\,dx$, $\int_1^\infty e^{-x}\,dx = e^{-1}$,

   Therefore $\int_1^\infty \dfrac{e^{-x}}{\sqrt{x}}\,dx$ converges.

19. (1) $\int_1^\infty \dfrac{1}{x^{3/2}}\,dx = 2$.

   (2) $\int_{-\infty}^{-1} \dfrac{3}{3x - x^2}\,dx = -2\ln 2$.

   (3) $\int_{-\infty}^\infty \dfrac{1}{e^x + e^{-x}}\,dx = \dfrac{\pi}{2}$.

   (4) $\int_0^1 \ln x\,dx = -1$.

   (5) $\int_{-1}^1 \dfrac{1}{x^{2/3}}\,dx = 6$.

20. (1) $\int_{\frac{\pi}{3}}^{\pi} \sin\left(x + \dfrac{\pi}{3}\right) dx = 0$.

   (2) $\int_0^{\frac{\pi}{2}} \sin x \cos^3 x\,dx = \dfrac{1}{4}$.

   (3) $\int_0^5 \dfrac{x^3}{x^2 + 1}\,dx = \dfrac{25}{2} - \dfrac{1}{2}\ln 26$.

(4) $\int_0^1 t\mathrm{e}^{-\frac{t^2}{2}}\mathrm{d}t = 1 - \mathrm{e}^{-\frac{1}{2}}$.

(5) $\int_1^{\mathrm{e}^2} \dfrac{\mathrm{d}x}{x\sqrt{1+\ln x}} = 2\sqrt{3} - 2$.

(6) $\int_0^{\sqrt{2}} \sqrt{2-x^2}\,\mathrm{d}x = \dfrac{\pi}{2}$.

21.(1) $\int_0^1 x\mathrm{e}^{-x}\mathrm{d}x = 1 - \dfrac{2}{\mathrm{e}}$.

(2) $\int_1^{\mathrm{e}} x\ln x\,\mathrm{d}x = \dfrac{1}{4}(\mathrm{e}^2 + 1)$.

(3) $\int_0^1 x\arctan x\,\mathrm{d}x = \dfrac{\pi}{4} - \dfrac{1}{2}$.

(4) $\int_0^5 \dfrac{x^3}{x^2+1}\mathrm{d}x = \dfrac{25}{2} - \dfrac{1}{2}\ln 26$.

Chapter 9 Applications of the Integral to Geometry(定积分的几何应用)

9.1 The Element Method of Definite Integrals（定积分的元素法）

回忆曲边梯形的面积：设 $y=f(x)\geqslant 0$ $(x\in[a,b])$，$A(x)=\int_a^x f(t)\mathrm{d}t$ 就是以 $[a,x]$ 为底的曲边梯形的面积，微分 $\mathrm{d}A(x)=f(x)\mathrm{d}x$ 表示将曲边梯形无限细分后以 $\mathrm{d}x$ 为宽的小曲边梯形面积 $f(x)\mathrm{d}x$，称为曲边梯形的面积元素.

以 $[a,b]$ 为底的曲边梯形的面积 A 就是以面积元素 $f(x)\mathrm{d}x$ 为被积表达式，以 $[a,b]$ 为积分区间的定积分，即

$$A=\int_a^b f(x)\mathrm{d}x.$$

一般情况下，为求某一个量 $U(x)$，先将此量分布在某一区间 $[a,b]$ 上，再将这一个量无限细分后求其元素 $\mathrm{d}U(x)$. 设 $\mathrm{d}U(x)=u(x)\mathrm{d}x$，则

$$U=\int_a^b u(x)\mathrm{d}x.$$

这种方法称为元素法.

9.2 Area between Two Curves（由两条曲线所围成的图形的面积）

9.2.1 Integration with Respect to x（对 x 积分求面积，竖切法）

Suppose that f and g are continuous functions on an interval $[a,b]$ and
$$f(x)\geqslant g(x) \text{ for } a\leqslant x\leqslant b.$$

[This means that the curve $y=f(x)$ lies above the curve $y=g(x)$ and that the two can touch but not cross.]

如图 9.1 所示，将曲边梯形进行分割后，面积元素是一个以 $(f(x)-g(x))$ 为高，以 $\mathrm{d}x$（或 $\mathrm{d}x=\Delta x$）为宽的长方形，即 $(f(x)-g(x))\mathrm{d}x$，所以由元素法可得：

The area A of the region bounded above by $y=f(x)$, below by $y=g(x)$, and on the sides by the lines $x=a$ and $x=b$, is

$$A = \int_a^b (f(x) - g(x))dx.$$

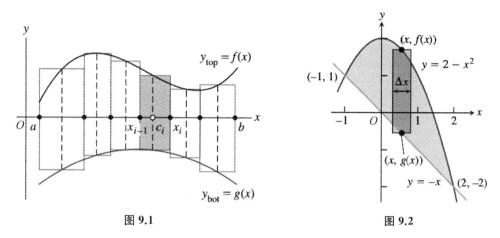

图 9.1　　　　　　　　　图 9.2

Example 9.1　竖切法求面积.

Find the area of the region enclosed by the parabola $y=2-x^2$ and the line $y=-x$.

Solution

如图 9.2 所示, 面积元素 $dA = [(2-x^2)-(-x)]\Delta x = [(2-x^2)-(-x)]dx$,

所以面积 $A = \int_{-1}^{2} dA = \int_{-1}^{2} [(1-x^2)-(-x)]dx$

$= \int_{-1}^{2} (2+x-x^2)dx = \left(2x + \frac{x^2}{2} - \frac{x^3}{3}\right)\Big|_{-1}^{2}$

$= \left(4 + \frac{4}{2} - \frac{8}{3}\right) - \left(-2 + \frac{1}{2} + \frac{1}{3}\right) = \frac{9}{2}.$

Example 9.2　竖切法求面积.

Find the area of the region in the first quadrant that is bounded above by $y=\sqrt{x}$ and below by the x-axis and the line $y=x-2$.

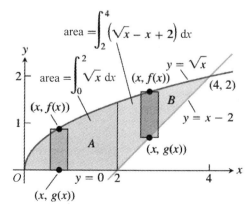

图 9.3

Solution

此例和 Example 9.1 的同之处在于，影部分的下曲线由 $y=0$ 和 $y=x-2$ 两个部分所构成（图 9.3），所以首要将阴影部分进行分割，然后分别求出各部分面积后相加.

$$\text{Total area} = \underbrace{\int_0^2 \sqrt{x}\,\mathrm{d}x}_{\text{area of A}} + \underbrace{\int_2^4 (\sqrt{x} - x + 2)\,\mathrm{d}x}_{\text{area of B}}$$

$$= \frac{2}{3} x^{3/2} \Big|_0^2 + \left(\frac{2}{3} x^{3/2} - \frac{x^2}{2} + 2x \right) \Big|_2^4$$

$$= \frac{2}{3} \times 2^{3/2} - 0 + \left(\frac{2}{3} \times 4^{3/2} - 8 + 8 \right) - \left(\frac{2}{3} \times 2^{3/2} - 2 + 4 \right)$$

$$= \frac{2}{3} \times 8 - 2$$

$$= \frac{10}{3}.$$

9.1.2 Integration with Respect to y（对 y 积分求面积，横切法）

If a region's bounding curves are described by functions of y, the approximating rectangles are horizontal instead of vertical and the basic formula has y in place of x.

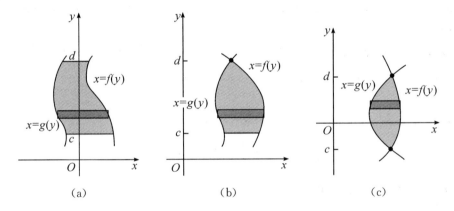

图 9.4

如图 9-4 所示，当横着对阴影部分进行分割时，面积元素是宽为 $(f(y)-g(y))$，高为 $\mathrm{d}y(\mathrm{d}y=\Delta y)$ 的小矩形的面积，面积元素 $\mathrm{d}A = (f(y)-g(y))\mathrm{d}y$，所以阴影部分的面积为

$$A = \int_c^d \mathrm{d}A = \int_c^d (f(y) - g(y))\,\mathrm{d}y.$$

In this equation, f always denotes the right-hand curve and g the left-hand curve, so $f(y) - g(y)$ is nonnegative.

Example 9.3 横切法求面积.

Find the area of the region in Example 9.2 by integrating with respect to y.

Chapter 9 Applications of the Integral to Geometry(定积分的几何应用)

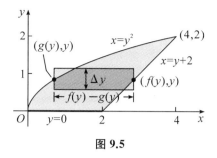

图 9.5

Solution
$$A = \int_0^2 (y+2-y^2)\mathrm{d}y = \left(2y+\frac{y^2}{2}-\frac{y^3}{3}\right)\Big|_0^2 = 4+\frac{4}{2}-\frac{8}{3}=\frac{10}{3}.$$

[注]Example 9.3 与 Example 9.2 的解法相比,用横切法时由于左、右曲线都为同一个表达式,所以不需要进行分割.

9.2.3 Area in Polar Coordinates(极坐标方程求面积)

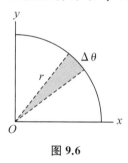

图 9.6

The area of a circular sector is exactly $\frac{1}{2}r^2\Delta\theta$(图 9.6).

Theorem 1 Area in Polar Coordinates.

如图 9.7 所示,用过极点的射线将阴影部分无限细分,分出的面积元素是以 $f(\theta)$ 为半径,以 $\mathrm{d}\theta(\mathrm{d}\theta=\Delta\theta)$ 为圆心角的扇形,即 $\mathrm{d}A=\frac{1}{2}(f(\theta))^2\mathrm{d}\theta$.

If $f(\theta)$ is a continuous function, then the area bounded by a curve in polar form $r=f(\theta)$ and the rays $\theta=\alpha$ and $\theta=\beta$ (with $\alpha<\beta$) is equal to
$$A=\frac{1}{2}\int_\alpha^\beta r^2\mathrm{d}\theta=\frac{1}{2}\int_\alpha^\beta (f(\theta))^2\mathrm{d}\theta$$

Theorem 2 Area between two curves.

如图 9.8 所示,用过极点的射线将阴影部分无限细分,分出的面积元素是分别以 $f_2(\theta)$ 和 $f_1(\theta)$ 为半径,以 $\mathrm{d}\theta$ 为圆心角的两个扇形面积之差,即
$$\mathrm{d}A=\frac{1}{2}[f_2(\theta)]^2\mathrm{d}\theta-\frac{1}{2}[f_1(\theta)]^2\mathrm{d}\theta=\frac{1}{2}((f_2(\theta))^2-(f_1(\theta))^2).$$

The area between two polar curves $r=f_1(\theta)$ and $r=f_2(\theta)$, for $\alpha\leqslant\theta\leqslant\beta$, is equal to:

Area between two curves $=\frac{1}{2}\int_\alpha^\beta((f_2(\theta))^2-(f_1(\theta))^2)\mathrm{d}\theta$.

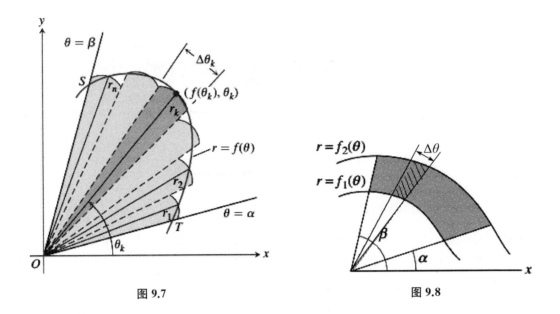

图 9.7　　　　　　　　　　图 9.8

Example 9.4　极曲线求面积.

Find the area of the region in the plane enclosed by the cardioid(心形线) $r=2(1+\cos\theta)$(如图 9.9).

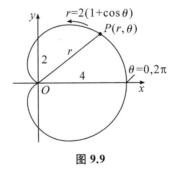

图 9.9

Solution

$$A = \frac{1}{2}\int_\alpha^\beta r^2 \, d\theta = \frac{1}{2}\int_0^{2\pi} 4(1+\cos\theta)^2 \, d\theta = \int_0^{2\pi} 2(1+2\cos\theta+\cos^2\theta) \, d\theta$$

$$= \int_0^{2\pi}\left(2+4\cos\theta+2\cdot\frac{1+\cos 2\theta}{2}\right)d\theta = \int_0^{2\pi}(3+4\cos\theta+\cos 2\theta)\,d\theta$$

$$= \left(3\theta+4\sin\theta+\frac{\sin 2\theta}{2}\right)\Big|_0^{2\pi} = 6\pi-0 = 6\pi.$$

Example 9.5　两条极曲线之间的面积.

Find the area of the region that lies inside the circle $r=1$ and outside the cardioid $r=1-\cos\theta$.

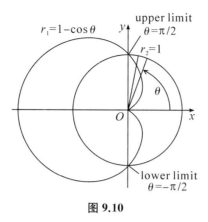

图 9.10

Solution

如图 9.10, area $=\dfrac{1}{2}\int_\alpha^\beta((f_2(\theta))^2-(f_1(\theta))^2)\mathrm{d}\theta = \dfrac{1}{2}\int_{-\pi/2}^{\pi/2}((f_2(\theta))^2-(f_1(\theta))^2)\mathrm{d}\theta$

$= 2\cdot\dfrac{1}{2}\int_0^{\pi/2}((f_2(\theta))^2-(f_1(\theta))^2)\mathrm{d}\theta$

$= \int_0^{\pi/2}(1-(1-2\cos\theta+\cos^2\theta))\mathrm{d}\theta$

$= \int_0^{\pi/2}\left(2\cos\theta-\left(\dfrac{1+\cos 2\theta}{2}\right)\right)\mathrm{d}\theta$

$= \left(2\sin\theta-\left(\dfrac{\theta}{2}+\dfrac{\sin 2\theta}{4}\right)\right)\Big|_0^{\pi/2} = 2-\dfrac{\pi}{4}.$

9.3 Volumes by Slicing(切片法求体积)

9.3.1 Solids with Known Cross Sections(截面面积已知的物体的体积)

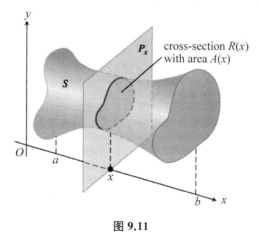

图 9.11

A cross-section of the solid S formed by intersecting S with a plane P_x perpendicular to the x-axis through the point x in the interval $[a,b]$.

如图 9.11 所示,垂直于 x 轴进行无限细分,分割得到的体积元素为

$$dV = A(x)dx.$$

The volume of a solid of known integrable cross-sectional area $A(x)$ is the integral of A from a to b,

$$V = \int_a^b A(x)dx (\text{垂直于 } x \text{ 轴的截面面积为 } A(x)).$$

Let $A(y)$ be the area of the horizontal cross section at height y of a solid body extending from $y=a$ to $y=b$, then

$$V = \int_a^b A(y)dy (\text{垂直于 } y \text{ 轴的截面面积为 } A(y)).$$

Example 9.6 垂直于 x 轴切片.

Compute the volume V of the solid lies between planes perpendicular to the x-axis at $x=-1$ and $x=1$. **The cross sections perpendicular to the x-axis** between these planes run from the semicircle to the semicircle $y=-\sqrt{1-x^2}$ to the semicircle $y=\sqrt{1-x^2}$.

a. The cross sections are circular disks with diameters in the xy-plane (如图 9.12).

b. The cross sections are squares with bases in the xy-plane (如图 9.13).

c. The cross sections are squares with diagonals in the xy-plane (如图 9.14). (The length of a square's diagonal is $\sqrt{2}$ times the length of its sides.)

d. The cross sections are equilateral triangles with bases in the xy-plane (如图 9.15).

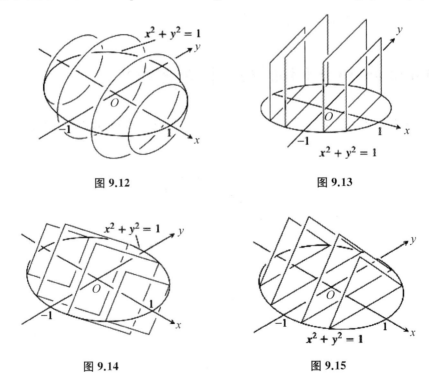

图 9.12　　　　　图 9.13

图 9.14　　　　　图 9.15

Solution

a. $V = \int_{-1}^{1} \pi(1-x^2)\,dx = \pi\left(x - \dfrac{x^3}{3}\right)\Big|_{-1}^{1} = \dfrac{4\pi}{3}$.

b. $V = \int_{-1}^{1} 4(1-x^2)\,dx = 4\left(x - \dfrac{x^3}{3}\right)\Big|_{-1}^{1} = \dfrac{16}{3}$.

c. $V = \int_{-1}^{1} 2(1-x^2)\,dx = 2\left(x - \dfrac{x^3}{3}\right)\Big|_{-1}^{1} = \dfrac{8}{3}$.

d. $V = \int_{-1}^{1} \dfrac{\sqrt{3}}{4}(2\sqrt{1-x^2})^2\,dx = \sqrt{3}\left(x - \dfrac{x^3}{3}\right)\Big|_{-1}^{1} = \dfrac{4\sqrt{3}}{3}$.

Example 9.7 垂直于 y 轴切片.

Compute the volume V of the solid, whose base is the region between the inverted parabola $y = 4 - x^2$ and the x-axis, and whose vertical **cross sections perpendicular to the y-axis** are semicircles.

Solution

To find a formula for the area $A(y)$ of the cross section, observe that $y = 4 - x^2$ can be written $x = \pm\sqrt{4-y}$. We see in the figure 9.16 that the cross section at y is a semicircle of radius $r = \sqrt{4-y}$. This semicircle has area:

$$A(y) = \dfrac{1}{2}\pi r^2 = \dfrac{\pi}{2}(4-y) \quad (\text{垂直于 } y \text{ 轴的截面是半径为 } x = \sqrt{4-y} \text{ 的半圆}).$$

Therefore

$$V = \int_0^4 A(y)\,dy = \dfrac{\pi}{2}\int_0^4 (4-y)\,dy = \dfrac{\pi}{2}\left(4y - \dfrac{1}{2}y^2\right)\Big|_0^4 = 4\pi.$$

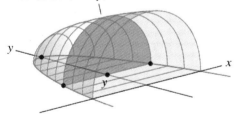

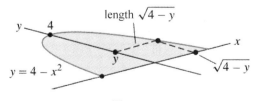

图 9.16

9.3.2 Solids of Revolution: Disks and Washers(旋转体的体积：圆盘法和垫圈法)

9.3.2.1 Solids of Revolution(旋转体)

A **solid of revolution** is a solid that is generated by revolving a plane region about a line that lies in the same plane as the region; the line is called the **axis of revolution**(旋转轴).

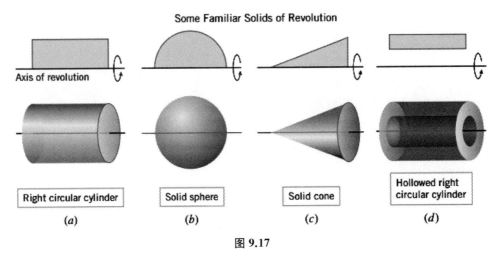

图 9.17

9.3.2.2 Disk 圆盘法

Let f be continuous and nonnegative on $[a,b]$, and let R be the region that is bounded above by $y=f(x)$, below by the x-axis, and on the sides by the lines $x=a$ and $x=b$ [图 9.18(a)]. Find the volume of the solid of revolution that is generated by revolving the region R about the x-axis.

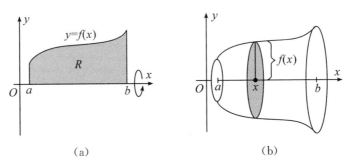

图 9.18

如图 9.18(b)所示，垂直于旋转轴作切面，将旋转体进行无限细分，分出的体积元素近似为底面半径是 $f(x)$，高为 $\mathrm{d}x$ 的圆柱体，即 $\mathrm{d}V=\pi\,[f(x)]^2\,\mathrm{d}x$.

$$V=\int_a^b \pi\,[f(x)]^2\,\mathrm{d}x.$$

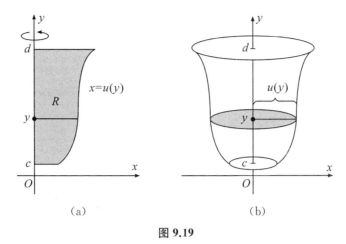

(a)   (b)

图 **9.19**

The volume of the solid of revolution that is generated by revolving the region R about the y-axis(如图 9.19) is

$$V = \int_c^d \pi [u(y)]^2 \mathrm{d}y.$$

Because the cross sections are disk shaped, the application of this formula is called the **method of disks**(圆盘法).

Example 9.8 以 x 轴为旋转轴的旋转体的体积.

Find the volume of the solid that is obtained when the region under the curve $y = \sqrt{x}$ over the interval $[1, 4]$ is revolved about the x-axis(图 9.20).

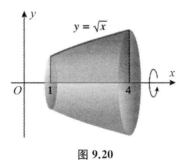

图 **9.20**

Solution

$$V = \int_a^b \pi [f(x)]^2 \mathrm{d}x = \int_1^4 \pi x \mathrm{d}x$$

$$= \frac{\pi x^2}{2} \Big|_1^4 = 8\pi - \frac{\pi}{2}$$

$$= \frac{15\pi}{2}.$$

Example 9.9 以 y 轴为旋转轴的旋转体的体积.

Find the volume of the solid generated when the region enclosed by $y = \sqrt{x}$, $y = 2$ and $x = 0$ is revolved about the y-axis(图 9.21).

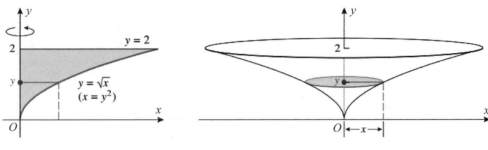

图 9.21

Solution

$$V = \int_c^d \pi [u(y)]^2 \, dy = \int_0^2 \pi y^4 \, dy = \frac{\pi y^5}{5}\bigg|_0^2 = \frac{32\pi}{5}.$$

9.3.2.3 Washer(垫圈法)

Let f and g be continuous and nonnegative on $[a,b]$, and suppose that $f(x) \geqslant g(x)$ for all x in the interval $[a,b]$. Let R be the region that is bounded above by $y = f(x)$, below by $y = g(x)$, and on the sides by the lines $x = a$ and $x = b$ [图 9.22(a)]. Find volume of the solid of revolution that is generated by revolving the region R about the x-axis [图 9.22(b)].

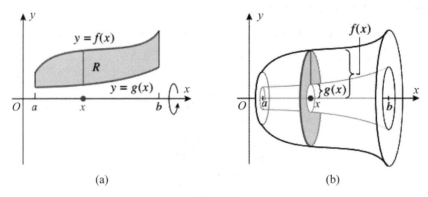

(a)   (b)

图 9.22

如图 9.22(b)所示，垂直于旋转轴作截面，将旋转体进行无限细分，分割出的体积元素是底面半径为 $f(x)$ 和 $g(x)$，高均为 dx 的两个圆柱体的体积之差，即

$$dV = \pi[f(x)]^2 dx - \pi[g(x)]^2 dx = \pi[(f(x))^2 - (g(x))^2] dx.$$

所以

$$V = \int_a^b \pi[f(x)^2 - g(x)^2] dx \,(\text{以 } x \text{ 轴为旋转轴的旋转体体积}).$$

The volume of the solid of revolution that is generated by revolving the region R about the y-axis(图 9.23) is

$$V = \int_a^b \pi[(w(y))^2 - (v(y))^2] dy \,(\text{以 } y \text{ 轴为旋转轴的旋转体体积}).$$

Because the cross sections are washer shaped, the application of this formula is called the **method of washers**(垫圈法).

Chapter 9 Applications of the Integral to Geometry(定积分的几何应用)

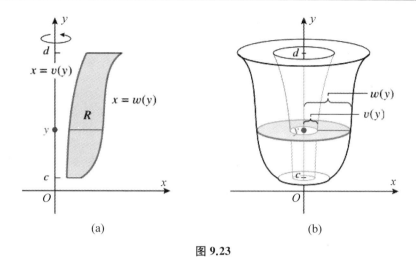

(a)　　　　　　　　　　　　　　(b)

图 9.23

Example 9.10 以 *x* 轴为旋转轴的旋转体体积.

Find the volume of the solid generated when the region between the graphs of the equations $f(x) = \frac{1}{2} + x^2$ and $g(x) = x$ over the interval $[0, 2]$ is revolved about the x-axis (图 9.24).

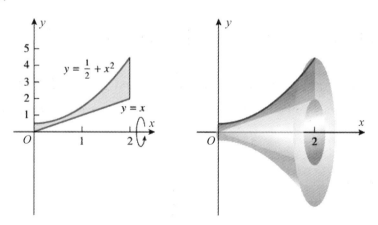

(a) Region defined by f and g. 　　(b) The resulting solid of revolution.

图 9.24 **Unequal scales on axes**

Solution

$$V = \int_a^b \pi([f(x)]^2 - [g(x)]^2) dx = \int_0^2 \pi[(\frac{1}{2} + x^2)^2 - x^2] dx$$
$$= \int_0^2 \pi(\frac{1}{4} + x^4) dx = \pi(\frac{x}{4} + \frac{x^5}{5}) \Big|_0^2$$
$$= \frac{69\pi}{10}.$$

Example 9.11 以 *y* 轴为旋转轴的旋转体体积.

Find the volume of the solid obtained by rotating the region under the graph of $f(x) =$

$9-x^2$ for $0 \leqslant x \leqslant 3$ about the vertical axis $x=-2$.

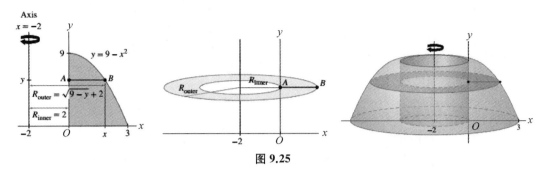

图 9.25

Solution

$$V = \pi \int_0^9 [(\sqrt{9-y}+2)^2 - 2^2] dy$$
$$= \pi \int_0^9 (9-y+4\sqrt{9-y}) dy$$
$$= \pi \left[9y - \frac{1}{2}y^2 - \frac{8}{3}(9-y)^{3/2} \right] \Big|_0^9$$
$$= \frac{225}{2}\pi.$$

9.4 Length of a Plan Curve(平面曲线的弧长)

9.4.1 Arc Length Formula in Cartesian coordinates(直角坐标下的弧长公式)

If $f'(x)$ continuous on $[a,b]$, we will say that $y=f(x)$ is a **smooth curve**(平滑的曲线) on $[a,b]$.

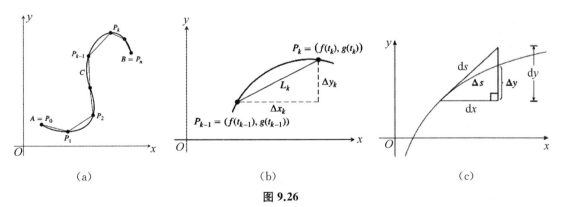

图 9.26

如图 9.26(a) 所示，将平滑曲线无限细分成无数段小弧线，这些小弧线可以用直线段近似代替[如图 9.26(b)]，当每段小弧线的长度都无限趋近于零时，弧长元素 $ds = \sqrt{(dx)^2 + (dy)^2}$，如图 9.26(c).

因此弧长为：
$$L = \int_a^b ds.$$

If $y=f(x)$ is a smooth curve on the interval $[a,b]$, then the arc length L of this

curve over $[a,b]$ is defined as
$$L = \int_a^b \sqrt{1+\left(\frac{dy}{dx}\right)^2}\,dx = \int_a^b \sqrt{1+[f'(x)]^2}\,dx \text{（自变量为 } x\text{）}.$$

Moreover, for a curve expressed in the form $x = g(y)$, where g is continuous on $[c, d]$, the arc length L from $y = c$ to $y = d$ can be expressed as
$$L = \int_a^b \sqrt{1+\left(\frac{dx}{dy}\right)^2}\,dy = \int_c^d \sqrt{1+[g'(y)]^2}\,dy \text{（自变量为 } y\text{）}.$$

Example 9.12 直角坐标方程求弧长.

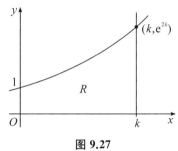

图 9.27

Let R be the region in the first quadrant bounded by the graph of $f(x) = e^{2x}$, the coordinate axes, and the vertical line $x = k$, where $k > 0$. The region R is shown in the figure above(见图 9.27).

Write, but do not evaluate, an expression involving an integral that gives the perimeter of R in terms of k.

Solution
$$\text{Perimeter} = 1 + k + e^{2k} + \int_0^k \sqrt{1+(2e^{2x})^2}\,dx.$$

9.4.2 Arc Length Formula for Parametric Curves(参数方程下的弧长公式)

If no segment of the curve represented by the parametric equations
$$x = x(t), \quad y = y(t) \quad (a \leqslant t \leqslant b)$$
is traced more than once as t increases from a to b, and if $\dfrac{dx}{dt}$ and $\dfrac{dy}{dt}$ are continuous functions for $a \leqslant t \leqslant b$, then the arc length L of the curve is given by
$$L = \int_a^b \sqrt{\left(\frac{dx}{dt}\right)^2 + \left(\frac{dy}{dt}\right)^2}\,dt.$$

Example 9.13 参数方程求弧长.

Find the circumference of a circle of radius a from the parametric equations
$$x = a\cos t, \quad y = a\sin t \quad (0 \leqslant t \leqslant 2\pi)$$

Solution
$$L = \int_a^b \sqrt{\left(\frac{dx}{dt}\right)^2 + \left(\frac{dy}{dt}\right)^2}\,dt = \int_0^{2\pi} \sqrt{(-a\sin t)^2 + (a\cos t)^2}\,dt = \int_0^{2\pi} a\,dt = at\,\Big|_0^{2\pi} = 2\pi a.$$

9.3.3 Arc Length of a Polar Curve(极坐标曲线求弧长)

If no segment of the polar curve $r=f(\theta)$ is traced more than once as θ increases from α to β, and if $\dfrac{dr}{d\theta}$ is continuous for $\alpha \leqslant \theta \leqslant \beta$, then the arc length L from $\theta=\alpha$ to $\theta=\beta$ is

$$L=\int_{\alpha}^{\beta}\sqrt{r^2+\left(\dfrac{dr}{d\theta}\right)^2}\,d\theta=\int_{\alpha}^{\beta}\sqrt{[f(\theta)]^2+[f'(\theta)]^2}\,d\theta.$$

Proof

$\because \begin{cases} x=r\cos\theta \\ y=r\sin\theta \end{cases} \Rightarrow \begin{cases} \dfrac{dx}{d\theta}=\dfrac{dr}{d\theta}\cdot\cos\theta+r\cdot(-\sin\theta) \\ \dfrac{dy}{d\theta}=\dfrac{dr}{d\theta}\cdot\sin\theta+r\cdot\cos\theta \end{cases}$

$ds=\sqrt{(dx)^2+(dy)^2}=\sqrt{\left(\dfrac{dx}{d\theta}\right)^2+\left(\dfrac{dy}{d\theta}\right)^2}=\sqrt{r^2+\left(\dfrac{dr}{d\theta}\right)^2}\,d\theta,$

$\therefore L=\int_{\alpha}^{\beta}\sqrt{r^2+\left(\dfrac{dr}{d\theta}\right)^2}\,d\theta.$

Practice Exercises(习题)

1. What is the area of the region enclosed by the graphs of $y=\dfrac{1}{1+x^2}$ and $y=x^2-\dfrac{1}{3}$?

   (A) 0.786　　　　(B) 0.791　　　　(C) 1.582　　　　(D) 1.837

2. The area of the region enclosed by the polar curve $r=1-\cos\theta$ is

   (A) π　　　　(B) $\dfrac{3\pi}{2}$　　　　(C) 2π　　　　(D) 3π

3. The area of the region between the graph of $y=4x^3+2$ and the x-axis from $x=1$ to $x=2$ is

   (A) 36　　　　(B) 23　　　　(C) 20　　　　(D) 17

4.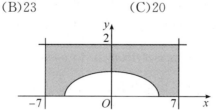

   The shaded region in the figure above is bounded by the graph of $y=\sqrt{\cos\left(\dfrac{\pi x}{10}\right)}$ and the lines $x=-7$, $x=7$, and $y=2$. What is the area of this region?

   (A) 20.372　　　　(B) 21.634　　　　(C) 24.634　　　　(D) 26.132

5. Which of the following gives the total area enclosed by the graph of the polar curve $r=\theta\sin 2\theta$ for $0 \leqslant \theta \leqslant 2\pi$?

   (A) $\displaystyle\int_0^{2\pi}\dfrac{1}{2}|\theta\sin 2\theta|\,d\theta$　　　　(B) $\displaystyle\int_0^{2\pi}|\theta\sin 2\theta|\,d\theta$

(C) $\int_0^{2\pi} \frac{1}{2}(\theta \sin 2\theta)^2 d\theta$ (D) $\int_0^{2\pi}(\theta \sin 2\theta)^2 d\theta$

6.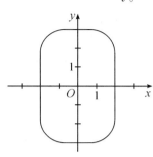

The base of solid is the region enclosed by the curve $\frac{x^4}{16}+\frac{y^4}{81}=1$ shown in the figure above. For the solid each cross section perpendicular to the x-axis is a semicircle. What is the volume of the solid?

(A) 12.356   (B) 22.249   (C) 24.712   (D) 49.425

7. Let R be the region in the first quadrant bounded above by the graph of $y=\ln(3-x)$, for $0 \leqslant x \leqslant 2$. R is the base of a solid for which each cross section perpendicular to the x-axis is a square. What is the volume of the solid?

(A) 0.442   (B) 1.029   (C) 1.296   (D) 3.233

8. Give the volume of the solid generated by revolving the region bounded by the graph of $y=\ln x$, the x-axis, the lines $x=1$ and $x=e$, about the y-axis.

(A) $\frac{1}{4}\pi(e^2+e)$ (B) $\frac{1}{4}\pi(e^2-e)$ (C) $\frac{1}{2}\pi(e^4+1)$ (D) $\frac{1}{2}\pi(e^2+1)$

9. A vase has the shape obtained by revolving the curve $y=2+\sin x$ from $x=0$ to $x=5$ about the x-axis, where x and y are measured in inches. What is the volume, in cubic inches, of the vase?

(A) 25.502   (B) 33.555   (C) 72.113   (D) 80.115

10. Which of the following gives the length of the curve $y=\sqrt{x}$ over the closed interval $[1,4]$?

(A) $\int_1^4 \sqrt{1+\frac{1}{2x}}\, dx$

(B) $\int_1^4 \sqrt{1-\frac{1}{4x}}\, dx$

(C) $\int_1^4 \sqrt{1+\frac{1}{4x}}\, dx$

(D) $\int_1^4 \sqrt{1+\frac{1}{4}x^2}\, dx$

11. Which of the following gives the length of the curve defined by the parametric equations $x(t)=\frac{t^2}{2}$ and $y(t)=\frac{t^3}{3}$ from $t=0$ to $t=1$?

(A) $\int_0^1 \sqrt{1+4t^2}\, dx$

(B) $\int_0^1 \sqrt{1+t^4}\, dx$

(C) $\int_0^1 \sqrt{t^2+t^4}\, dx$

(D) $\int_0^1 \sqrt{\frac{t^4}{4}+\frac{t^6}{9}}\, dx$

12.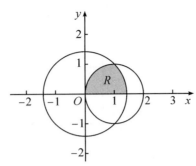

The figure above shows the graphs of the circles $x^2+y^2=2$ and $(x-1)^2+y^2=1$. The graphs intersect at the points $(1,1)$ and $(1,-1)$. Let R be shaded region in the first quadrant bounded by the two circles and the x-axis.

a. Set up an expression involving one or more integrals with respect to x that represents the area of R.

b. Set up an expression involving one or more integrals with respect to y that represents the area of R.

c. The polar equations of the circles are $r=\sqrt{2}$ and $r=2\cos\theta$, respectively. Set up an expression involving one or more integrals with respect to the polar angle θ that represents the area of R.

13. A curved wedge is cut from a cylinder of radius 3 by two planes. One plane is perpendicular to the axis of the cylinder. The second plane crosses the first plane at a 45° angle at the center of the cylinder. Find the volume of the wedge.

14.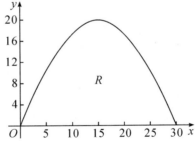

A baker is creating a birthday cake. The base of the cake is the region R in the first quadrant under the graph of $y=f(x)$ for $0 \leqslant x \leqslant 30$, where $f(x)=20\sin\left(\dfrac{\pi x}{30}\right)$. Both x and y are measured in centimeters. The region R is shown in the figure above. The derivative of f is $f'(x)=\dfrac{2\pi}{3}\cos\left(\dfrac{\pi x}{30}\right)$.

a. The region R is cut out of a 30-centimeter by 20-centimeter rectangular sheet of cardboard, and the remaining cardboard is discarded. Find the area of the discarded cardboard.

b. The cake is a solid with base R. Cross sections of the cake perpendicular to the x-axis are semicircles. If the baker uses 0.05 gram of unsweetened chocolate for each cubic centimeter of cake, how many grams of unsweetened chocolate will be in the cake?

c. Find the perimeter of the base of the cake.

习题参考答案

1.C 2.B 3.D 4.A 5.C 6.D 7.B 8.D 9.D 10.C 11.C

12. a. Area $=\int_0^1 \sqrt{1-(x-1)^2}\,dx + \int_1^{\sqrt{2}} \sqrt{2-x^2}\,dx$ or

    Area $=\dfrac{1}{4}(\pi \cdot 1^2) + \int_1^{\sqrt{2}} \sqrt{2-x^2}\,dx$.

   b. Area $=\int_0^1 [\sqrt{2-y^2} - (1-\sqrt{1-y^2})]\,dy$.

   c. Area $=\int_0^{\frac{\pi}{4}} \dfrac{1}{2}(\sqrt{2})^2\,d\theta + \int_{\frac{\pi}{4}}^{\frac{\pi}{2}} \dfrac{1}{2}(2\cos\theta)^2\,d\theta$ or

    Area $=\dfrac{\pi}{8}(\sqrt{2})^2 + \int_{\frac{\pi}{4}}^{\frac{\pi}{2}} \dfrac{1}{2}(2\cos\theta)^2\,d\theta$.

13.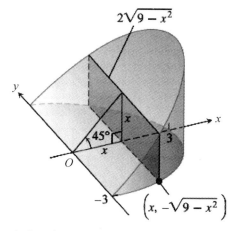

We draw the wedge and sketch a typical cross-section perpendicular to the x-axis. The cross-section at x is a rectangle of area:

$$A(x) = \text{height} \times \text{width} = x \cdot 2\sqrt{9-x^2} = 2x\sqrt{9-x^2}.$$

The rectangles run from $x=0$ to $x=3$, so we have

$$V = \int_a^b A(x)\,dx = \int_0^3 2x\sqrt{9-x^2}\,dx$$

$$= -\dfrac{2}{3}(9-x^2)^{3/2}\Big|_0^3$$

$$= 0 + \dfrac{2}{3} \times 9^{3/2} = 18.$$

14. a. Area $= 30 \times 20 - \int_0^{30} f(x)\,dx = 218.028 \text{ cm}^2$.

b. Volume $= \int_0^{30} \dfrac{\pi}{2}\left(\dfrac{f(x)}{2}\right)^2 \mathrm{d}x = 2356.194 \text{ cm}^3$.

Therefore, the baker needs $2356.194 \times 0.05 = 117.809$ or 117.810 grams of chocolate.

c. Perimeter $= 30 + \int_0^{30} \sqrt{1+(f'(x))^2}\, \mathrm{d}x = 81.803$ or 81.804 cm.

Chapter 10 Differential Equations(微分方程)

10.1 Definitions of Differential Equations(微分方程的相关概念)

Example 10.1 引例.

If a particle move alone the x-axis, the velocity function is $v(t) = \dfrac{\mathrm{d}x}{\mathrm{d}t} = 4t^3 - 3t^2$. (10.1)

So position function is $x(t) = \int (4t^3 - 3t^2)\mathrm{d}t = t^4 - t^3 + C$ (C is a arbitrary constant). (10.2)

Initial position (at time $t=0$) of the particle is $x=3$, that is $x(0) = 0 - 0 + C = 3$.
The solution to the initial-value problem is
$$x(t) = t^4 - t^3 + 3. \tag{10.3}$$

例 10.1 是在已知质点的速度函数的条件下,求质点的位置函数的运动问题.在现实生活中,我们有时知道的是函数及其导数之间的关系,这样的关系式就是微分方程.通过例 10.1 我们可以理解微分方程的相关概念.

10.1.1 Differential Equation(微分方程)

A differential equation is an equation that involves an unknown function $y = y(x)$ and its first or higher derivatives.

含有未知函数的一阶或高阶导数的方程称为微分方程.微分方程中所出现的未知函数的最高阶导数的阶数叫作微分方程的阶数,(10.1)式为一阶微分方程.

10.1.2 General Solution(微分方程的通解)

A general solution is a function $y = f(x)$ satisfying the differential equation. Solutions usually depend on one or more arbitrary constants.

如果微分方程的解中含有任意常数,且任意常数的个数与微分方程的阶数相同,则这样的解叫作微分方程的通解,如(10.2)式为微分方程(10.1)的通解.

10.1.3 Particular Solution(微分方程的特解)

不含任意常数的微分方程的解称为微分方程的特解,如(10.3)式为微分方程(10.1)的一个特解.

10.1.4 Initial Condition and Initial-Value Problem(初始条件与初值问题)

确定通解中的任意常数的值的条件称为初始条件,求满足初始条件的微分方程的特解的问题称为初值问题.

10.2 Separable Differential Equations(可分离变量的微分方程)

如果一个一阶微分方程能写成
$$g(y)\mathrm{d}y = f(x)\mathrm{d}x$$
的形式,也就是通过恒等变形能使微分方程的两个变量 x 和 y 分别位于等式的两边,那么原方程就称为可分离变量的微分方程.

Example 10.2 求可分离变量微分方程的解.

Consider the differential equation $\dfrac{\mathrm{d}y}{\mathrm{d}x} = y^2(2x+2)$. Let $y = f(x)$ be the particular solution to the differential equation with initial condition $f(0) = -1$, find $y = f(x)$.

Solution

$\dfrac{\mathrm{d}y}{\mathrm{d}x} = y^2(2x+2)$ 是可分离变量的微分方程.

第一步:分离变量,将含有 y 和 x 的表达式通过恒等变形分别位于等号的左右两边.

$$\frac{\mathrm{d}y}{y^2} = (2x+2)\mathrm{d}x$$

第二步:两边分别积分,求通解.

$$\int \frac{\mathrm{d}y}{y^2} = \int (2x+2)\mathrm{d}x.$$

$-\dfrac{1}{y} + C_1 = x^2 + 2x + C_2$(其中 C_1, C_2 为任意常数).

C_1, C_2 可以合并为任意常数 C,即

$-\dfrac{1}{y} = x^2 + 2x + C$(微分方程的隐式通解)

第三步:由初始条件,确定任意常数 C 的值,求出特解.

$f(0) = -1$,

$-\dfrac{1}{-1} = 0^2 + 2 \times 0 + C \Rightarrow C = 1.$

$-\dfrac{1}{y} = x^2 + 2x + 1$(微分方程的隐式特解).

$y = -\dfrac{1}{x^2 + 2x + 1} = -\dfrac{1}{(x+1)^2}$(微分方程的显式特解).

[注]解可分离变量的微分方程的基本步骤:

第一步:分离变量,将方程写成 $g(y)\mathrm{d}y = f(x)\mathrm{d}x$ 的形式;

第二步:两端积分,$\int g(y)\mathrm{d}y = \int f(x)\mathrm{d}x$,设积分后得 $G(y) = F(x) + C$,其中 $G(y) = $

$F(x)+C$ 称为隐式(通)解.

第三步:由初始条件,确定任意常数 C 的值,求出特解.

Example 10.3 解可分离变量的微分方程.

Find the particular solution $W=W(t)$ to the differential equation $\dfrac{dW}{dt}=\dfrac{1}{25}(W-300)$ with initial condition $W(0)=1400$.

Solution

$\dfrac{dW}{dt}=\dfrac{1}{25}(W-300)$

$\Rightarrow \dfrac{dW}{(W-300)}=\dfrac{1}{25}dt$

$\Rightarrow \displaystyle\int \dfrac{dW}{(W-300)}=\int \dfrac{1}{25}dt$

$\Rightarrow \ln|W-300|=\dfrac{1}{25}t+C.$

$\ln|400-300|=\dfrac{1}{25}(0)+C \Rightarrow \ln 1100=C.$

$\because t=0, W=1400,$

$\therefore |W-300|=W-300=1100e^{\frac{1}{25}t}$(由于初始条件满足微分方程的特解,所以考试中根据初始条件决定去掉绝对值的符号,不要求对绝对值符号进行讨论).

特解为:$W(t)=300+1100e^{\frac{1}{25}t}.$

10.3 Numerical and Graphical Methods(微分方程的数值和图像解法)

10.3.1 Euler's Method(欧拉方法)

欧拉方法是利用微分方程和初始条件,对函数值进行逐步切线近似的一种求函数近似值的方法.

To approximate the solution of the initial-value problem

$$\dfrac{dy}{dx}=f(x,y), \quad f(x_0)=y_0$$

proceed as follows:

Step 1 Choose a nonzero number Δx to serve as an **increment**(增量) or **step size**(步长) along the x-axis, and let

$$x_1=x_0+\Delta x, x_2=x_1+\Delta x, x_3=x_2+\Delta x, \cdots$$

Step 2 Compute successively:

$$f(x_1)\approx f(x_0)+f'(x_0,y_0)\Delta x$$
$$f(x_2)\approx f(x_1)+f'(x_1,y_1)\Delta x$$
$$f(x_3)\approx f(x_2)+f'(x_2,y_2)\Delta x$$
$$\vdots$$

$$f(x_{n+1}) \approx f(x_n) + f'(x_n, y_n)\Delta x$$
$$\vdots$$

Example 10.4 欧拉方法.

Consider the differential equation $\dfrac{dy}{dx} = y^2(2x+2)$. Let $y = f(x)$ be the particular solution to the differential equation with initial condition $f(0) = -1$.

Use Euler's method, starting at $x = 0$ with two steps of equal size, to approximate $f\left(\dfrac{1}{2}\right)$.

Solution

这里,步长 $\Delta x = \dfrac{\frac{1}{2} - 0}{2} = \dfrac{1}{4}$.

$$f\left(\frac{1}{4}\right) \approx f(0) + f'(0)\left(\frac{1}{4}\right) = -1 + 2 \times \frac{1}{4} = -\frac{1}{2}.$$

$$f\left(\frac{1}{2}\right) \approx f\left(\frac{1}{4}\right) + f'\left(\frac{1}{4}\right)\left(\frac{1}{4}\right) = -\frac{1}{2} + \left(-\frac{1}{2}\right)^2\left(2 \times \frac{1}{4} + 2\right) \times \frac{1}{4} = -\frac{11}{32}.$$

Example 10.5 欧拉方法.

Let f be the function whose graph goes through the point $(3,6)$ and whose derivative is given by $f'(x) = \dfrac{1+e^x}{x^2}$.

a. Write an equation of the line tangent to the graph of f at $x = 3$ and use it to approximate $f(3.1)$.

b. Use Euler's method, starting at $x = 3$ with a step size of 0.05, to approximate $f(3.1)$. Use f'' to explain why this approximation is less than $f(3.1)$.

Solution

a. $f'(3) = \dfrac{1+e^3}{9} = 2.342$ or 2.343.

$y - 6 = \dfrac{1+e^3}{9}(x-3)$.

$y = \dfrac{1+e^3}{9}(x-3) + 6$.

$f(3.1) \approx \dfrac{1+e^3}{9}(3.1-3) + 6 = 6.234$.

b. $f(3.05) \approx f(3) + f'(3) \times 0.05 = 6 + 0.11714 = 6.11714$.

$f(3.1) \approx 6.11714 + f'(3.05) \times 0.05 = 6.11714 + 2.37735 \times 0.05 = 6.236$.

$f''(x) = \dfrac{x^2 e^x - 2x(1+e^x)}{x^4} = \dfrac{(x-2)e^x - 2}{x^3}$.

For $x \geq 3$, $f''(x) > \dfrac{e^x - 2}{x^3} > 0$ and the graph of f is concave upward on $(3, 3.1)$.

Therefore, the approximation lines at 3 and 3.05 lie below the graph.

10.3.2 Slope Field and Solution Curve(斜率场与解曲线)

对于一阶微分方程 $y'=f(x,y)$,根据导数的几何意义 y' 代表切线的斜率,则微分方程对应的某个过点 (x,y) 的特解的图像便是一条**解曲线(Solution Curve)**.由于过 (x,y) 的切线的斜率为 $f(x,y)$,所以用过点 (x,y) 的斜率为 $f(x,y)$ 的小线段可以直观表示该点的导数值(图 10.1、表 10.1).

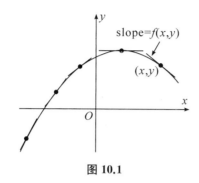

图 10.1

对微分方程 $\dfrac{\mathrm{d}y}{\mathrm{d}x}=y-x$ 可以列表求出各点的导数值,逐个用斜率为导数值的小线段表示,得到的图像称为斜率场(图 10.2),当分点越多时得到的斜率场就越密(图 10.3).

表 10.1 微分方程 $\dfrac{\mathrm{d}y}{\mathrm{d}x}=y-x$ 各点对应的导数值

	$y=-3$	$y=-2$	$y=-1$	$y=0$	$y=1$	$y=2$	$y=3$
$x=-3$	0	1	2	3	4	5	6
$x=-2$	-1	0	1	2	3	4	5
$x=-1$	-2	-1	0	1	2	3	4
$x=0$	-3	-2	-1	0	1	2	3
$x=1$	-4	-3	-2	-1	0	1	2
$x=2$	-5	-4	-3	-2	-1	0	1
$x=3$	-6	-5	-4	-3	-2	-1	0

微分方程 $\quad \dfrac{\mathrm{d}y}{\mathrm{d}x}=y-x \quad$ (10-4)

有通解: $\quad y=x+1+C\mathrm{e}^x \quad$ (10-5)

将微分方程的通解绘制成函数图像,得到的曲线称为解曲线.显然,由于通解中含有任意常数 C,所以通解对应的解曲线是一簇曲线.若已知解曲线过点 $(0,2)$,则从 (10-5) 式可以得到微分方程 (10-4) 的特解:

$$y=x+1+\mathrm{e}^x \quad (10-6)$$

(10-6) 式作为微分方程的特解,所对应的解曲线就是一条曲线.斜率场中当我们取的分点越来越密,小线段越来越短,所绘出的斜率场就越来越接近解曲线(图 10.4).所以,斜率场是求微分方程解曲线的一种直观的方法.

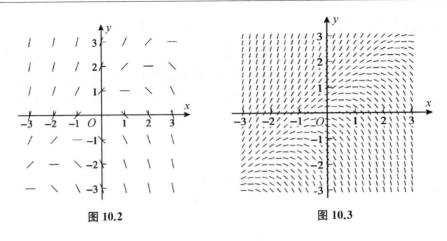

图 10.2　　　　　　　图 10.3

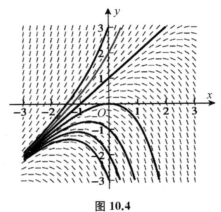

图 10.4

Example 10.6　绘制斜率场.

Consider the differential equation given by $\dfrac{dy}{dx} = x(y-1)^2$.

a. On the axes provided, sketch a slope field for the given differential equation at the eleven points indicated.

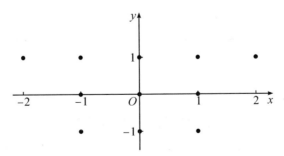

b. Use the slope field for the given differential equation to explain why a solution could not have the graph shown below.

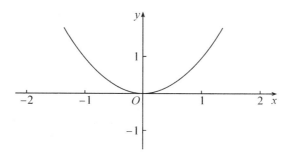

Solution:

a.图中格点对应的斜率列表计算如下：

	$y=-1$	$y=0$	$y=1$
$x=-2$	—	—	0
$x=-1$	-4	-1	0
$x=0$	0	0	0
$x=1$	4	1	0
$x=2$	—	—	0

根据表中结果画出斜率场：

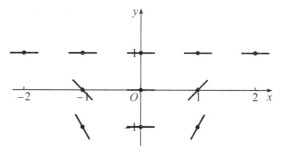

b. The graph does not have slope 0 where $y=1$.

or

The slope field shown suggests that solutions are asymptotic to $y=1$ from below, but the graph does not exhibit this behavior.

Example 10.7 由斜率场绘制解曲线.

Consider the logistic differential equation $\dfrac{dy}{dt} = \dfrac{y}{8}(6-y)$. Let $y=f(t)$ be the particular solution to the differential equation with $f(0)=8$. A slope field for this differential equation is given below. Sketch possible solution curves through the points (3, 2) and (0,8).

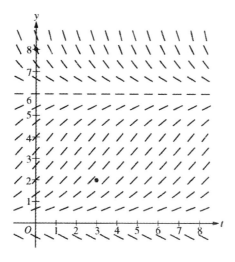

Solution

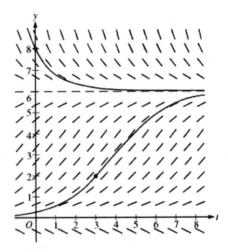

[注]根据斜率场作解曲线时要注意以下几点：

（1）解曲线形状和斜率场中小线段的走势相同.

（2）斜率场中的小线段与解曲线凡相交处必相切.

（3）考试中一些关键点处的斜率线段和解曲线一定要画准确，如与坐标轴的交点、过初始条件的点.

Example 10.8 *由微分方程确定斜率场.*

In parts (a)~(f), match the differential equation with the differential equation with the slope field, and explain your reasoning.

(a) $y'=\dfrac{1}{x}$; (b) $y'=\dfrac{1}{y}$; (c) $y'=e^{-x^2}$; (d) $y'=y^2-1$; (e) $y'=\dfrac{x+y}{x-y}$; (f) $y'=(\sin x)(\sin y)$.

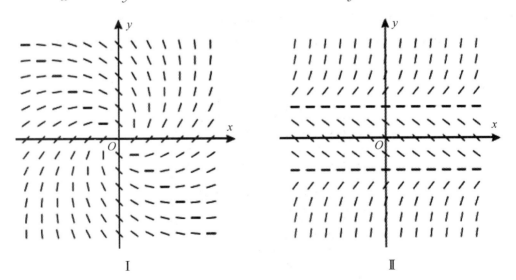

I 　　　　　　　　　II

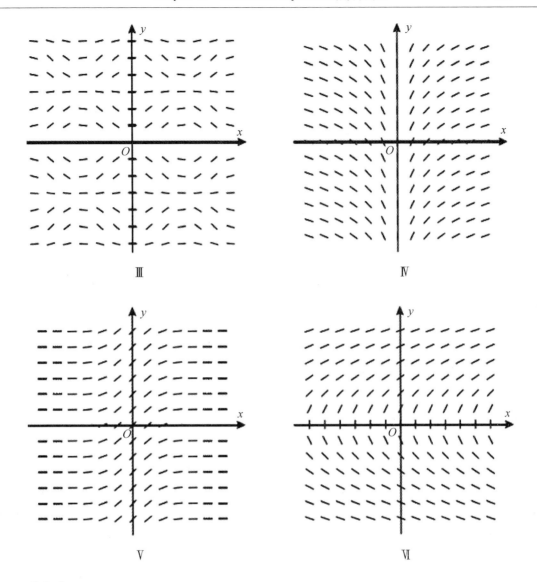

Solution

(a) 微分方程 $y'=\dfrac{1}{x}$ 在 $x=0$ 处 $y'=\pm\infty$，所以斜率场在 y 轴有垂直的小线段.同时 $x>0$ 时 $y'>0$，$x<0$ 时 $y'<0$，所以斜率场在第 I、IV 象限的小线段的斜率为正，在第 II、III 象限的小线段的斜率为负，故 $y'=\dfrac{1}{x}$ 对应的斜率场为图 IV.

(b) 微分方程 $y'=\dfrac{1}{y}$ 在 $y=0$ 处 $y'=\pm\infty$，所以斜率场在 x 轴有垂直切线，故 $y'=\dfrac{1}{y}$ 对应的斜率场为图 VI.

(c) 微分方程 $y'=\mathrm{e}^{-x^2}$ 在 $x=0$ 处有 $f'(0)=1$，所以斜率场在 y 轴有斜率为 1 的平行小线段，故图 V 与微分方程 $y'=\mathrm{e}^{-x^2}$ 相匹配.

(d) 微分方程 $y'=y^2-1$ 在 $y=\pm 1$ 时有 $y'=0$，所以斜率场在 $y=\pm 1$ 处有水平的小线段，故图 II 与微分方程 $y'=y^2-1$ 相匹配.

(e) 微分方程 $y'=\dfrac{x+y}{x-y}$ 在 $y=x$ 处 $y'=\pm\infty$，在 $y=-x$ 处 $y'=0$，所以斜率场在 $y=x$ 处为垂直的小线段，在 $y=-x$ 处为水平的小线段，故图 I 与微分方程 $y'=\dfrac{x+y}{x-y}$ 相匹配.

(f) 微分方程 $y'=(\sin x)(\sin y)$ 在 $x=0$ 和 $y=0$ 处有 $y'=0$，所以斜率场在 x 轴和 y 轴有水平的小线段，所以微分方程 $y'=(\sin x)(\sin y)$ 与图Ⅲ相匹配.

[注]斜率场与微分方程的匹配问题的解题方法：

(1) 观察 x 轴上小线段的切线斜率方向，看是否有对称或平行关系.

(2) $\dfrac{dy}{dx}=0$ 处的斜率场有水平的小线段.

(3) $\dfrac{dy}{dx}=\pm\infty$ 处的斜率场有垂直的小线段.

(4) 观察各象限对应的导数的符号，确定各个象限的斜率场的小线段的斜率为正还是为负.

10.4 Applications of First-Order Differential Equations(一阶微分方程的应用)

10.4.1 Exponential Growth and Decay(指数增长和指数衰减)

10.4.1.1 Definition(定义)

A quantity $y=y(t)$ is said to have an **exponential growth** if it increases at a rate that is proportional to the amount of the quantity present, and it is said to have an **exponential decay** if it decreases at a rate that is proportional to the amount of the quantity present. Thus, for an exponential growth model, the quantity $y(t)$ satisfies an equation of the form

指数增长的微分方程为： $\dfrac{dy}{dt}=ky\,(k>0)$

and for an exponential decay model, the quantity $y(t)$ satisfies an equation of the form

指数衰减的微分方程为： $\dfrac{dy}{dt}=-ky\,(k>0)$

The constant k is called the **growth constant**(增长常数)or the **decay constant**(衰减常数), as appropriate.

指数增长：

$$\dfrac{dy}{dt}=ky\,(k>0)$$

$\Rightarrow\quad \dfrac{1}{y}dy=k\,dt$

$\Rightarrow\quad \int \dfrac{1}{y}dy=\int k\,dt$

$\Rightarrow\quad \ln|y|=kt+C_1$

$\Rightarrow \quad y = Ce^{kt}$ (general solution)

Initial condition: $y = y_0$ when $t = 0$

$\Rightarrow \quad y = y_0 e^{kt}$ (指数增长的特解)

指数衰减:

$$\frac{dy}{dt} = -ky \ (k > 0)$$

$\Rightarrow \quad \dfrac{1}{y} dy = -k\, dt$

$\Rightarrow \quad \displaystyle\int \dfrac{1}{y} dy = -\int k\, dt$

$\Rightarrow \quad \ln|y| = -kt + C_1$

$\Rightarrow \quad y = Ce^{-kt}$ (general solution)

Initial condition: $y = y_0$ when $t = 0$

$\Rightarrow \quad y = y_0 e^{-kt}$ (指数衰减的特解)

10.4.1.2 Doubling Time and Half-Life(双倍时间和半衰期)

If a quantity y has an exponential growth model, then the time required for the original size to double is called the doubling time, and if y has an exponential decay model, then the time required for the original size to reduce by half is called the half-life.

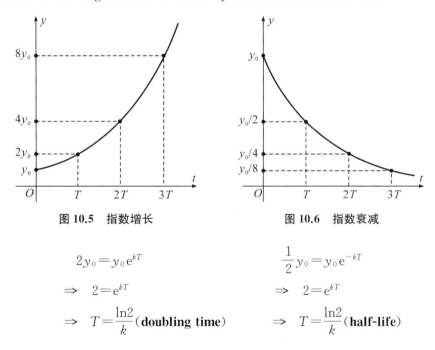

图 10.5 指数增长　　　　　图 10.6 指数衰减

$2y_0 = y_0 e^{kT}$　　　　　　　　$\dfrac{1}{2} y_0 = y_0 e^{-kT}$

$\Rightarrow \quad 2 = e^{kT}$　　　　　　　　$\Rightarrow \quad 2 = e^{kT}$

$\Rightarrow \quad T = \dfrac{\ln 2}{k}$ (**doubling time**)　　$\Rightarrow \quad T = \dfrac{\ln 2}{k}$ (**half-life**)

Example 10.9 碳-14 年代测定法.

It is a fact of physics that radioactive elements disintegrate spontaneously in a process called **radioactive decay**. Every radioactive element has a specific half-life. For example, the half-life of radioactive carbon−14 is about 5730 years.

In 1988 the Vatican authorized the British Museum to date a cloth relic known as the

Shroud of Turin, possibly the burial shroud of Jesus of Nazareth. This cloth, which first surfaced in 1356, contains the negative image of a human body that was widely believed to be that of Jesus. The report of the British Museum showed that the fibers in the cloth contained between 92% and 93% of their original carbon$-$14. Use this information to estimate the age of the shroud.

Solution

$$T = \frac{\ln 2}{k} \Rightarrow k = \frac{\ln 2}{5730} \approx 0.000121$$

This implies that if there are y_0 units of carbon$-$14 present at time $t=0$, then the number of units present after t years will be approximately

$$y(t) = y_0 e^{-0.000121t}$$

$$\Rightarrow \frac{y(t)}{y_0} = e^{-0.000121t} \Rightarrow t = -\frac{1}{0.000121} \ln\left(\frac{y(t)}{y_0}\right)$$

Thus, taking $\dfrac{y(t)}{y_0}$ to be 0.93 and 0.92, we obtain

$$t = -\frac{1}{0.000121} \ln(0.93) \approx 600 \quad \text{and} \quad t = -\frac{1}{0.000121} \ln(0.92) \approx 689$$

This means that when the test was done in 1988, the shroud was between 600 and 689 years old, thereby placing its origin between 1299 A.D. and 1388 A.D. Thus, if one accepts the validity of carbon$-$14 dating, the Shroud of Turin cannot be the burial shroud of Jesus of Nazareth.

10.4.1.3 Continuously Compounded Interest(连续复利)

Differential equation: $\dfrac{\mathrm{d}A}{\mathrm{d}t} = rA$

Initial condition: $A(0) = A_0$

The amount of money in the account after t years is then

$$A(t) = A_0 e^{rt}$$

Interest paid according to this formula is said to be compounded continuously. The number r is the continuous interest rate.

假定以固定的年利率 r(用小数表示)投资 A_0 美元。若一年内 k 次把利息加入账目，则 t 年后的现金总额是

$$A(t) = A_0 \left(1 + \frac{r}{k}\right)^{kt}$$

利息可以每月($k=12$)、每周($k=52$)、每天($k=365$)，或甚至更频繁地，每小时或每分钟加入(银行家称为"复利")。

如果利息不是离散的区间加入，而是连续地以正比于账户现金的速率把利息加入账目，就可以用一个初值问题为账目的增长建模。

微分方程为: $\dfrac{\mathrm{d}A}{\mathrm{d}t} = rA$

初始条件：$\qquad A(0)=A_0$

t 年后账户的现金总额为：
$$A(t)=A_0 e^{rt}$$

按这个公式付的利息称为连续复利，数 r 是连续利率．

Example 10.10

假定你在一个账户以 6.3% 的年利息存款 800 美元．8 年后你将有多少美元？如果利息是

a.连续复利息．

b.季度复利息．

Solution

这里 $A_0=800, r=0.063$．

8 年后，账户中的金额精确到分是：

a.连续复利息：
$$A(8)=800 e^{0.063\times 8}=1324.26.$$

b.季度复利息：
$$A(8)=800\times \left(1+\frac{0.063}{4}\right)^{4\times 8}=1319.07.$$

10.4.2 Restrict Growth and Decay(约束增长和约束衰减)

A quantity $y=y(t)$ whose rate of change is proportional to the difference $y-b$. Differential equation:
$$\frac{dy}{dt}=k(y-b)(k>0) \quad \text{or} \quad \frac{dy}{dt}=-k(y-b)(k>0)$$

10.4.2.1 养老金问题的微分方程：$\dfrac{dy}{dt}=k(y-b)(k>0)$

通解为：$y(t)=b+Ce^{kt}$．

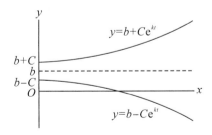

图 10.7 养老金问题的解曲线

从解曲线可以观察出，如果初始值大于 b，那么 $\lim\limits_{t\to\infty}y(t)=\infty$，即养老金可以永远提取；若初始值小于 b，那么 $\lim\limits_{t\to\infty}y(t)=-\infty$，即养老金在有限时间内会被提取完．

Example 10.11 养老金问题．

An annuity earns interest at the rate $r=0.07$, and withdrawals are made continuously at a rate of $N = \$500/\text{year}$.

a. When will the annuity run out of money if the initial deposit is $P(0) = \$5000$?

b. Show that the balance increases indefinitely if $P(0) = \$9000$.

Solution

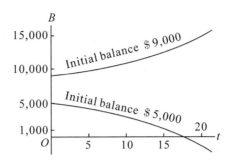

We have $N/r = \dfrac{500}{0.07} \approx 7143$(若初期存入 7143 美元,则每年的利息刚好是 500 美元),so $P(t) = 7143 + Ce^{0.07t}$.

a. If $P(0) = 5000 = 7143 + Ce^0$, then $C = -2143$ and $P(t) = 7143 - 2143e^{0.07t}$.

The account runs out of money when $P(t) = 7143 - 2143e^{0.07t} = 0$, or

$$e^{0.07t} = \frac{7143}{2143} \Rightarrow 0.07t = \ln\left(\frac{7143}{2143}\right) \approx 1.2$$

The annuity money runs out at time $t = \dfrac{1.2}{0.07} \approx 17$ years.

b. If $P(0) = 9000 = 7143 + Ce^0$, then $C = 1857$ and

$$P(t) = 7143 + 1857e^{0.07t}$$

Since the coefficient $C = 1857$ is positive, the account never runs out of money. In fact, $P(t)$ increases indefinitely as $t \to \infty$.

10.4.2.2 牛顿冷却定律的微分方程:$\dfrac{dy}{dt} = -k(y - b)(k > 0)$

通解为:$y(t) = b + Ce^{-kt}$.

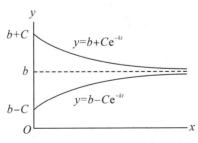

图 10.8 牛顿冷却定律的解曲线

从解曲线可以观察出,不管初始温度是否大于室温 b,$\lim\limits_{t \to \infty} y(t) = b$ 说明最终物体会恢复成室温.

Example 10.12 牛顿冷却定律.

A hot metal bar with cooling constant $k = 2.1$ min^{-1} is submerged in a large tank of water held at temperature $T_0 = 10°C$. Let $y(t)$ be the bar's temperature at time t (in minutes).

a. Find the differential equation satisfied by $y(t)$ and find its general solution.

b. What is the bar's temperature after 1 min if its initial temperature was 180°C?

c. What was the bar's initial temperature if it cooled to 80°C in 30 s?

Solution

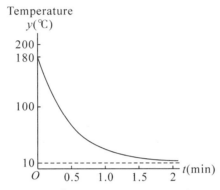

a. Since $k = 2.1$ min^{-1}, $y(t)$ (with t in minutes) satisfies

$$\frac{dy}{dt} = -2.1(y-10) \text{ (因为} \frac{dy}{dt} < 0, \text{且} \frac{dy}{dt} \text{与温差}(y-10)\text{成正比)}$$

$$\Rightarrow y(t) = 10 + Ce^{-2.1t}$$

b. $y(0) = 10 + C = 180$

$\Rightarrow C = 170$

$\Rightarrow y(1) = 10 + 170e^{-2.1(1)} \approx 30.8°C$

c. If the temperature after 30 s is 80°C, then $y(0.5) = 80$, and we have

$$10 + Ce^{-2.1 \times 0.5} = 80 \Rightarrow Ce^{-2.1 \times 0.5} = 70 \Rightarrow C = 70e^{1.05} \approx 200$$

$$\Rightarrow y(0) = 10 + 200e^{-2.1 \times 0} = 10 + 200 = 210°C.$$

10.4.3 The Logistic Equation(逻辑斯蒂方程)

10.4.3.1 Definition(定义)

The rate of change of a quantity (for example, a population) may be proportional both to the amount (size) of the quantity and to the difference between a fixed constant A and its amount(size).

Logistic differential equation: $\quad \dfrac{dy}{dt} = ky(A-y) \ (k>0, A>0)$

The general solution: $\quad y = \dfrac{A}{1+Ce^{-Akt}}$

Here $k > 0$ is the **growth constant**, and $A > 0$ is a constant called the **carrying capacity** (承载能力).

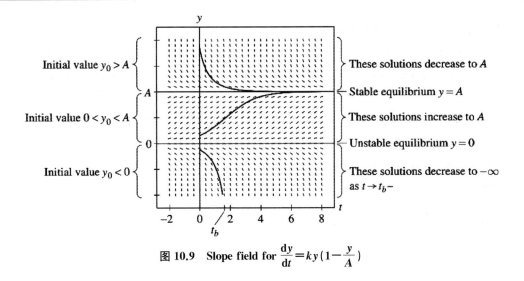

图 10.9　Slope field for $\dfrac{dy}{dt}=ky\left(1-\dfrac{y}{A}\right)$

10.4.3.2　Logistic Curve 的性质

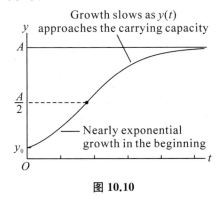

图 10.10

当初始值 $0<y_0<A$ 时，Logistic 方程的解曲线为一条 S 形曲线，这条曲线称为 Logistic Curve. 从曲线的图像（图 10.10）可见：

(1) 当 $t\to\infty$ 时，$y\to A$，此时数量无限接近于承载能力 A.

(2) y 的增长速度 $\dfrac{dy}{dt}$ 从 $t=0$ 开始越来越快，在 $y=\dfrac{A}{2}$ 处达到最快，然后 y 的增长速度逐渐放缓，$t\to\infty$ 时 $\dfrac{dy}{dt}\to 0$.

(3) $y=\dfrac{A}{2}$ 处为 Logistic Curve 的 inflection.

Example 10.13　逻辑斯蒂方程.

The growth rate of a population P of bears in a newly established wildlife preserve is modeled by the differential equation $\dfrac{dP}{dt}=0.008P(100-P)$, where t is measured in years.

a. What is the carrying capacity for bears in this wildlife preserve?

b. What is the bear population when the population is growing the fastest?

c. What is the rate of change of the population when it is growing the fastest?

Solution

a. The carrying capacity is 100 bears.

b. The bear population is growing the fastest when it is half the carrying capacity, 50 bears.

c. When $P=50$, $\dfrac{dP}{dt}=0.008\times 50\times(100-50)=20$ bears per year. Although the derivative represents the instantaneous growth rate, it is reasonable to say that the population grows by about 20 bears that year.

Practice Exercises(习题)

1. During a certain epidemic, the number of people that are infected at any time increases at a rate proportional to the number of people that are infected at that time. If 1000 people are infected when the epidemic is first discovered, and 1200 are infected 7 days later, how many people are infected 12 days after the epidemic is first discovered?

   (A) 343　　　(B) 1343　　　(C) 1367　　　(D) 1400

2. Bacteria in a certain culture increase at a rate proportional to the number present. If the number of bacteria doubles in three hours, in how many hours will the number of bacteria triple?

   (A) $\dfrac{3\ln 3}{\ln 2}$　　　(B) $\dfrac{2\ln 3}{\ln 2}$　　　(C) $\dfrac{\ln 3}{\ln 2}$　　　(D) $\ln\left(\dfrac{27}{2}\right)$

3. The population $P(t)$ of a species satisfies the logistic differential equation $\dfrac{dP}{dt}=P\left(2-\dfrac{P}{5000}\right)$, where the initial population $P(0)=3000$ and t is the time in years. What is $\lim\limits_{t\to\infty} P(t)$?

   (A) 2500　　　(B) 5000　　　(C) 10000　　　(D) 12000

4. The number of moose in a national park is modeled by the function M that satisfies the logistic differential equation $\dfrac{dM}{dt}=0.6M\left(1-\dfrac{M}{200}\right)$, where t is the time in years and $M(0)=50$. What is $\lim\limits_{t\to\infty} M(t)$?

   (A) 50　　　(B) 200　　　(C) 500　　　(D) 1000

5.

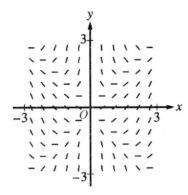

Shown above is a slope field for which of the following differential eqations?

(A) $\dfrac{dy}{dx}=\dfrac{x^2-y^2}{x}$ (B) $\dfrac{dy}{dx}=\dfrac{x^2-y^2}{y}$

(C) $\dfrac{dy}{dx}=\dfrac{x^2+y^2}{x}$ (D) $\dfrac{dy}{dx}=x^2-y^2$

6. Let f be the function satisfying $f'(x)=-3xf(x)$, for all real numbers x, with $f(1)=4$ and $\lim\limits_{x\to\infty}f(x)=0$.

(a) Use Euler's method, starting at $x=1$ with a step size of 0.5, to approximate $f(2)$.

(b) Write an expression for $y=f(x)$ by solving the differential equation $\dfrac{dy}{dx}=-3xy$ with the initial condition $f(1)=4$.

习题参考答案

1. C 2. A 3. C 4. B 5. A

6. (a) $f(1.5)\approx f(1)+f'(1)\times 0.5=4-3\times 1\times 4\times 0.5=-2$.

$f(2)\approx -2+f'(1.5)\times 0.5=-2-3\times 1.5\times(-2)\times 0.5=2.5$.

(b) $\dfrac{1}{y}dy=-3xdx$

$\ln y=-\dfrac{3}{2}x^2+k$

$y=Ce^{-\frac{3}{2}x^2}$

$4=Ce^{-\frac{3}{2}x^2}\Rightarrow C=4e^{\frac{3}{2}}$

$y=4e^{\frac{3}{2}}e^{-\frac{3}{2}x^2}$.

Chapter 11 Sequences and Series(序列和级数)

11.1 Sequences(序列)

11.1.1 Definition(定义)

A sequence can be thought of as a list of numbers written in a definite order:
$$a_1, a_2, a_3, \cdots, a_n, \cdots$$
The number a_1 is called the first term, a_2 is the second term, and in general a_n is the nth term.

A sequence is a function whose domain is a set of integers.

序列(国内教材也称作数列)是按一定顺序排列的无穷多个数,是定义域为正整数的特殊函数:$a_n = f(n), n = 1, 2, 3\cdots$. 其中 a_n 称为第 n 项或一般项.

序列一般表示为:$a_1, a_2, a_3, \cdots, a_n, \cdots$ 或 $\{a_n\}$ 或 $\{a_n\}_{n=1}^{\infty}$.

Example 11.1

Some sequences can be defined by giving a formula for the nth term. In the following examples we give three descriptions of the sequence: one by using the preceding notation, another by using the defining formula, and a third by writing out the terms of the sequence. Notice that doesn't have to start at 1.

a. $\left\{\dfrac{n}{n+1}\right\}_{n=1}^{\infty}$ $\qquad a_n = \dfrac{n}{n+1} \qquad \left\{\dfrac{1}{2}, \dfrac{2}{3}, \dfrac{3}{4}, \cdots, \dfrac{n}{n+1}, \cdots\right\}$

b. $\left\{\dfrac{1}{2^n}\right\}_{n=1}^{\infty}$ $\qquad a_n = \dfrac{1}{2^n} \qquad \left\{\dfrac{1}{2}, \dfrac{1}{4}, \dfrac{1}{8}, \cdots, \dfrac{1}{2^n}, \cdots\right\}$

c. $\left\{(-1)^{n+1}\dfrac{n}{n+1}\right\}_{n=1}^{\infty}$ $\qquad a_n = (-1)^{n+1}\dfrac{n}{n+1} \qquad \left\{\dfrac{1}{2}, -\dfrac{2}{3}, \dfrac{3}{4}, \cdots, (-1)^{n+1}\dfrac{n}{n+1}, \cdots\right\}$

d. $\{2n-1\}_{n=3}^{\infty}$ $\qquad a_n = 2n-1 \qquad \{5, 7, 9, \cdots, 2n-1, \cdots\}$

11.1.2 Sequence converges or diverges(序列收敛或发散)

A sequence $\{a_n\}$ has the limit L and we write

$$\lim_{n\to\infty} a_n = L$$

If $\lim\limits_{n\to\infty} a_n$ exists, we say the sequence converges (or is convergent). Otherwise, we say the sequence diverges (or is divergent).

若序列一般项的极限存在，即 $\lim\limits_{n\to\infty} a_n = L$（$L$ 为实数），则称序列 $\{a_n\}$ 收敛；反之则称序列发散．

Example 11.2 *序列收敛与发散.*

Are the following sequences converge or diverge?

a. $\left\{\dfrac{5n}{2n-1}\right\}$; b. $\left\{\dfrac{e^n}{n}\right\}$; c. $\left\{\dfrac{e^n}{1+e^n}\right\}$.

Solution

a. $\because \lim\limits_{x\to\infty}\dfrac{5x}{2x-1}=\dfrac{5}{2}$, $\therefore \lim\limits_{x\to\infty}\dfrac{5n}{2n-1}=\dfrac{5}{2}$, so $\left\{\dfrac{5n}{2n-1}\right\}$ converges.

b. $\because \lim\limits_{n\to\infty}\dfrac{e^n}{n}=\lim\limits_{x\to\infty}\dfrac{e^x}{x}=\lim\limits_{x\to\infty}\dfrac{e^x}{1}=\infty$, $\therefore \left\{\dfrac{e^n}{n}\right\}$ diverges.

c. $\because \lim\limits_{n\to\infty}\dfrac{e^n}{1+e^n}=\lim\limits_{x\to\infty}\dfrac{e^x}{1+e^x}=1$, $\therefore \left\{\dfrac{e^n}{1+e^n}\right\}$ converges.

[注] 常见极限：

(1) $\lim\limits_{n\to\infty}\sqrt[n]{n}=1$;

(2) $\lim\limits_{n\to\infty}\sqrt[n]{x}=1\,(x>0)$;

(3) $\lim\limits_{n\to\infty} x^n = 0\,(|x|<0)$;

(4) $\lim\limits_{n\to\infty}\sqrt[n]{n}=1=\lim\limits_{n\to\infty} n^{1/n}=\lim\limits_{n\to\infty} e^{(1/n)\ln n}=e^0=1$.

11.2　Series（级数）

11.2.1　Definition（定义）

给定一个序列 $a_1, a_2, a_3, \cdots, a_n, \cdots$，则由此序列构成的表达式 $a_1+a_2+a_3+\cdots+a_n+\cdots$ 叫作**常数项无穷级数（Infinite Series）**，简称级数，记为 $\sum\limits_{n=1}^{\infty} a_n$，即

$$\sum_{n=1}^{\infty} a_n = a_1 + a_2 + a_3 + \cdots + a_n + \cdots$$

其中第 n 项 a_n 叫作级数的一般项．

Given a series $\sum\limits_{n=1}^{\infty} a_n = a_1+a_2+a_3+\cdots+a_n+\cdots$, let S_n denote its **n th partial sum（n 项部分和）**:

$$S_n = \sum_{i=1}^{n} a_i = a_1 + a_2 + a_3 + \cdots + a_n$$

If the sequence $\{S_n\}$ is convergent and $\lim\limits_{n\to\infty} S_n = S$ exists as a real number, then the series $\sum_{n=1}^{\infty} a_n$ is called convergent and we write

$$\sum_{n=1}^{\infty} a_n = a_1 + a_2 + a_3 + \cdots + a_n + \cdots = S$$

The number S is called the sum of the series. If the sequence $\{S_n\}$ is divergent, then the series is called divergent.

级数 $\sum_{n=1}^{\infty} a_n$ 的部分和序列 $\{S_n\}$ 收敛，则级数 $\sum_{n=1}^{\infty} a_n$ 收敛.

级数 $\sum_{n=1}^{\infty} a_n$ 的部分和序列 $\{S_n\}$ 发散，则级数 $\sum_{n=1}^{\infty} a_n$ 发散.

11.2.2 几种常见的级数

11.2.2.1 Geometric Series(几何级数)

$$\sum_{n=0}^{\infty} ar^n = a + ar + ar^2 + \cdots + ar^n + \cdots \quad (r \neq 0)$$

Such series are called **geometric series**, and the number r is called the **ratio** for the series.

Here are some examples:

$1 + 2 + 4 + \cdots + 2^k + \cdots$	$a = 1, r = 2$
$\dfrac{3}{10} + \dfrac{3}{10^2} + \dfrac{3}{10^3} + \cdots + \dfrac{3}{10^i} + \cdots$	$a = \dfrac{3}{10}, r = \dfrac{1}{10}$
$\dfrac{1}{2} - \dfrac{1}{4} + \dfrac{1}{8} - \cdots + (-1)^{n+1} \dfrac{1}{2^n} + \cdots$	$a = \dfrac{1}{2}, r = -\dfrac{1}{2}$
$1 + 1 + 1 + \cdots + 1 + \cdots$	$a = 1, r = 1$
$1 - 1 + 1 - \cdots + (-1)^{k+1} + \cdots$	$a = 1, r = -1$
$1 + x + x^2 + \cdots + x^k + \cdots$	$a = 1, r = x$

Theorem(定理)

A geometric series

$$\sum_{n=0}^{\infty} ar^n = a + ar + ar^2 + \cdots + ar^n + \cdots \quad (r \neq 0 \text{ is the } \textbf{common ratio})$$

converges if $|r| < 1$ and **diverges** if $|r| \geq 1$.

If the series converges, then the sum is

$$\sum_{n=0}^{\infty} ar^n = \frac{a}{1-r}$$

Proof

If $r = 1$, then $S_n = a + a + a + \cdots + a = na \to \pm \infty$. Since $\lim_{n \to \infty} S_n$ doesn't exist, the geometric series diverges in this case.

If $r \neq 1$, we have

$$S_n = a + ar + ar^2 + \cdots + ar^{n-1}$$

and

$$rS_n = ar + ar^2 + ar^3 + \cdots + ar^n$$

Subtracting these equations, we get
$$S_n - rS_n = a - ar^n$$
$$S_n = \frac{a(1-r^n)}{1-r}$$

If $-1 < r < 1$, $\lim\limits_{n \to \infty} S_n = \lim\limits_{n \to \infty} \frac{a(1-r^n)}{1-r} = \frac{a}{1-r} - \frac{a}{1-r} \lim\limits_{n \to \infty} r^n = \frac{a}{1-r}$.

If $|r| \geq 1$, $\lim\limits_{n \to \infty} S_n$ does not exist. Therefore the geometric series diverges.

Example 11.3

Find the sum of the geometric series
$$5 - \frac{10}{3} + \frac{20}{9} - \frac{40}{27} + \cdots$$

Solution

The first term is $a = 5$ and the common ratio is $r = \dfrac{-\frac{10}{3}}{5} = -\dfrac{2}{3}$. Since $|r| = \dfrac{2}{3} < 1$, the series is convergent and its sum is
$$5 - \frac{10}{3} + \frac{20}{9} - \frac{40}{27} + \cdots = \frac{5}{1 - \left(-\frac{2}{3}\right)} = \frac{5}{\frac{5}{3}} = 3.$$

11.2.2.2 p-Series(p—级数)

p-series $\sum\limits_{n=1}^{\infty} \dfrac{1}{n^p} = 1 + \dfrac{1}{2^p} + \dfrac{1}{3^p} + \cdots + \dfrac{1}{n^p} + \cdots (p > 0)$

converges if $p > 1$ and diverges if $0 < p \leq 1$.

Example 11.4

Give a value of p such that $\sum\limits_{n=1}^{\infty} \dfrac{1}{n^p}$ diverges, but $\sum\limits_{n=1}^{\infty} \dfrac{1}{n^{2p}}$ converges. Give reasons why your value of p is correct.

Solution

Some p such that $\dfrac{1}{2} < p \leq 1$ because the p-series $\sum\limits_{n=1}^{\infty} \dfrac{1}{n^p}$ diverges for $0 < p \leq 1$ and the p-series $\sum\limits_{n=1}^{\infty} \dfrac{1}{n^{2p}}$ converges for $2p > 1$.

11.2.2.3 Harmonic Series(调和级数)

Harmonic series $\sum\limits_{n=1}^{\infty} \dfrac{1}{n} = 1 + \dfrac{1}{2} + \dfrac{1}{3} + \cdots + \dfrac{1}{n} + \cdots$ diverges.(调和级数即 $p=1$ 的 $p-$级数,发散)

11.2.3 收敛级数的性质

性质 1

如果级数 $\sum_{n=1}^{\infty} a_n$ 和 $\sum_{n=1}^{\infty} b_n$ 分别收敛于和 A 和 B,则 $\forall \alpha, \beta \in \mathbf{R}$,级数 $\sum_{n=1}^{\infty}(\alpha a_n \pm \beta b_n)$ 也收敛,且其和为 $\alpha A + \beta B$.

性质 2

在级数中去掉、加上或改变有限多项,不会改变级数的收敛性.

比如:由级数 $\frac{1}{1 \times 2} + \frac{1}{2 \times 3} + \frac{1}{3 \times 4} + \cdots + \frac{1}{n(n+1)} + \cdots$ 是收敛的,可知:数 $10000 + \frac{1}{1 \times 2} + \frac{1}{2 \times 3} + \frac{1}{3 \times 4} + \cdots + \frac{1}{n(n+1)} + \cdots$ 与级数 $\frac{1}{3 \times 4} + \frac{1}{4 \times 5} + \cdots + \frac{1}{n(n+1)} + \cdots$ 也收敛.

由于级数 $\sum_{n=1}^{\infty} \frac{1}{n} = 1 + \frac{1}{2} + \frac{1}{3} + \cdots + \frac{1}{n} + \cdots$ 发散,则级数 $\sum_{n=2}^{\infty} \frac{1}{n} = \frac{1}{2} + \frac{1}{3} + \cdots + \frac{1}{n} + \cdots$ 也发散.

性质 3

如果级数 $\sum_{n=1}^{\infty} a_n$ 收敛,则对该级数的各项任意加括号后所成的新级数仍收敛,且其和不变.

[注意] 如果加括号后所成的级数收敛,则不能断定去括号后原来的级数也收敛.例如,级数 $(1-1)+(1-1)+\cdots$ 收敛于零,但级数 $1-1+1-1+\cdots$ 却是发散的.

[推论] 如果加括号后所成的级数发散,则原来的级数也发散.

性质 4 *n*th Term Test (The Divergence Test)

(a) If $\lim_{n \to \infty} a_n \neq 0$, then the series $\sum a_n$ diverges.

(b) If $\lim_{n \to \infty} a_n = 0$, then the series $\sum a_n$ may either converge or diverge.

Example 11.5

Prove the divergence of $\sum_{n=1}^{\infty} \frac{n}{4n+1}$.

Solution

$\lim_{n \to \infty} a_n = \lim_{n \to \infty} \frac{n}{4n+1} = \frac{1}{4}$.

The *n*th term a_n does not converge to zero, so the series diverges by *n*th term test.

11.2.4 Tests for Convergence of Nonnegative Series(非负项级数判别法)

11.2.4.1 Definition(定义)

若 $\sum a_n$ 中各项均有 $a_n \geqslant 0$,则称 $\sum a_n$ 为 **非负项级数(Nonnegative Series)**.

1.2.4.2 The Integral Test(积分判别法)

Let $a_n = f(n)$, where $f(x)$ is positive, decreasing, and continuous for $x \geqslant 1$.

(a) If $\int_1^\infty f(x)\mathrm{d}x$ converges, then $\sum_{n=1}^\infty a_n$ converges.

(b) If $\int_1^\infty f(x)\mathrm{d}x$ diverges, then $\sum_{n=1}^\infty a_n$ diverges.

Example 11.6 积分判别法.

Proof the p-series $\sum_{n=1}^\infty \dfrac{1}{n^p} = 1 + \dfrac{1}{2^p} + \dfrac{1}{3^p} + \cdots \dfrac{1}{n^p} + \cdots (p>0)$ converges if $p>1$ and diverges if $0 < p \leqslant 1$.

Proof

If $p \leqslant 0$, then the general term n^{-p} does not tend to zero, so the series diverges by nth term test.

If $p > 0$, then $f(x) = x^{-p} = \dfrac{1}{x^p}$ is positive and decreasing, so the Integral Test applies.

$$\int_1^\infty \frac{1}{x^p}\mathrm{d}x = \begin{cases} \dfrac{1}{p-1}, & \text{if } p > 1 \\ \infty, & \text{if } 0 < p \leqslant 1 \end{cases}$$

Therefore, $\sum_{n=1}^\infty \dfrac{1}{n^p}$ converges for $p > 1$ and diverges for $p \leqslant 1$.

11.2.4.3 The Comparison Test(比较判别法)

Assume that there exists $M > 0$ such that $0 \leqslant a_n \leqslant b_n$ for $n \geqslant M$.

(a) If $\sum_{n=1}^\infty b_n$ converges, then $\sum_{n=1}^\infty a_n$ also converges.

(b) If $\sum_{n=1}^\infty a_n$ diverges, then $\sum_{n=1}^\infty b_n$ also diverges.

Example 11.7 比较判别法.

Does $\sum_{n=1}^\infty \dfrac{1}{\sqrt{n(n+1)}}$ converge or diverge?

Solution

Since $\dfrac{1}{\sqrt{n(n+1)}} > \dfrac{1}{\sqrt{(n+1)^2}} = \dfrac{1}{n+1}$, and the Harmonic series $\sum_{n=1}^\infty \dfrac{1}{n}$ diverges, so $\sum_{n=1}^\infty \dfrac{1}{n+1}$ diverges.

$\sum_{n=1}^\infty \dfrac{1}{\sqrt{n(n+1)}}$ diverges by the Comparison Test.

11.2.4.4 Comparison Test in Limit Form(比较判别法的极限形式)

Suppose $\sum a_n$ and $\sum b_n$ are series with positive terms. If

$$\lim_{n \to \infty} \frac{a_n}{b_n} = L$$

Where L is a finite number and $L>0$, then either both series converge or both diverge.

Example 11.8 比较判别法的极限形式.

Determine whether $\sum\limits_{n=3}^{\infty}\dfrac{1}{\sqrt{n^2+4}}$ converges.

Solution

Apply the Limit Comparison Test with $a_n=\dfrac{1}{\sqrt{n^2+4}}$ and $b_n=\dfrac{1}{n}$.

Then

$$L=\lim_{n\to\infty}\frac{a_n}{b_n}=\lim_{n\to\infty}\frac{n}{\sqrt{n^2+4}}=1$$

Since $\sum\limits_{n=3}^{\infty}\dfrac{1}{n}$ diverges and $L>0$, the series $\sum\limits_{n=3}^{\infty}\dfrac{1}{\sqrt{n^2+4}}$ also diverges.

11.2.4.5 The Ratio Test(比值判别法)

Let $\sum a_n$ be a series with positive terms and suppose that:

$$\rho=\lim_{n\to\infty}\frac{a_{n+1}}{a_n}$$

(a) If $\rho<1$, then $\sum a_n$ converges.

(b) If $\rho>1$ or $\rho=\infty$, then $\sum a_n$ diverges.

(c) If $\rho=1$, then $\sum a_n$ may converges or diverges, so that another test must be tried.

Example 11.9 比值判别法.

Each of the following series has positive terms, so the ratio test applies. In each part, use the ratio test to determine whether the following series converge or diverge.

(a) $\sum\limits_{n=1}^{\infty}\dfrac{n}{2^n}$ (b) $\sum\limits_{n=1}^{\infty}\dfrac{n^n}{n!}$ (c) $\sum\limits_{n=3}^{\infty}\dfrac{(2n)!}{4^n}$ (d) $\sum\limits_{n=1}^{\infty}\dfrac{1}{2n-1}$

Solution

(a) $\sum\limits_{n=1}^{\infty}\dfrac{n}{2^n}$ converges, since

$$\rho=\lim_{n\to\infty}\frac{a_{n+1}}{a_n}=\lim_{n\to\infty}\frac{n+1}{2^{n+1}}\cdot\frac{2^n}{n}=\frac{1}{2}\lim_{n\to\infty}\frac{n+1}{n}=\frac{1}{2}<1.$$

(b) $\sum\limits_{n=1}^{\infty}\dfrac{n^n}{n!}$ diverges, since

$$\rho=\lim_{n\to\infty}\frac{a_{n+1}}{a_n}=\lim_{n\to\infty}\frac{(n+1)^{n+1}}{(n+1)!}\cdot\frac{n!}{n^n}=\lim_{n\to\infty}\frac{(n+1)^n}{n^n}=\lim_{n\to\infty}\left(1+\frac{1}{n}\right)^n=\mathrm{e}>1.$$

(c) $\sum\limits_{n=3}^{\infty}\dfrac{(2n)!}{4^n}$ diverges, since

$$\rho=\lim_{n\to\infty}\frac{a_{n+1}}{a_n}=\lim_{n\to\infty}\frac{[2(n+1)]!}{4^{n+1}}\cdot\frac{4^n}{(2n)!}=\frac{1}{4}\lim_{n\to\infty}(2n+2)(2n+1)=\infty.$$

(d) The ratio test is of no help since
$$\rho = \lim_{n \to \infty} \frac{a_{n+1}}{a_n} = \lim_{n \to \infty} \frac{1}{2(n+1)-1} \cdot \frac{2n-1}{1} = \lim_{n \to \infty} \frac{2n-1}{2n+1} = 1.$$

However, the integral test proves that the series diverges since
$$\int_1^\infty \frac{1}{2x-1} \, dx = \lim_{b \to \infty} \int_1^b \frac{1}{2x-1} \, dx = \lim_{b \to \infty} \frac{1}{2} \ln(2x-1) \Big|_1^b = \infty.$$

11.2.4.6 The nth Root Test(根值判别法)

Let $\sum a_n$ be a series with positive terms and suppose that:
$$\rho = \lim_{n \to \infty} \sqrt[n]{a_n}$$

(a) If $\rho < 1$, then $\sum a_n$ converges.

(b) If $\rho > 1$ or $\rho = \infty$, then $\sum a_n$ diverges.

(c) If $\rho = 1$, then $\sum a_n$ may converges or diverges, so that another test must be tried.

Example 11.10 根值判别法.

Determine whether $\sum \left(\frac{n}{2n+1}\right)^n$ converges.

Solution

The series $\sum \left(\frac{n}{2n+1}\right)^n$ converges by the nth Root Test, since
$$\lim_{n \to \infty} \sqrt[n]{\left(\frac{n}{2n+1}\right)^n} = \lim_{n \to \infty} \frac{n}{2n+1} = \frac{1}{2}.$$

11.2.5 Alternating Series and Absolute Convergence(交错级数和绝对收敛)

11.2.5.1 Definition(定义)

A series in which the terms are alternately positive and negative is an alternating series.

称各项正负交错出现的级数为交错级数.

若 $a_n > 0 (n=1,2,3,\cdots)$,称 $\sum_{n=1}^\infty (-1)^{n+1} a_n$ 或 $\sum_{n=1}^\infty (-1)^n a_n$ 为交错级数.

$$\sum_{n=1}^\infty (-1)^{n+1} a_n = a_1 - a_2 + a_3 - a_4 + \cdots + (-1)^{n+1} a_n + \cdots$$

$$\sum_{n=1}^\infty (-1)^n a_n = -a_1 + a_2 - a_3 + a_4 + \cdots + (-1)^n a_n + \cdots$$

11.2.5.2 The Alternating Series Test(交错级数判别法)

Assume that $\{a_n\}$ is a positive sequence that is decreasing and converges to 0:
$$a_1 > a_2 > a_3 > a_4 > \cdots > 0, \quad \lim_{n \to \infty} a_n = 0$$

Then the following alternating series converges:
$$S = \sum_{n=1}^\infty (-1)^{n+1} a_n = a_1 - a_2 + a_3 - a_4 + \cdots$$

Furthermore, $S \leqslant a_1$.

11.2.5.3　The Alternating Series Estimation Theorem(交错级数估计定理)

Let $S = \sum_{n=1}^{\infty} (-1)^{n+1} a_n$, $S_n = \sum_{k=1}^{n} (-1)^{k+1} a_k = a_1 - a_2 + a_3 - a_4 + \cdots + (-1)^{n+1} a_n$, where $\{a_n\}$ is a positive decreasing sequence that converges to 0. Then
$$|S - S_n| < a_{n+1}$$

In other words, the error committed when we approximate S by S_n is less than the size of the first omitted term a_n.

Example 11.11　交错级数判别法及交错级数估计定理.

Consider the infinite series $\sum_{n=3}^{\infty} (-1)^{n+1} a_n = \frac{1}{3\ln 3} - \frac{1}{4\ln 4} + \frac{1}{5\ln 5} - \cdots$. Identify properties of this series that guarantee the series converges. Explain why the sum of this series is less than $\frac{1}{3}$.

Solution

The terms in this alternating series decrease in absolute value and $\lim_{n \to \infty} \frac{1}{n \ln n} = 0$. Therefore, the Alternating Series Test guarantees that this series converges.

Furthermore, $\text{sum} < \frac{1}{3\ln 3} < \frac{1}{3}$.

Therefore, the sum of series is less than $\frac{1}{3}$.

11.2.6　Absolute and Conditional Convergence(绝对收敛和条件收敛)

11.2.6.1　Absolute Convergence(绝对收敛)

The series $\sum a_n$ converges absolutely if $\sum |a_n|$ converges.

11.2.6.2　Absolute Convergence Implies Convergence(绝对收敛则收敛)

If $\sum |a_n|$ converges, then $\sum a_n$ also converges.

11.2.6.3　Conditional Convergence(条件收敛)

An infinite series $\sum a_n$ converges conditionally if $\sum a_n$ converges but $\sum |a_n|$ diverges.

Example 11.12　条件收敛.

a. Show that $\sum_{n=1}^{\infty} \frac{(-1)^{n+1}}{n}$ converges conditionally.

b. Show that $|S - S_6| < \frac{1}{7}$.

c. Find an n such that S_n approximates S with an error less than 10^{-3}.

Solution

a. The terms $a_n = \dfrac{1}{n}$ are positive and decreasing, and $\lim\limits_{n\to\infty} a_n = 0$. Therefore, S converges by the AST. The harmonic series $\sum\limits_{n=1}^{\infty} \dfrac{1}{n}$ diverges, so S converges conditionally but not absolutely.

b.
$$|S - S_n| < a_{n+1} = \dfrac{1}{n+1}$$

For $n = 6$, we obtain $|S - S_6| < a_7 = \dfrac{1}{7}$.

c. We can make the error less than 10^{-3} by choosing n so that
$$\dfrac{1}{n+1} \leqslant 10^{-3} \implies n+1 \geqslant 10^3 \implies n \geqslant 999$$

Using a computer algebra system, we find that $S_{999} \approx 0.69365$.

[注]根据交错级数判别法，$\sum\limits_{n=1}^{\infty} \dfrac{(-1)^{n+1}}{n^p}(p > 0)$ 收敛.

(1) $0 < p \leqslant 1$ 时，$\sum\limits_{n=1}^{\infty} \dfrac{(-1)^{n+1}}{n^p}$ 条件收敛;

(2) $p > 1$ 时，$\sum\limits_{n=1}^{\infty} \dfrac{(-1)^{n+1}}{n^p}$ 绝对收敛.

Example 11.13

Give a value of p such that $\sum\limits_{n=1}^{\infty} \dfrac{(-1)^n}{n^p}$ converges, but $\sum\limits_{n=1}^{\infty} \dfrac{1}{n^{2p}}$ converges. Give reasons why your value of p is correct.

Solution

Some p such that $0 < p \leqslant \dfrac{1}{2}$ because $\sum\limits_{n=1}^{\infty} \dfrac{(-1)^n}{n^p}$ converges by The Alternating Series Test (AST), but the p-series $\sum\limits_{n=1}^{\infty} \dfrac{1}{n^{2p}}$ diverges for $2p \leqslant 1$.

11.3　Power Series(幂级数)

11.3.1　Definition(定义)

11.3.1.1　函数项级数

给定一个定义在区间 I 上的函数列 $\{u_n(x)\}$，由该函数列构成的表达式
$$u_1(x) + u_2(x) + u_3(x) + \cdots + u_n(x) + \cdots$$

称为定义在区间 I 上的(函数项)级数，记为 $\sum\limits_{n=1}^{\infty} u_n(x)$.

11.3.1.2 幂级数(Power Series)

函数项级数中简单而常见的一类级数就是各项都为幂函数的函数项级数,这种形式的级数称为幂级数,它的形式是

$$\sum_{n=1}^{\infty} a_n x^n = a_0 + a_1 x + a_2 x^2 + \cdots + a_n x^n + \cdots \textbf{(Power Series in } x\textbf{)}$$

式中常数 $a_0, a_1, a_2, \cdots, a_n, \cdots$ 叫作幂级数的系数.

幂级数的例子: $1 + x + x^2 + x^3 + \cdots + x^n + \cdots$

$$1 + x + \frac{1}{2!}x^2 + \cdots + \frac{1}{n!}x^n + \cdots$$

幂级数的一般形式如下:

$$\sum_{n=1}^{\infty} a_n (x-x_0)^n = a_0 + a_1(x-x_0) + a_2(x-x_0)^2 + \cdots + a_n(x-x_0)^n + \cdots \textbf{(Power Series in } (x-x_0)\textbf{)}$$

11.3.2 Radius of Convergence and Interval of Convergence(收敛半径与收敛区间)

If a numerical value is substituted for x in a power series $\sum_{n=1}^{\infty} a_n x^n$, then the resulting series of numbers may either converge or diverge. This leads to the problem of determining the set of x-values for which a given power series converges; this is called its **convergence set**(收敛集).

11.3.2.1 Theorem

For any power series in x, exactly one of the following is true:

(a) The series converges only for $x=0$(级数仅在 $x=0$ 处收敛).

(b) The series converges absolutely (and hence converges) for all real values of x(级数对一切实数收敛).

(c) The series converges absolutely (and hence converges) for all x in some finite open interval $(-R, R)$ and diverges if $x < -R$ or $x > R$. At either of the values $x = R$ or $x = -R$, the series may converge absolutely, converge conditionally, or diverge, depending on the particular series [级数在区间 $(-R, R)$ 收敛,在区间端点 $x = R$ 或 $x = -R$ 可能收敛也可能发散,在 $(-\infty, -R)$ 与 (R, ∞) 发散].

11.3.2.2 Radius of Convergence and Interval of Convergence(收敛半径与收敛区间)

如图 11.1, the convergence set of a power series in x is called the **interval of convergence**. In the case where the convergence set is the single value $x = 0$ we say that the series has **radius of convergence** 0(若级数仅在 $x=0$ 处收敛,则收敛半径 $R=0$), in the case where the convergence set is $(-\infty, \infty)$ we say that the series has **radius of convergence** ∞ [若级数仅在 $(-\infty, \infty)$ 收敛,则收敛半径 $R = \infty$].

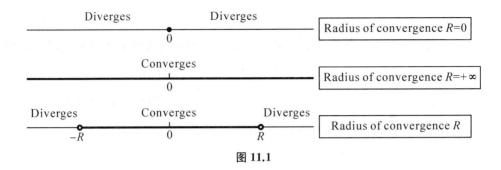

图 11.1

Example 11.14 求级数的收敛半径和收敛区间.

Find the interval of convergence and radius of convergence of the following power series.

a. $\sum_{n=0}^{\infty} x^n$;

b. $\sum_{n=0}^{\infty} \frac{x^n}{n!}$;

c. $\sum_{n=0}^{\infty} n! x^n$;

d. $\sum_{n=0}^{\infty} \frac{(-1)^n x^n}{3^n (n+1)}$.

Solution

a. Applying the ratio test for absolute convergence to the given series, we obtain

$$\rho = \lim_{n \to \infty} \left| \frac{u_{n+1}}{u_n} \right| = \lim_{n \to \infty} \left| \frac{x^{n+1}}{x^n} \right| = \lim_{n \to \infty} |x| = |x|$$

So the series converges absolutely if $\rho = |x| < 1$ and diverges if $\rho = |x| > 1$. The test is inconclusive if $|x| = 1$ (i.e., if $x = 1$ or $x = -1$), which means that we will have to investigate convergence at these values separately. At these values, the series becomes

$$\sum_{n=0}^{\infty} 1^n = 1 + 1 + 1 + 1 + \cdots (x = 1)$$

$$\sum_{n=0}^{\infty} (-1)^n = 1 - 1 + 1 - 1 + \cdots (x = -1)$$

Both of which diverge; thus, the interval of convergence for the given power series is $(-1, 1)$, and the radius of convergence is $R = 1$.

b. Applying the ratio test for absolute convergence to the given series, we obtain

$$\rho = \lim_{n \to \infty} \left| \frac{u_{n+1}}{u_n} \right| = \lim_{n \to \infty} \left| \frac{x^{n+1}}{(n+1)!} \cdot \frac{n!}{x^n} \right| = \lim_{n \to \infty} \left| \frac{x}{n+1} \right| = 0$$

Since $\rho < 1$ for all x, the series converges absolutely for all x.

Thus, the interval of convergence is $(-\infty, \infty)$ and the radius of convergence is $R = \infty$.

c. If $x \neq 0$, then the ratio test for absolute convergence yields

$$\rho = \lim_{n \to \infty} \left| \frac{u_{n+1}}{u_n} \right| = \lim_{n \to \infty} \left| \frac{(n+1)! \, x^{n+1}}{n! \, x^n} \right| = \lim_{n \to \infty} |(n+1)x| = \infty$$

Therefore, the series diverges for all nonzero values of x. Thus, the interval of convergence is the single value $x=0$ and the radius of convergence is $R=0$.

d. Since $|(-1)^n|=|(-1)^{n+1}|=1$, we obtain

$$\rho = \lim_{n\to\infty}\left|\frac{u_{n+1}}{u_n}\right| = \lim_{n\to\infty}\left|\frac{x^{n+1}}{3^{n+1}(n+2)} \cdot \frac{3^n(n+1)}{x^n}\right| = \frac{|x|}{3}\lim_{n\to\infty}\left|\frac{n+1}{n+2}\right| = \frac{|x|}{3}$$

The ratio test for absolute convergence implies that the series converges absolutely if $|x|<3$ and diverges if $|x|>3$. The ratio test fails to provide any information when $|x|=3$, so the cases $x=-3$ and $x=3$ need separate analyses. Substituting $x=-3$ in the given series yields

$$\sum_{n=0}^{\infty}\frac{(-1)^n(-3)^n}{3^n(n+1)} = \sum_{n=0}^{\infty}\frac{1}{n+1}$$

Which is divergent harmonic series $1+\frac{1}{2}+\frac{1}{3}+\frac{1}{4}+\cdots$

Substituting $x=3$ in the given series yields

$$\sum_{n=0}^{\infty}\frac{(-1)^n(3)^n}{3^n(n+1)} = \sum_{n=0}^{\infty}\frac{(-1)^n}{n+1} = 1-\frac{1}{2}+\frac{1}{3}-\frac{1}{4}+\cdots$$

Which is the conditionally convergent alternating harmonic series. Thus, the interval of convergence for the given series is $(-3,3]$ and the radius of convergence is $R=3$.

11.3.2.3 Computation of Power Series(幂级数的运算)

Theorem 1

Let $f(x)=\sum_{n=0}^{\infty}a_n x^n$ and $g(x)=\sum_{n=0}^{\infty}b_n x^n$, k is any real number and N is any positive integer.

(a) $f(kx) = \sum_{n=0}^{\infty}a_n(kx)^n = \sum_{n=0}^{\infty}a_n k^n x^n$.

(b) $f(x^k) = \sum_{n=0}^{\infty}a_n(x^k)^n = \sum_{n=0}^{\infty}a_n x^{nk}$.

(c) $f(x) \pm g(x) = \sum_{n=0}^{\infty}a_n x^n \pm \sum_{n=0}^{\infty}b_n x^n = \sum_{n=0}^{\infty}(a_n \pm b_n)x^n$.

However the interval of convergence for the series on the right side may differ from the one on the left side.

Theorem 2 Term-by-Term Differentiation and Integration(逐项求导与逐项积分)

Assume that $F(x) = \sum_{n=0}^{\infty}a_n(x-c)^n$

has radius of convergence $R>0$. Then $F(x)$ is differentiable on $(c-R,c+R)$ (or for all x if $R=\infty$).

Furthermore, we can integrate and differentiate term by term. For $x \in (c-R,c+R)$,

$$F'(x) = \sum_{n=0}^{\infty}[a_n(x-c)^n]' = \sum_{n=0}^{\infty}na_n(x-c)^{n-1}$$

$$\int_0^x F(t)\,dt = \int_0^x \sum_{n=0}^{\infty} a_n (t-c)^n\,dt = \sum_{n=0}^{\infty} (\int_0^x a_n (t-c)^n\,dt) = \sum_{n=0}^{\infty} \frac{a_n}{n+1}(x-c)^{n+1}$$

Example 11.15 级数的运算.

a. $\dfrac{1}{1-x} = 1 + x + x^2 + \cdots + x^n + \cdots \ (-1 < x < 1)$, prove that for $-1 < x < 1$,

$$\frac{1}{(1-x)^2} = 1 + 2x + 3x^2 + 4x^3 + 5x^4 + \cdots$$

b. Prove that for $-1 < x < 1$, $\tan^{-1} x = \sum_{n=0}^{\infty} \dfrac{(-1)^n x^{2n+1}}{2n+1} = x - \dfrac{x^3}{3} + \dfrac{x^5}{5} - \dfrac{x^7}{7} + \cdots$

Solution

a. $\dfrac{1}{1-x} = 1 + x + x^2 + \cdots + x^n + \cdots \ (-1 < x < 1)$

$\Rightarrow \dfrac{d}{dx}\left(\dfrac{1}{1-x}\right) = \dfrac{d}{dx}(1 + x + x^2 + \cdots + x^n + \cdots) \ (-1 < x < 1)$

$\Rightarrow \dfrac{1}{(1-x)^2} = 1 + 2x + 3x^2 + 4x^3 + 5x^4 + \cdots \ (-1 < x < 1)$

b. $\dfrac{1}{1-x} = 1 + x + x^2 + \cdots + x^n + \cdots \ (-1 < x < 1)$

$\Rightarrow \dfrac{1}{1+x^2} = 1 - x^2 + x^4 - x^6 + \cdots \ (-1 < x < 1)$

$\Rightarrow \tan^{-1} x = \int_0^x (1 - t^2 + t^4 - t^6 + \cdots)\,dt = x - \dfrac{x^3}{3} + \dfrac{x^5}{5} - \dfrac{x^7}{7} + \cdots$

11.4 Taylor Series and Maclaurin Series(泰勒级数与麦克劳林级数)

11.4.1 Taylor Series and Maclaurin Series(泰勒级数与麦克劳林级数)

If $f(x)$ has derivatives of all orders at a, then we call the series

$$\sum_{n=0}^{\infty} \frac{f^{(n)}(a)}{n!}(x-a)^n = f(a) + f'(a)(x-a) + \frac{f''(a)}{2!}(x-a)^2 + \cdots + \frac{f^{(n)}(a)}{n!}(x-a)^n + \cdots$$

the **Taylor series for f about $x = a$**. In the special case where $a = 0$, this series becomes

$$\sum_{n=0}^{\infty} \frac{f^{(n)}(0)}{n!} x^n = f(0) + f'(0)x + \frac{f''(0)}{2!}x^2 + \cdots + \frac{f^{(n)}(0)}{n!}x^n + \cdots$$

in which case we call it the **Maclaurin series for $f(x)$**.

11.4.2 常见的 Maclaurin Series

$$\frac{1}{1-x} = 1 + x + x^2 + \cdots + x^n + \cdots \ (-1 < x < 1)$$

$$e^x = 1 + x + \frac{1}{2!}x^2 + \cdots + \frac{1}{n!}x^n + \cdots \ (-\infty < x < +\infty)$$

$$\sin x = x - \frac{x^3}{3!} + \frac{x^5}{5!} - \cdots + (-1)^{n-1}\frac{x^{2n-1}}{(2n-1)!} + \cdots \ (-\infty < x < +\infty)$$

$$\cos x = 1 - \frac{x^2}{2!} + \frac{x^4}{4!} - \cdots + (-1)^n \frac{x^{2n}}{(2n)!} + \cdots \ (-\infty < x < +\infty)$$

$$\ln(1+x) = x - \frac{x^2}{2} + \frac{x^3}{3} - \frac{x^4}{4} + \cdots + (-1)^n \frac{x^{n+1}}{n+1} + \cdots \quad (-1 < x \leqslant 1)$$

11.4.3 Taylor and Maclaurin Polynomials(泰勒多项式与麦克劳林多项式)

If f can be differentiated n times at a, then we define the **nth Taylor polynomial for f about $x=a$** to be

$$P_n(x) = \sum_{k=0}^{n} \frac{f^{(k)}(a)}{k!}(x-a)^k$$
$$= f(a) + f'(a)(x-a) + \frac{f''(a)}{2!}(x-a)^2 + \cdots + \frac{f^{(n)}(a)}{n!}(x-a)^n$$

The **nth Maclaurin polynomial for f about $x=0$** to be

$$P_n(x) = \sum_{k=0}^{n} \frac{f^{(k)}(0)}{k!} x^k = f(0) + f'(0)x + \frac{f''(0)}{2!}x^2 + \cdots + \frac{f^{(n)}(0)}{n!}x^n$$

Note that the nth Taylor and Maclaurin polynomials are the nth partial sums for the corresponding Maclaurin and Taylor series.

11.4.4 Lagrange remainder(拉格朗日余项)

11.4.4.1 The nth Order Remainder n(阶余项)

It will be convenient to have a notation for the error in the approximation $f(x) \approx P_n(x)$. Accordingly, we will let $R_n(x)$ denote the difference between $f(x)$ and its nth Taylor polynomial, that is,

$$R_n(x) = f(x) - P_n(x) = f(x) - \sum_{k=0}^{n} \frac{f^{(k)}(a)}{k!}(x-a)^k$$

The function $R_n(x)$ is called the **nth order remainder or the error term** for the approximation of f by $P_n(x)$. It is also called the **Lagrange form** of the remainder, and bounds on $R_n(x)$ found using this form are **Lagrange error bounds**. $R_n(x)$称为n阶拉格朗日余项.

11.4.4.2 The Remainder Estimation Theorem(余项估计定理)

If the function $f(x)$ can be differentiated $n+1$ times on an interval containing the number $x=a$, for all x in the interval, then

$$|R_n(x)| \leqslant \frac{\max\limits_{c \text{ is between } x \text{ and } a} |f^{(n+1)}(c)|}{(n+1)!} |x-a|^{n+1}$$

for all x in the interval.

Example 11.16 余项估计定理.

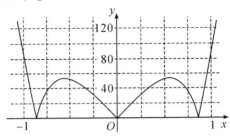

Graph of $y=|f^{(5)}(x)|$

Let $f(x)=\sin(x^2)+\cos x$. The graph of $y=|f^{(5)}(x)|$ is shown above.

a. Write the first four nonzero terms of the Taylor series for $\sin x$ about $x=0$, and write the first four nonzero terms of the Taylor series for $\sin(x^2)$ about $x=0$.

b. Write the first four nonzero terms of the Taylor series for $\cos x$ about $x=0$. Use this series and the series for $\sin(x^2)$, found in part (a), to write the first four nonzero terms of the Taylor series for f about $x=0$.

c. Find the value of $f^{(6)}(0)$.

d. Let $P_4(x)$ be the fourth-degree Taylor polynomial for f about $x=0$. Using information from the graph of $y=|f^{(5)}(x)|$ shown above, show that $\left|f\left(\frac{1}{4}\right)-P_4\left(\frac{1}{4}\right)\right|<\frac{1}{3000}$.

Solution

a. $\sin x = x - \dfrac{x^3}{3!} + \dfrac{x^5}{5!} - \dfrac{x^7}{7!} + \cdots$

$\sin(x^2) = x^2 - \dfrac{x^6}{3!} + \dfrac{x^{10}}{5!} - \dfrac{x^{14}}{7!} + \cdots$

b. $\cos x = 1 - \dfrac{x^2}{2!} + \dfrac{x^4}{4!} - \dfrac{x^6}{6!} + \cdots$

$f(x) = 1 + \dfrac{x^2}{2} + \dfrac{x^4}{4!} - \dfrac{121 x^6}{6!} + \cdots$

c. $\dfrac{f^{(6)}(0)}{6!}$ is the coefficient of x^6 in the Taylor series for f about $x=0$. Therefore $f^{(6)}(0) = -121$.

d. The graph of $y=|f^{(5)}(x)|$ indicates that $\max\limits_{0 \leqslant c \leqslant \frac{1}{4}} |f^{(5)}(c)| < 40$.

Therefore,

$$\left|f\left(\frac{1}{4}\right)-P_4\left(\frac{1}{4}\right)\right| \leqslant \frac{\max\limits_{0 \leqslant c \leqslant \frac{1}{4}} |f^{(5)}(c)|}{5!}\left(\frac{1}{4}\right)^5 < \frac{40}{120 \times 4^5} = \frac{1}{3072} < \frac{1}{3000}.$$

Practice Exercises(习题)

1. $\sum_{k=n}^{\infty} \left(\frac{1}{3}\right)^k =$

   (A) $\frac{3}{2} - \left(\frac{1}{3}\right)^n$ (B) $\frac{3}{2}\left[1-\left(\frac{1}{3}\right)^n\right]$ (C) $\frac{3}{2}\left(\frac{1}{3}\right)^n$ (D) $\frac{3}{2}\left(\frac{1}{3}\right)^{n+1}$

2. The sum of the infinite geometric series $\frac{3}{2} + \frac{9}{16} + \frac{27}{128} + \frac{81}{1024} + \cdots$ is

   (A) 2.40 (B) 2.45 (C) 2.50 (D) 3.11

3. Which of the following series are convergent?

   I. $1 + \frac{1}{2^2} + \frac{1}{3^2} + \cdots + \frac{1}{n^2} + \cdots$

   II. $1 + \frac{1}{2} + \frac{1}{3} + \cdots + \frac{1}{n} + \cdots$

   III. $1 - \frac{1}{3} + \frac{1}{3^2} - \cdots + \frac{(-1)^{n+1}}{3^{n-1}} + \cdots$

   (A) I only (B) II only (C) I and III (D) I, II and III

4. The infinite series $\sum_{k=1}^{\infty} a_k$ has nth partial sum $S_n = \frac{n}{3n+1}$ for $n \geqslant 1$. What is the sum of the series $\sum_{k=1}^{\infty} a_k$?

   (A) $\frac{1}{3}$ (B) $\frac{1}{2}$ (C) 1 (D) $\frac{3}{2}$

5. Which of the following series converge?

   I. $\sum_{n=1}^{\infty} (-1)^{n+1} \frac{1}{2n+1}$

   II. $\sum_{n=1}^{\infty} \frac{1}{n} \left(\frac{3}{2}\right)^n$

   III. $\sum_{n=2}^{\infty} \frac{1}{n \ln n}$

   (A) I only (B) II only (C) I and II (D) I, II and III

6. Which of the following series converge?

   I. $\sum_{n=3}^{\infty} \frac{2}{n^2+1}$

   II. $\sum_{n=1}^{\infty} \left(\frac{6}{7}\right)^n$

   III. $\sum_{n=2}^{\infty} \frac{(-1)^n}{n}$

   (A) I only (B) II only

(C) Ⅰ and Ⅱ   (D) Ⅰ, Ⅱ and Ⅲ

7. Which of the following series converge?

Ⅰ. $\sum_{n=1}^{\infty} \dfrac{1}{n\sqrt{n}}$

Ⅱ. $\sum_{n=1}^{\infty} \left(\dfrac{1}{7}\right)^n$

Ⅲ. $\sum_{n=2}^{\infty} \dfrac{1}{n\ln n}$

(A) Ⅰ only   (B) Ⅱ only

(C) Ⅰ and Ⅱ   (D) Ⅰ, Ⅱ and Ⅲ

8. The power series $\sum_{n=1}^{\infty} \dfrac{(x-5)^n}{2^n n^2}$ has radius of convergence 2. At which of the following values of x can the alternating series test be used with this series to verify convergence at x?

(A) 6   (B) 4   (C) 2   (D) 0

9. For $x>0$, the power series $1-\dfrac{x^2}{3!}+\dfrac{x^4}{5!}-\dfrac{x^6}{7!}+\cdots+(-1)^n \dfrac{x^{2n}}{(2n+1)!}+\cdots$ converges to which of the following?

(A) $\cos x$ (B) $\sin x$ (C) $\dfrac{\sin x}{x}$ (D) $e^x - e^{x^2}$

10. The power series $\sum_{n=1}^{\infty} a_n (x-3)^n$ converges at $x=5$. Which of the following must be true?

(A) The series diverges at $x=0$.

(B) The series diverges at $x=1$.

(C) The series converges at $x=1$.

(D) The series converges at $x=2$.

11. Let f be a function having derivatives of all orders for $x>0$ such that $f(3)=2$, $f'(3)=-1$, $f''(3)=6$, and $f'''(3)=12$. Which of the following is the third-degree Taylor polynomial for f about $x=3$?

(A) $2-x+6x^2+12x^3$

(B) $2-(x-3)+3(x-3)^2+2(x-3)^3$

(C) $2-(x-3)+3(x-3)^2+4(x-3)^3$

(D) $2-(x-3)+6(x-3)^2+4(x-3)^3$

12. If the series $\sum_{n=1}^{\infty} a_n$ converges and $a_n>0$ for all n, which of the following must be true?

(A) $\lim_{n\to\infty} \left|\dfrac{a_{n+1}}{a_n}\right|=0$ (B) $|a_n|<1$ for all n

(C) $\sum_{n=1}^{\infty} na_n$ diverges   (D) $\sum_{n=1}^{\infty} \frac{a_n}{n}$ converges

13. The Maclaurin series for the function f is given by $f(x) = \sum_{n=2}^{\infty} \frac{(-1)^n (2x)^n}{n-1}$ on its interval of convergence.

    (a) Find the interval of convergence for the Maclaurin series of f. Justify your answer.

    (b) Show that $y = f(x)$ is a solution to the differential equation $xy' - y = \frac{4x^2}{1+2x}$ for $|x| < R$, where R is the radius of convergence from part (a).

14. The function f is defined by the power series
$$f(x) = \sum_{n=0}^{\infty} \frac{(-1)^n x^{2n}}{(2n+1)!} = 1 - \frac{x^2}{3!} + \frac{x^4}{5!} - \frac{x^6}{7!} + \cdots + \frac{(-1)^n x^{2n}}{(2n+1)!} + \cdots$$
for all real numbers x.

    (a) Find $f'(0)$ and $f''(0)$. Determine whether f has a local maximum, a local minimum, or neither at $x = 0$. Give a reason for your answer.

    (b) Show that $1 - \frac{1}{3!}$ approximates $f(1)$ with error less than $\frac{1}{100}$.

15. Let f be the function given by $f(x) = \frac{2x}{1+x^2}$.

    (a) Write the first four nonzero terms and the general term of the Taylor series for f about $x = 0$.

    (b) Does the series found in part (a), when evaluated at $x = 1$, converge to $f(1)$? Explain why or why not.

    (c) The derivative of $\ln(1+x^2)$ is $\frac{2x}{1+x^2}$. Write the first four nonzero terms of the Taylor series for $\ln(1+x^2)$ about $x = 0$.

    (d) Use the series found in part (c) to find a rational number A such that $\left| A - \ln\left(\frac{5}{4}\right) \right| < \frac{1}{100}$. Justify your answer.

习题参考答案

1.C 2.A 3.C 4.A 5.A 6.D 7.C 8.B 9.C 10.D 11.B 12.D

13. (a) $\lim_{n \to \infty} \left| \frac{\frac{(2x)^{n+1}}{(n+1)-1}}{\frac{(2x)^n}{n-1}} \right| = \lim_{n \to \infty} \left| 2x \cdot \frac{n-1}{n} \right| = |2x|$

$|2x| < 1$ for $|x| < \frac{1}{2}$

Therefore the radius of convergence is $\frac{1}{2}$.

When $x=-\frac{1}{2}$, the series is $\sum_{n=2}^{\infty}\frac{(-1)^n(-1)^n}{n-1}=\sum_{n=2}^{\infty}\frac{1}{n-1}$. This is the alternating harmonic series, which diverges.

When $x=\frac{1}{2}$, the series $\sum_{n=2}^{\infty}\frac{(-1)^n}{n-1}$ is the alternating harmonic series, which converges.

The interval of convergence for the Maclaurin series of f is $\left(-\frac{1}{2},\frac{1}{2}\right]$.

(b) $y=\frac{(2x)^2}{1}-\frac{(2x)^3}{2}+\frac{(2x)^4}{3}-\cdots+\frac{(-1)^n(2x)^n}{n-1}+\cdots$

$=4x^2-4x^3+\frac{16}{3}x^4-\cdots+\frac{(-1)^n(2x)^n}{n-1}+\cdots$

$y'=8x-12x^2+\frac{64}{3}x^3-\cdots+\frac{(-1)^n n(2x)^n\cdot 2}{n-1}+\cdots$

$xy'=8x^2-12x^3+\frac{64}{3}x^4-\cdots+\frac{(-1)^n n(2x)^{n+1}}{n-1}+\cdots$

$xy'-y=4x^2-8x^3+16x^4-\cdots+(-1)^n(2x)^n+\cdots$

$=4x^2(1-2x+4x^2-\cdots+(-1)^n(2x)^{n-2}+\cdots)$

The series $1-2x+4x^2-\cdots+(-1)^n(2x)^{n-2}+\cdots=\sum_{n=1}^{\infty}(-2x)^n$ is a geometric series that converges to $\frac{1}{1+2x}$ for $|x|<\frac{1}{2}$.

Therefore $xy'-y=\frac{4x^2}{1+2x}$ for $|x|<\frac{1}{2}$.

14. (a) $f'(0)=$ coefficient of x term $=0$

$f''(0)=2$ coefficient of x^2 term $=-\frac{1}{3}$

f has a local maximum at $x=0$ because $f'(0)=0$ and $f''(0)<0$

(b) $f(1)=1-\frac{1}{3!}+\frac{1}{5!}-\frac{1}{7!}+\cdots+\frac{(-1)^n}{(2n+1)!}+\cdots$

This is an alternating series whose terms decrease in absolute value with limit 0. Thus, the error is less than the first omitted term, so $\left|f(1)-\left(1-\frac{1}{3!}\right)\right|\leqslant\frac{1}{5!}=\frac{1}{120}<\frac{1}{100}$.

15. (a) $\frac{1}{1-u}=1+u+u^2+\cdots+u^n+\cdots$

$\frac{1}{1+x^2}=1-x^2+x^4-x^6+\cdots+(-x^2)^n+\cdots$

$\frac{2x}{1+x^2}=2x-2x^3+2x^5-2x^7+\cdots+(-1)^n 2x^{2n+1}+\cdots$

(b) No, the series does not converge when $x=1$ because when $x=1$, the terms of the series do not converge to 0.

(c) $\ln(1+x^2) = \int_0^x \dfrac{2t}{1+t^2} dt = \int_0^x (2t - 2t^3 + 2t^5 - 2t^7 + \cdots) dt$

$\qquad = x^2 - \dfrac{1}{2}x^4 + \dfrac{1}{3}x^6 - \dfrac{1}{4}x^8 + \cdots$

(d) $\ln\left(\dfrac{5}{4}\right) = \ln\left(1 + \dfrac{1}{4}\right) = \left(\dfrac{1}{2}\right)^2 - \dfrac{1}{2}\left(\dfrac{1}{2}\right)^4 + \dfrac{1}{3}\left(\dfrac{1}{2}\right)^6 - \dfrac{1}{4}\left(\dfrac{1}{2}\right)^8 + \cdots$

Let $A = \left(\dfrac{1}{2}\right)^2 - \dfrac{1}{2}\left(\dfrac{1}{2}\right)^4 = \dfrac{7}{32}$.

Since the series is a converging alternating series and the absolute values of the individual terms decrease to 0,

$$\left| A - \ln\left(\dfrac{5}{4}\right) \right| < \left| \dfrac{1}{3}\left(\dfrac{1}{2}\right)^6 \right| = \dfrac{1}{3} \times \dfrac{1}{64} < \dfrac{1}{100}.$$

Chapter 12　图形计算器的使用

在 AP 微积分的考试过程中,图形计算器只允许在多项选择题的 B 部分(共 15 道题)和自由问答的 A 部分(共两道大题)中使用.一般情况下,每次考试中大约有 5～8 道题属于必须在图形计算器辅助下才能解决的问题.美国大学理事会指出,在 AP 微积分考试过程中,学生要具备以下四项图形计算器的使用能力.

(1) Plot the graph of a function within an arbitrary viewing window.

在任意视窗下作函数图像.

(2) Find the zeros of functions (solve equations numerically).

求函数的零点(用数字方法解方程和方程组).

(3) Numerically calculate the derivative of a function.

求函数在某一点的导数值.

(4) Numerically calculate the value of a definite integral.

求定积分.

下面通过几道例题说明图形计算器在以上四个方面的具体操作.

[提醒]在考试前,请将计算器按以下要求设置.

(1)按 [AC/ON] 开机,在菜单 [MENU] 窗口,按数字键 [1],进入 **"Run－Matrix"** 窗口(见图 12.1).

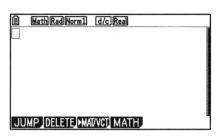

图 12.1

2.按 [SHIFT] [MENU](SET UP),进入设置窗口(见图 12.2),按照图 12.2 和图 12.3 进行图形计算器的基本设置,在使用计算器的过程中要注意图 12.1 中的第一行状态栏,特别注意计算器是处于弧度制(Rad)还是角度制(Deg).

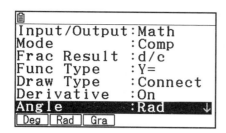

图 12.2

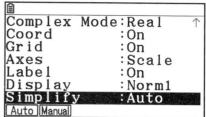

图 12.3

3.在菜单(MENU)窗口移动光标找到 ![System] 图标,按下 EXE 进入"System"窗口,对计算器的系统进行设置(见图 12.4).

图 12.4

图 12.5

其中,在图 12.4 中按下 F3(Language)可以更改计算器的显示语言,在图 12.4 中按 F5(Reset)可以将计算器恢复到出厂设置,在图 12.4 中按 F6(▷)可以设置用户名等(见图 12.5).

【说明】

(1)图 12.2 中"Input/Output"设置为"Math",设置计算器采用数学格式显示计算结果(结果包括 $\sqrt{}$ 或 π),可以通过按 S↔D 转化为小数形式.

(2)国外的考试中反三角函数一般表示为 $\sin^{-1}$,$\cos^{-1}$,$\tan^{-1}$,分别对应为:$\sin^{-1}=\arcsin$,$\cos^{-1}=\arccos$,$\tan^{-1}=\arctan$.

(3)一定要记住考试中将图形计算器设置为弧度制!

例 12.1 Finding the Derivative of the Parametric Functions(参数方程求点导数).

The position of a particle moving in the xy-plane is given by the parametric function $x(t)=t\sin t$ and $y(t)=6\mathrm{e}^{-3t}$. What is the slope of the line tangent to the path of the particle at $t=2$?

(A)0.490 (B)-1.726 (C)1.819 (D)-1.912

【题目大意】质点在 xy-平面运动的位置函数为:$x(t)=t\sin t$,$y(t)=6\mathrm{e}^{-3t}$.求 $t=2$ 时运动路径的切线斜率.

解:使用"Run-Matrix"功能辅助求解,步骤如下:

(1)在主菜单窗口(MENU),按数字键 **1**,进入"Run-Matrix"窗口.

(2)按 ▤ F4(Math F4 X,θ,T sin X,θ,T ▷ 2 ▽ F4 6 SHIFT ln (-) 3 X,θ,T ▷ ▷ 2 EXE 输入算式(见图 12.6),故答案为 B(**注意**:计算器要处于"弧度制").

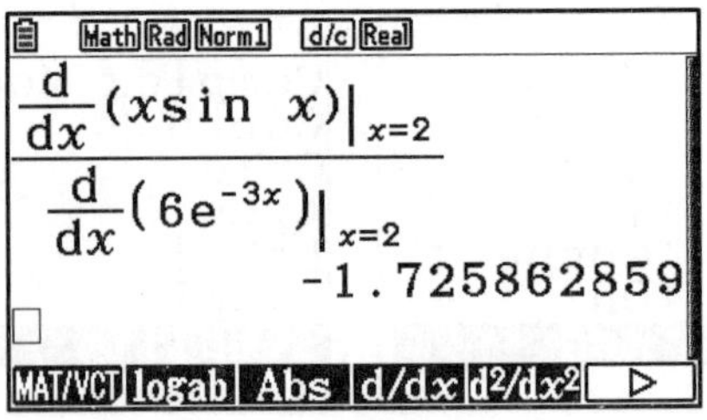

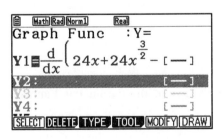

图 12.6

例 12.2 Graphing the Derivative Function(作导函数的图形).

The height h, in meters, of an object at time t is given by $h(t)=24t+24t^{\frac{3}{2}}-16t^2$. What is the height of the object at the instant when it reaches its maximum upward velocity?

(A)2.545 meters　　(B)10.263 meters　　(C)34.125 meters　　(D)54.889 meters

【题目大意】物体在 t 时刻的高度为 $h(t)=24t+24t^{\frac{3}{2}}-16t^2$ 米,求该物体的速度达到最大值时的高度为多少.

分析:高度为 $h(t)=24t+24t^{\frac{3}{2}}-16t^2$,则速度函数为 $v(t)=\dfrac{dh(t)}{dt}$,只需作出速度函数的图像求出速度为正时的最大值对应的自变量的值,即为物体速度达到最大值的时刻,代入高度函数即可求出物体的高度.

解:使用图形计算器的"Graph"功能辅助求解,步骤如下:

(1)在主菜单窗口(MENU),按数字键 5,进入"Graph"窗口.

(2)依次按 F1 2 4 X,θ,T + 2 4 X,θ,T ∧ ▤ 3 ▽ 2 ▷ ▷ − 1 6 X,θ,T x^2 ▷ X,θ,T EXE,输入"$\dfrac{d}{dx}(24x+24x^{\frac{3}{2}}-16x^2)\Big|_{x=x}$"表示速度函数 $v(x)=\dfrac{dh(x)}{dx}$(见图 12.7).

图 12.7

(3)按 F6(DRAW)得图 12.8(视窗设为标准窗"INITIAL"),虽然此时无法从图像中看到速度的最大值,但是按 F5(G−Solv) F2(MAX),还是可以得到速度为最大值时的时间 $t=0.316406$(见图 12.9).

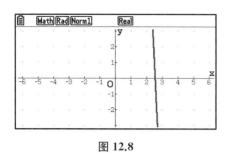

图 12.8

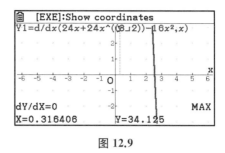

图 12.9

此时可将 $t=0.316406$ 代入高度函数 $h(t)=24t+24t^{\frac{3}{2}}-16t^2$，得到结果大约为 10.263 米，故答案为 B．

【说明】

（1）在步骤（3）中得到图 12.8 后，按 F2（Zoom）F5（AUTO），计算器将使函数图像尽量充满整个屏幕，得到图 12.10．

使用图形计算器的"**Graph**"模式研究函数图像时，在不知道函数的取值范围时，使用标准窗"INITIAL"或自动"AUTO"通常可以显示出题目需要的函数图像，通过 F2（Zoom）中的 F3（放大"IN"）或 F4（缩小"OUT"）结合光标的移动，使图像变得美观（见图 12.11）．

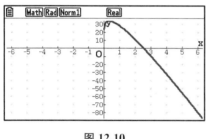

图 12.10

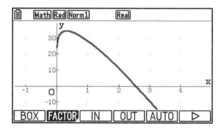

图 12.11

（2）依次按 F5（G－Solv）F2（MAX），可以得到速度为最大值时的时间 $t=0.316406$（见图 12.12）．

由于当 $t=0$ 时 $h(0)=0$，依次按 F5（G－Solv）F6（▷）F3（$\int dx$）F1（$\int dx$）0 EXE（输入积分下限"Lower Bound"0）0 · 3 1 6 4 EXE（输入定积分的上限"Upper Bound"0.3164），得到定积分的结果（见图 12.13）．

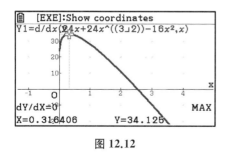

图 12.12

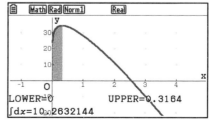

图 12.13

图 12.13 的结果显示从 $t=0$ 时刻开始到物体的速度达到最大时，物体的位移约为 10.263米，由于 $t=0$ 时刻物体的高度为 0，所以此时物体的高度为 10.263米，故答案为 B．

例 12.3 Finding the Maximum of the Accumulation Function(求变上限积分的最大值).

Let f be the function given by $f(x)=\int_{\frac{1}{3}}^{x}\cos(\frac{1}{t^2})dt$ for $\frac{1}{3}\leqslant x \leqslant 1$. At which of the following values of x does f attain a relative maximum?

(A)0.357 and 0.798　(B)0.4 and 0.564　(C)0.4 only　(D)0.461

【题目大意】 $f(x)=\int_{\frac{1}{3}}^{x}\cos\left(\frac{1}{t^2}\right)dt\left(\frac{1}{3}\leqslant x\leqslant 1\right)$，当 x 取下列哪些值时，f 达到极大值？

分析： 因为 $f'(x)=\left(\int_{\frac{1}{3}}^{x}\cos\left(\frac{1}{t^2}\right)dt\right)'=\cos\left(\frac{1}{x^2}\right)$，使用图形计算器的图形模块作 $f'(x)$ 的函数图像，然后用一阶导数判别法即可.

解： 使用图形计算器的"Graph"功能辅助求解，步骤如下：

(1)在主菜单窗口(MENU)，按数字键 5，进入"Graph"窗口.

(2)依次按 COS ▤ 1 ▽ X,θ,T x² ▶ , SHIFT + ▤ 1 ▽ 3 ▶ , 1 SHIFT − EXE，输入函数表达式及考察的定义区间 $Y1=\cos\left(\frac{1}{x^2}\right),\left[\frac{1}{3},1\right]$（见图 12.14）.

(3)按 F6 (DRAW)（见图 12.14）[通过 F3 (V−Window)设置"Xmin=0.1, Xmax=1.2, Xscale=0.1, Ymin=−2, Ymax=2, Yscale=0.2"].从图 12.15 观察出，$f'(x)$ 仅第二个零点满足函数的取值由正变负，由一阶导数判别法知：该零点为函数 $f(x)$ 在 $\frac{1}{3}\leqslant x \leqslant 1$ 时取得的极大值点.

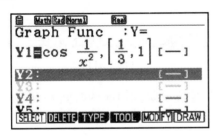

图 12.14

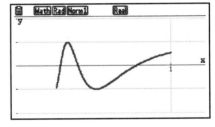

图 12.15

(4)依次按 F5 (G−Solv) F1 (ROOT) ▶（见图 12.16），故答案为 D.

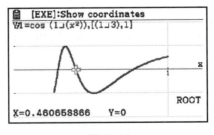

图 12.16

例 12.4 Finding the Net Change With a Negative Rate of Change(求变化率为负时的净增量).

The rate of the altitude of a hot-air balloon is given by $r(t)=t^3-4t^2+6$ for $0\leqslant t\leqslant 8$.

Which of the following expressions gives the change in altitude of the balloon during the time the altitude is decreasing?

(A) $\int_{1.572}^{3.514} r(t)\mathrm{d}t$ (B) $\int_{0}^{3} r(t)\mathrm{d}t$ (C) $\int_{0}^{2.667} r(t)\mathrm{d}t$ (D) $\int_{1.572}^{3.514} r'(t)\mathrm{d}t$

【题目大意】热气球高度的变化率为函数 $r(t)=t^3-4t^2+6, 0\leqslant t\leqslant 8$. 下列哪个表达式表示高度降低过程中的高度的改变量.

分析：令 $h(t)$ 表示热气球在时刻 t 的高度，则 $h'(t)=r(t)$.

解法一：使用图形计算器的"Graph"功能辅助求解，步骤如下：

(1) 在主菜单窗口(MENU)，按数字键 5，进入"Graph"窗口.

(2) 依次按 [X,θ,T] [∧] [3] [▶] [−] [4] [X,θ,T] [x²] [+] [6] [,] [SHIFT] [+] [0] [,] [8] [SHIFT] [−] [EXE]，输入函数表达式 $Y1=x^3-4x^2+6,[0,8]$（见图 12.17）.

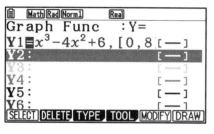

图 12.17

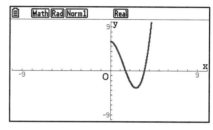

图 12.18（视窗为标准窗"STANDRD"）

(3) 按 [F6](DRAW) 得到高度变化率的图像（见图 12.18）.

(4) 依次按 [F5](G−Solv) [F1](ROOT) 得到第一个零点 $x=1.571993268$（见图 12.19），按 [▶] 得到第二个零点 $x=3.514136929$（见图 12.20），故答案为 A.

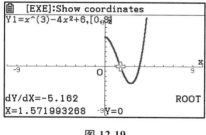

图 12.19

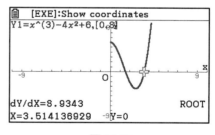

图 12.20

【说明】在图 12.20 中，依次按 [F5](G−Solv) [F6]([▶]) [F3]($\int \mathrm{d}x$) [F2](ROOT)（见图 12.21），按 [EXE] 后绿色虚线变成绿色实线（设置为积分下限），按 [▶] [EXE] 后绿色虚线变成绿色实线（设置为积分上限），得到两个零点之间的定积分（见图 12.22）.

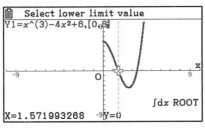

图 12.21

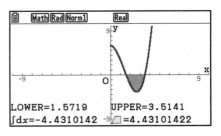

图 12.22

图 12.22 中显示函数 $Y1=x^3-4x^2+6$,$[0,8]$,当 $x\in[1.5719,3.5141]$ 时,函数图像在 x 轴下方,说明高度的变化率 $r(t)=t^3-4t^2+6$ 在此时为负值,所以高度是降低的.

高度的净增量(net change)$=h(3.514)-h(1.572)=\int_{1.572}^{3.514}r(t)\mathrm{d}t=\int_{1.572}^{3.514}(t^3-4t^2+6)\mathrm{d}t=\int_{1.572}^{3.514}(x^3-4x^2+6)\mathrm{d}x\approx-4.431$,说明高度下降了 4.431 个单位.

解法二:使用图形计算器的"Equation"功能辅助求解,步骤如下:

(1)在主菜单窗口(MENU),按X,θ,T,进入"Equation"窗口或使用光标移动到"Equation"模式后按EXE进入"Equation"窗口.

(2)按F2(Polynomial)F2(3),并输入方程 $x^3-4x^2+6=0$ 的各项系数(见图 12.23).

(3)按EXE得到方程的三个实根(见图 12.24).

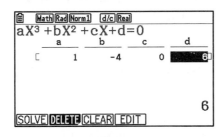

图 12.23

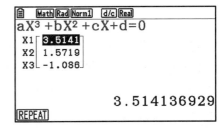

图 12.24

t	$[0,1.572)$	1.572	$(1.572,3.514)$	3.514	$(3.514,8]$
$r(t)$	+	0	−	0	+

$r(t)$ 的值为负的区间为 $t\in(1.572,3.514)$,故高度的变化量为 $\int_{1.572}^{3.514}r(t)\mathrm{d}t$,故答案为 A.

例 12.5 Finding the Points When the Particle Changes Direction(求直线运动的转向点).

A particle moves on the x-axis with velocity given by $v(t)=3t^4-11t^2+9t-2$ for $-3\leqslant t\leqslant 3$. How many times does the particle change direction as t increases from -3 to 3?

(A) zero   (B) one   (C) two   (D) three

【题目大意】一个质点沿 x 轴运动,其速度函数为 $v(t)=3t^4-11t^2+9t-2$,$-3\leqslant t\leqslant 3$,当时刻 t 从 -3 到 3 时,质点改变了几次运动方向?

分析:质点改变运动方向的时刻满足两个条件:$v(t)=0$,且在该时刻前后 $v(t)$ 异号.

解法一:使用图形计算器的"Graph"功能辅助求解,步骤如下:

(1)在主菜单(MENU)窗口,按数字键5,进入"Graph"窗口.

（2）依次按 3 X,θ,T ^ 4 ▶ − 1 1 X,θ,T x² + 9 X,θ,T − 2 , SHIFT + (−) 3 , 3 SHIFT − EXE,输入速度函数及题目需要考察的定义区间：$Y1=3x^4-11x^2+9x-2$，$[-3,3]$.

(3)按 F6 (DRAW)得到图 12.25.

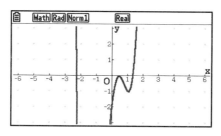

图 12.25 （视窗为"INITIAL"）

（4）按 F5 (G−Solv) F1 (ROOT)，得到速度函数在区间$[-3,3]$的第一个零点（见图 12.26），然后按 ▶ 得到第二个零点（见图 12.27）.由于速度函数是多项式函数，所以速度函数在区间$[-3,3]$共有两个零点，而在零点两侧速度函数异号，故答案为 C.

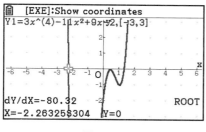

图 12.26

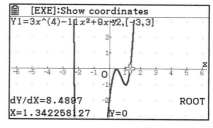

图 12.27

解法二：使用图形计算器的"Equation"功能辅助求解，步骤如下：

(1)在主菜单窗口(MENU)，按 X,θ,T，进入"Equation"窗口.

(2)按 F2 (POLY) F3 (4)，参考图 12.28 依次输入方程 $3x^4-11x^2+9x-2=0$ 各项系数（见图 12.28）.

(3)按 F1 (SOLVE)得到方程 $3x^4-11x^2+9x-2=0$ 的两个实根（见图 12.29）.

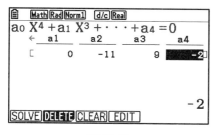

图 12.28

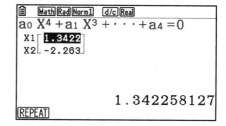

图 12.29

由于多项式方程 $3x^4-11x^2+9x-2=0$ 在$[-3,3]$共有两个实根，由标根法可知：速度函数在这两个根两侧一定异号，故答案为 C.

【说明】

在图12.28中依次按 SHIFT MENU 进入设置窗口,将复数模式"Complex Mode"按 F2 $(a+bi)$ 设置(见图12.30).

依次按 EXE F1 (SOLVE)得到多项式方程 $3x^4-11x^2+9x-2=0$ 的4个根(见图12.31).

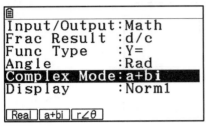

图12.30

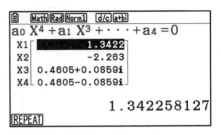

图12.31

例12.6 Finding Inflections(求拐点).

Let f be the function with derivative defined by $f'(x)=\sin(x^3)$ on the interval $-1.8<x<1.8$. How many points of inflection does the graph of f have on this interval?

(A)two　　　　(B)three　　　　(C)four　　　　(D)five

【题目大意】当 $-1.8<x<1.8$ 时,$f'(x)=\sin(x^3)$,则 f 的图形在该区间有多少个拐点?

分析:拐点是函数凹凸区间的分界点,由 $f'(x)$ 的函数图像判断拐点有两种方法:拐点为 $f'(x)$ 图像的极值点或 $f'(x)$ 的图像改变单调性的点.

解:使用图形计算器的"Graph"功能辅助求解,步骤如下:

(1)在主菜单窗口(MENU),按数字键 5,进入"Graph"窗口.

(2) 依次按 sin (X,θ,T ^ 3 ▶) , SHIFT + (-) 1 . 8 , 1 . 8 SHIFT - EXE,输入函数表达式 $Y1=\sin(x^3)$,$[-1.8,1.8]$(见图12.32),然后按 F6 (DRAW)得到图12.33.

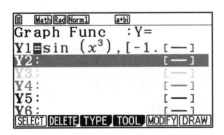

图12.32

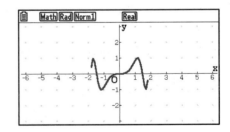

图12.33(视窗为"INITIAL")

从图12.33观察出当 $-1.8<x<1.8$ 时,$y'=(x)$ 的图像有4个极值点或有4个点的两侧 $y'=(x)$ 的单调性发生了改变,故答案为C.

【说明】在图12.33中依次按 F5 (G-Solv) F2 (MAX),在图12.34中显示该区间内的第一个极大值点的坐标,按 ▶ 显示第二个极大值点的坐标.在图12.33中按 F5 (G-Solv) F3 (MIN)可以显示出该区间的极小值(见图12.35),当 $-1.8<x<1.8$ 时,$y'=x$ 的图像有2个极大值点和2个极小值点.这4个极值点为 $y=f(x)$ 的拐点.

Chapter 12 图形计算器的使用

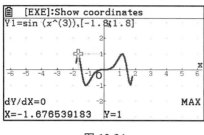

图 12.34

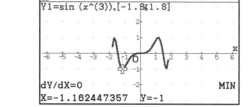

图 12.35

例 12.7 Finding the Area of a Plane Region(求曲边梯形面积).

What is the area enclosed by the curves $y=x^3-8x^2+18x-5$ and $y=x+5$?
(A)10.667　　　　(B)11.833　　　　(C)14.583　　　　(D)21.333

【题目大意】曲线 $y=x^3-8x^2+18x-5$ 与 $y=x+5$ 围成的区域的面积为多少?

解:使用图形计算器的"Graph"功能辅助求解,步骤如下:

(1)在主菜单(MENU)窗口,按数字键 5,进入"Graph"窗口.

(2)依次按 X,θ,T ∧ 3 ▶ − 8 X,θ,T x^2 + 1 8 X,θ,T − 5 EXE 输入第一条曲线的表达式 $Y1=x^3-8x^2+18x-5$,依次按 X,θ,T + 5 EXE 输入第二条曲线的表达式 $Y2=x+5$(见图 12.36).

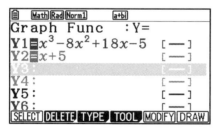

图 12.36

(3)按 F6 (DRAW)(见图 12.37),按 SHIFT (见图 12.38).

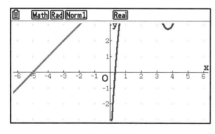

图 12.37 (视窗为"INITIAL")

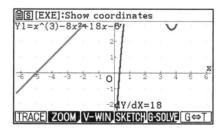

图 12.38

在图 12.37 和图 12.38 中,按 F1 F2 F3 F4 F5 F6 功能是一样的.按 F1 (TRACE)可以显示图像在光标处的坐标.按 F2 (ZOOM)可以设定视窗范围,也可以对视窗进行缩放等.按 F3 (V−WIN)可以进行视窗的设置.在图 12.38 中,按 F3 (V−Window) F3 (标准窗"STANDRD") EXE F6 (DRAW)得到的图像更有利于解题(见图 12.39),通过光标的移动可以得到图 12.40.按 F4 (草图"SKETCH")可以作切线、法线等草图(AP 微积分考试一般不涉及).按 F5(图解"G-SOLVE")可以进行图像解析.

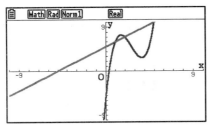

图 12.39 （视窗为"STANDRD"）

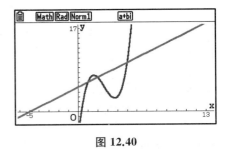

图 12.40

(4)在图 12.40 中依次按 F5(G-Solv) F6(▷) F3($\int dx$) F3(INTSECT)，图像出现最左边的一个交点(见图 12.41)，按 EXE 将交点(1,6)的横坐标值 $x=1$ 设为积分下限，然后按 ▷ ▷ 光标移动至第三个交点，按 EXE 得到定积分结果和两条曲线在两个交点之间所围成的曲边梯形的面积(见图 12.42).

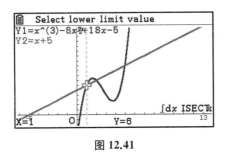

图 12.41

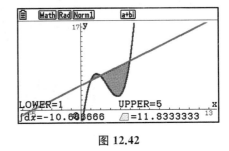

图 12.42

图 12.42 显示曲线 $y=x^3-8x^2+18x-5$ 与 $y=x+5$ 围成的区域的面积约为 11.833，故答案为 B.

例 12.8 Free-Response 1(自由问答 1).

For $t \geqslant 0$, a particle is moving along a curve so that its position at time t is $(x(t), y(t))$. At time $t=2$, the particle is at position $(1,5)$. It is known that $\dfrac{dx}{dt}=\dfrac{\sqrt{t+2}}{e^t}$ and $\dfrac{dy}{dt}=\sin^2 t$.

(a) Is the horizontal movement of the particle to the left or to the right at time $t=2$? Explain your answer. Find the slope of the path of the particle at time $t=2$.

(b) Find the x-coordinate of the particle's position at time $t=4$.

(c) Find the speed of the particle at time $t=4$. Find the acceleration vector of the particle at time $t=4$.

(d) Find the distance traveled by the particle from time $t=2$ to $t=4$.

【题目大意】一个沿曲线运动的质点在时刻 t 的位置为 $(x(t),y(t))$，$t \geqslant 0$ 且 $\dfrac{dx}{dt}=\dfrac{\sqrt{t+2}}{e^t}$，$\dfrac{dy}{dt}=\sin^2 t$. 当 $t=2$ 时，质点的位置为 $(1,5)$.

(a)质点在 $t=2$ 时是水平向左还是水平向右移动？解释你的结果并求质点在 $t=2$ 时刻运动路径的斜率.

(b)求质点在 $t=4$ 时位置的横坐标.

(c)求质点在 $t=4$ 时的速率和加速度向量.

(d)求质点从 $t=2$ 到 $t=4$ 运动的距离.

解：(a)质点在时刻 t 的位置为 $(x(t),y(t))$，则其速度向量为 $(x'(t),y'(t))$.

因为 $t=2$ 时，$x'(2)=\dfrac{\sqrt{t+2}}{\mathrm{e}^t}\bigg|_{t=2}=\dfrac{\sqrt{2+2}}{\mathrm{e}^2}=\dfrac{2}{\mathrm{e}^2}>0$，所以此时质点向右运动.

$t=2$ 时运动路径的斜率为：$\dfrac{\mathrm{d}y}{\mathrm{d}x}\bigg|_{t=2}=\dfrac{\dfrac{\mathrm{d}y}{\mathrm{d}t}\bigg|_{t=2}}{\dfrac{\mathrm{d}x}{\mathrm{d}t}\bigg|_{t=2}}=\dfrac{\sin^2 2}{\dfrac{2}{\mathrm{e}^2}}\approx 3.0547.$

使用图形计算器操作如下：

(1)在主菜单窗口(MENU)，按数字**1**，进入"**Run－Matrix**"窗口.

(2)依次按 (sin 2) x^2 ═ ═ 2 ▽ SHIFT ln 2 EXE，输入计算表达式得到计算结果(见图 12.43).

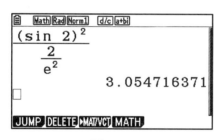

图 12.43

(b)由题意可知，质点在 $t=4$ 时，位置的横坐标为：

$$x(4)=x(2)+\int_2^4 x'(t)\mathrm{d}t=1+\int_2^4\dfrac{\sqrt{t+2}}{\mathrm{e}^t}\mathrm{d}t\approx 1.2530.$$

使用图形计算器操作如下：

在图 12.43 中依次按 **1** ＋ F4 ▷ F1 ═ SHIFT x^2 X,θ,T ＋ 2 ▽ SHIFT ln X,θ,T ▷ ▷ ▷ 2 △ 4 EXE，输入计算表达式得到计算结果(见图 12.44).

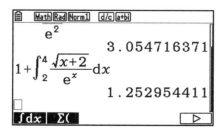

图 12.44

(c)质点在 $t=4$ 时的速率为：

$$\sqrt{(x'(t))^2+(y'(t))^2}\Big|_{t=4}=\sqrt{(x'(4))^2+(y'(4))^2}$$

$$=\sqrt{(\frac{\sqrt{6}}{e^4})^2+(\sin^2 4)^2}=\sqrt{\frac{6}{e^8}+\sin^4 4}\approx 0.5745.$$

质点在 $t=4$ 时的加速度向量为：$(x''(t),y''(t))=(x''(4),y''(4))=$

$(\dfrac{d(\dfrac{\sqrt{t+2}}{e^t})}{dt}\Big|_{t=4},\dfrac{d(\sin^2 t)}{dt}\Big|_{t=4})\approx(-0.0411,0.9894).$

使用图形计算器操作如下：

(1)在图 12.44 中依次按 [SHIFT] [x²] [6] [▭] [SHIFT] [ln] [8] [▶] [▶] [+] [(] [sin] [4] [)] [∧] [4] [EXE]，输入计算表达式得到计算结果(见图 12.45).

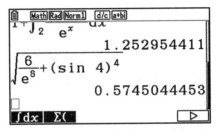

图 12.45

（2）按 [EXIT] [F4]（MATH）[F4]（d/dx）[▭] [SHIFT] [x²] [X,θ,T] [+] [2] [▼] [SHIFT] [ln] [X,θ,T] [▶] [▶] [▶] [4] [EXE]，计算得到 $t=4$ 时的加速度向量的 X 坐标(见图 12.46).

(3)按 [F4] (d/dx) [(] [sin] [X,θ,T] [)] [x²] [▶] [4] [EXE] 计算得到 $t=4$ 时的加速度向量的 y 坐标(见图 12.47).

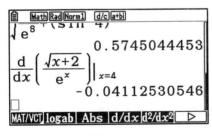

图 12.46

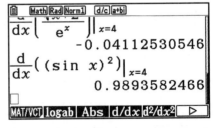

图 12.47

(d)根据距离公式,质点从 $t=2$ 到 $t=4$ 时刻运动的距离为：

$$\int_2^4\sqrt{(x'(t))^2+(y'(t))^2}\,dt=\int_2^4\sqrt{\frac{t+2}{e^{2t}}+(\sin t)^4}\,dt\approx 0.6510.$$

使用图形计算器操作如下：

在图 12.47 中依次按 [F6]（[▶]）[F1]（∫dx）[SHIFT] [x²] [▭] [X,θ,T] [+] [2] [▼] [SHIFT] [ln] [2] [X,θ,T] [▶] [▶] [+] [(] [sin] [X,θ,T] [)] [∧] [4] [▶] [▶] [▶] [2] [▲] [4] [EXE]，计算得到质点从 $t=2$ 到 $t=4$ 时刻运动的距离(见图 12.48).

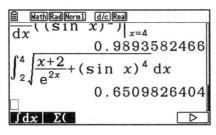

图 12.48

例 12.9 Free-Response 2(自由问答 2).

Grass clippings are placed in a bin, where they decompose. For $0 \leqslant t \leqslant 30$, the amount of grass clippings remaining in the bin is modeled by $A(t)=6.687(0.931)^t$, where $A(t)$ is measured in pounds and t is measured in days.

(a) Find the average rate of change of $A(t)$ over the interval $0 \leqslant t \leqslant 30$. Indicate units of measure.

(b) Find the value of $A'(15)$. Using correct units, interpret the meaning of the value in the context of the problem.

(c) Find the time t for which the amount of grass clippings in the bin is equal to the average amount of grass clippings in the bin over the interval $0 \leqslant t \leqslant 30$.

(d) For $t>30$, $L(t)$, the linear approximation to A at $t=30$, is a better model for the amount of grass clippings remaining in the bin. Use $L(t)$ to predict the time at which there will be 0.5 pound of grass clippings remaining in the bin. Show the work that leads to your answer.

【题目大意】剪下的草被放在一个箱子里分解.设 t 天后($0 \leqslant t \leqslant 30$),箱子中剩下的草量为 $A(t)=6.687(0.931)^t$.

(a)求 $A(t)$ 在区间 $0 \leqslant t \leqslant 30$ 的平均变化率,并标明单位.

(b)求 $A'(15)$ 的值,并解释其在上下文中的含义(要求使用准确的单位).

(c)求箱子中所剩草量等于 $0 \leqslant t \leqslant 30$ 时的平均草量的时刻 t.

(d)令 $L(t)$ 表示当 $t>30$ 时 $A(t)$ 在 $t=30$ 的线性近似,$L(t)$ 是当 $t>30$ 时较 $A(t)$ 更优的模型.用 $L(t)$ 预测当箱子中剩下的草量为 0.5 磅时的时间,并写出解题过程.

解:(a)求 $A(t)$ 在区间 $0 \leqslant t \leqslant 30$ 的平均变化率为:$\dfrac{A(30)-A(0)}{30-0}=\dfrac{6.687\times(0.931)^{30}-6.687}{30}=$ -0.197 磅/天.

使用图形计算器操作如下:

(1)在主菜单窗口(**MENU**),按数字 **1**,进入"**Run－Matrix**"窗口.

(2) 依次按 ▦ 6 · 6 8 7 (0 · 9 3 1) ∧ 3 0 ▶ − 6 · 6 8 7 ▼ 3 0 **EXE**,得到 $A(t)$ 在区间 $0 \leqslant t \leqslant 30$ 的平均变化率的计算结果(见图 12.49).

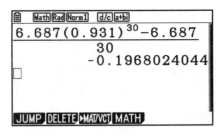

图 12.49

(b) $A'(t)$ 表示 $A(t)$ 的瞬时变化率, $A'(15) = \dfrac{d(6.687(0.931)^t)}{dt}\bigg|_{t=15} \approx -0.164$,
$A'(15) = -0.164$, 表示第 15 天时箱子中所剩草量以 0.164 磅/天的速度减少.

使用图形计算器操作如下:

在图 12.49 中继续输入 [F4](MATH)[F4](d/dx)[6][·][6][8][7][(][0][·][9][3][1][)] [∧][X,θ,T][▶][▶][1][5][EXE], 计算结果见图 12.50.

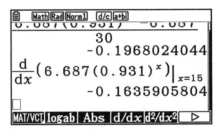

图 12.50

(c) 根据连续函数在区间的平均数的计算公式得:

当 $0 \leqslant t \leqslant 30$ 时, 平均草量为 $\dfrac{\int_0^{30} A(t)\,dt}{30-0} = \dfrac{\int_0^{30} 6.687(0.931)^t\,dt}{30} \approx 2.752635$.

求箱子中所剩草量等于 $0 \leqslant t \leqslant 30$ 时的平均草量的时间 t, 即求方程 $A(t) = 6.687(0.931)^t = 2.752635$ 在区间 $0 \leqslant t \leqslant 30$ 的根.

解法一:使用"Graph"功能辅助求解,步骤如下:

(1) 在主菜单窗口([MENU]), 按数字键[5], 进入"Graph"窗口.

(2) 依次按 [6][·][6][8][7][(][0][·][9][3][1][)][∧][X,θ,T][▶][,][SHIFT][+][0][,][3][0][SHIFT][−][EXE], 输入函数表达式 $Y1 = 6.687(0.931)^x$, $[0, 30]$, 按[F6](DRAW)画出图像 (见图 12.51).

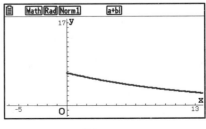

图 12.51

然后依次按 F5(G—Solv) F6(▷) F2(X—CAL),输入 Y:2.752635(见图 12.52),按 EXE 得出结果(见图 12.53).

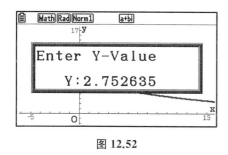

图 12.52

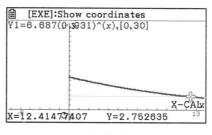

图 12.53

从图像可见:$Y1=6.687(0.931)^x$,$[0,30]$是一个单调递减的函数,当 $x=12.41477407$ 时,$y=2.752635$,且方程在区间$[0,30]$只有唯一的根,即 $t=12.415$ 天时箱子中所剩草量等于 $0 \leqslant t \leqslant 30$ 的平均草量.

解法二:使用图形计算器的解方程命令"**SolveN**"辅助求解,操作步骤如下:

(1)在主菜单窗口(MENU),按数字键 1,进入"**Run－Matrix**"窗口.

(2)按 OPTN F4(CALC) F5(SolveN),调用"SolveN"函数命令,按 6 · 6 8 7 (0 · 9 3 1) ∧ X,θ,T ▷ SHIFT · 2 · 7 5 2 6 3 5 EXE 输入方程表达式后出现图 12.54,按 EXIT 得到方程的根(见图 12.55).

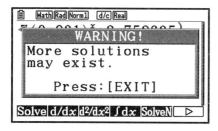

图 12.54

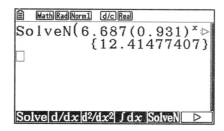

图 12.55

解法三:使用图形计算器的"**Equation**"模式辅助求解,操作步骤如下:

(1)在主菜单窗口(MENU),按 X,θ,T,进入"**Equation**"窗口,按 F3(SOLVER)(见图 12.56),此时采用牛顿法求方程的解.

(2)在图 12.56 中输入 6 · 6 8 7 (0 · 9 3 1) ∧ X,θ,T ▷ SHIFT · 2 · 7 5 2 6 3 5 EXE EXE(见图 12.57),得到方程的一个根 $x=12.41477407$.

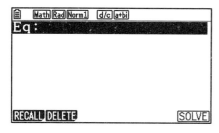

图 12.56

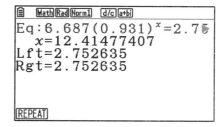

图 12.57

可以重新设置 x 的下界"Lower"和上界"Upper"（见图 12.58），求其他范围内的解，按 [F6](SOLVE)（见图 12.59），说明原方程在 $x \in [13,30]$ 无解．

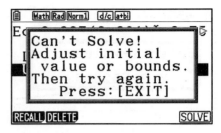

图 12.58　　　　　　图 12.59

(d) 当 $t > 30$ 时，$L(t) = A(30) + A'(30)(t-30)$．

用图形计算器可以计算出：$A(30) = 6.687 \times (0.931)^{30} \approx 0.7829$，$A'(30) = \left. \dfrac{\mathrm{d}(6.687(0.931)^t)}{\mathrm{d}t} \right|_{t=30} \approx -0.056$，则剩下的草量为 0.5 磅的时间可以由方程 $L(t) = 0.7829 - 0.056(t-30) = 0.5$ 求出，解得 $t \approx 35.0518$．

Chapter 13　AP 微积分模拟试题及解析

试题和试题自由问答的答案均来源于 College Board,考试题型及时间安排见下表：

Section I	**Total Time**：1 hour 45 minutes **Number of Questions**：45 * **Percent of Total Score**：50% **Writing Instrument**：Pencil required * The number of questions may vary slightly depending on the form of the exam.	**Part A**： **Number of Questions**：30 **Time**：1 hour No calculator allowed	**Part B**： **Number of Questions**：15 **Time**：45 minutes Graphing calculator required
Section II	**Total Time**：1 hour 30 minutes **Number of Questions**：6 **Percent of Total Score**：50% **Writing Instrument**：Either pencil or pen with black or dark blue ink **Note**：For Section II, if students finish Part A before the end of the timed 30 minutes for Part A, they cannot begin working on Part B. Students must wait until the beginning of the timed 1 hour for Part B. However, during the timed portion for Part B, students may work on the problems in Part A without the use of a calculator.	**Part A**： **Number of Questions**：2 **Time**：30 minutes **Percent of Section II Score**：33.33% Graphing calculator required	**Part B**： **Number of Questions**：4 **Time**：1 hour **Percent of Section II Score**：66.67% No calculator allowed

CALCULUS BC
SECTION Ⅰ, Part A
Time—1 hour
Number of questions—30

NO CALCULATOR IS ALLOWED FOR THIS PART OF THE EXAM.

（第一部分为30道单项选择题，不允许使用计算器，答题时间为1个小时）

Directions: Solve each of the following problems, using the available space for scratch work. After examining the form of the choices, decide which is the best of the choices given and fill in the corresponding circle on the answer sheet. No credit will be given for anything written in this exam booklet. Do not spend too much time on any one problem.

In this exam:

(1) Unless otherwise specified, the domain of a function f is assumed to be the set of all real numbers x for which $f(x)$ is a real number.

(2) The inverse of a trigonometric function f may be indicated using the inverse function notation f^{-1} or with the prefix "arc"(e.g., $\sin^{-1} x = \arcsin x$).

1. If $f(x) = \cos^2(3x-5)$, then $f'(x) =$

   (A) $6\cos(3x-5)$　　　　　　　　(B) $-3\sin^2(3x-5)$

   (C) $-3\sin(3x-5)\cos(3x-5)$　　(D) $-6\sin(3x-5)\cos(3x-5)$

【答案】D

本题的考点是 The Chain Rule(链式法则)，$f(x) = (\cos(3x-5))^2$ 是三层函数的复合.
$f'(x) = 2\cos(3x-5) \cdot [-\sin(3x-5)] \times 3 = -6\sin(3x-5)\cos(3x-5)$

2. $\displaystyle\int \frac{1}{t\sqrt{t}} dt =$

   (A) $-2t^{-1/2} + C$　　　　　　　(B) $-\dfrac{3}{2} t^{-5/2} + C$

   (C) $-\dfrac{2}{5} t^{-5/2} + C$　　　　　(D) $2t^{1/2} \ln t + C$

【答案】A

本题的考点是利用基本积分法求不定积分：

$$\int \frac{1}{t\sqrt{t}} dt = \int t^{-\frac{3}{2}} dt = \frac{t^{-\frac{3}{2}+1}}{-\frac{3}{2}+1} + C = -2t^{-\frac{1}{2}} + C$$

3.If $f(x)=\dfrac{5-x}{x^3+2}$,then $f'(x)=$

(A)$\dfrac{-4x^3+15x^2-2}{(x^3+2)^2}$ (B)$\dfrac{-2x^3+15x^2+2}{(x^3+2)^2}$

(C)$\dfrac{2x^3-15x^2-2}{(x^3+2)^2}$ (D)$\dfrac{4x^3-15x^2-2}{(x^3+2)^2}$

【答案】C

本题的考点是求导的除法公式：
$$f'(x)=\dfrac{-(x^3+2)-(5-x)\cdot 3x^2}{(x^3+2)^2}=\dfrac{2x^3-15x^2-2}{(x^3+2)^2}$$

4.The position of a particle moving in the xy-plane is given by the vector $\langle 4t^3, y(2t)\rangle$, where y is a twice-differentiable function of t. At time $t=\dfrac{1}{2}$, what is the acceleration vector of the particle?

(A)$\langle 3,2y''(1)\rangle$ (B)$\langle 6,4y''(1)\rangle$ (C)$\langle 12,2y''(1)\rangle$ (D)$\langle 12,4y''(1)\rangle$

【答案】D

题目大意：质点在 xy-平面运动，其位置向量为 $\langle 4t^3,y(2t)\rangle$，$y$ 二阶可导，求 $t=\dfrac{1}{2}$ 时的加速度向量.

本题的考点是平面运动问题由位置向量求加速度向量.

$\overrightarrow{R(t)}=\langle x(t),y(t)\rangle$
$=\langle 4t^3,y(2t)\rangle$

$\overrightarrow{a(t)}=\langle \dfrac{d^2x}{dt^2},\dfrac{d^2y}{dt^2}\rangle$

$\overrightarrow{a(\dfrac{1}{2})}=\langle(12t^2)',2y'(2t)\rangle=\langle 24t,4y''(2t)\rangle$

5.To what number does the series $\sum\limits_{k=0}^{\infty}\left(\dfrac{-e}{\pi}\right)^k$ converge?

(A)0 (B)$\dfrac{-e}{\pi+e}$

(C)$\dfrac{\pi}{\pi+e}$ (D)The series does not converge

【答案】C

本题的考点是几何级数 $\sum\limits_{n=0}^{\infty}ar^n=\dfrac{a}{1-r}(|r|<1)$.

$\sum\limits_{k=0}^{\infty}\left(\dfrac{-e}{\pi}\right)^k=1-\dfrac{e}{\pi}+\cdots$这是公比为 $r=-\dfrac{e}{\pi}$，$|r|<1$，首项 $a=1$ 的几何级数，所以收敛，且有

$$\sum_{k=0}^{\infty}\left(\dfrac{-e}{\pi}\right)^k=\dfrac{a}{1-r}=\dfrac{1}{1+\dfrac{e}{\pi}}=\dfrac{\pi}{\pi+e}.$$

6.

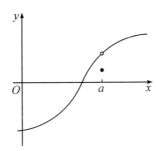

The graph of $y=f(x)$ is shown above. Which of the following is true?

(A) $\lim\limits_{h\to\infty}\dfrac{f(a+h)-f(a)}{h}$ exists

(B) $\lim\limits_{x\to a^+}f(x)\neq \lim\limits_{x\to a^-}f(x)$

(C) $\lim\limits_{x\to a}f(x)\neq f(a)$

(D) $\lim\limits_{x\to a}f(x)$ does not exist

【答案】C

本题是由图像判断导数的概念及可导与连续的关系的综合题.

由于函数 $f(x)$ 在 $x=a$ 处不连续,所以 $f(x)$ 在 $x=a$ 处不可导,故 $\lim\limits_{h\to\infty}\dfrac{f(a+h)-f(a)}{h}$ does not exists,排除 A 选项.

从函数 $f(x)$ 在 $x=a$ 两侧的函数图像观察出 $\lim\limits_{x\to a^+}f(x)=\lim\limits_{x\to a^-}f(x)$,说明 $\lim\limits_{x\to a}f(x)$ 存在,但是 $\lim\limits_{x\to a}f(x)\neq f(a)$,所以 B,D 选项错误,C 选项正确.

7. If $\displaystyle\int_4^{-10}g(x)\mathrm{d}x=-3$ and $\displaystyle\int_4^6 g(x)\mathrm{d}x=5$, then $\displaystyle\int_{-10}^6 g(x)\mathrm{d}x=$

(A)-8 (B)-2 (C)2 (D)8

【答案】D

本题的考点是定积分对积分区间的可加性.

$\displaystyle\int_{-10}^6 g(x)\mathrm{d}x=\int_{-10}^4 g(x)\mathrm{d}x+\int_4^6 g(x)\mathrm{d}x=3+5=8.$

8. The length of the curve $y=\sin(3x)$ from $x=0$ to $x=\dfrac{\pi}{6}$ is given by

(A) $\displaystyle\int_0^{\pi/6}(1+9\cos^2(3x))\mathrm{d}x$ (B) $\displaystyle\int_0^{\pi/6}\sqrt{1+\sin^2(3x)}\,\mathrm{d}x$

(C) $\displaystyle\int_0^{\pi/6}\sqrt{1+3\cos(3x)}\,\mathrm{d}x$ (D) $\displaystyle\int_0^{\pi/6}\sqrt{1+9\cos^2(3x)}\,\mathrm{d}x$

【答案】D

本题的考点是直角坐标方程求弧长 $L=\displaystyle\int_a^b\sqrt{1+\left(\dfrac{\mathrm{d}y}{\mathrm{d}x}\right)^2}\,\mathrm{d}x$

这里,$\dfrac{\mathrm{d}y}{\mathrm{d}x}=3\cos(3x)$,

所以弧长 $L=\displaystyle\int_0^{\frac{\pi}{6}}\sqrt{1+\left(\dfrac{\mathrm{d}y}{\mathrm{d}x}\right)^2}\,\mathrm{d}x=\int_0^{\pi/6}\sqrt{1+9\cos^2(3x)}\,\mathrm{d}x.$

9. The slope of the line tangent to the graph of $y=xe^x$ at $x=\ln 2$ is

(A)$2\ln 2$ (B)$2\ln 2+2$ (C)$e^2(\ln 2)+e^2$ (D)$2+\dfrac{2\ln 2}{e}$

【答案】 B

本题的考点是直角坐标方程求过切点的切线的斜率.

$y'|_{x=\ln 2}=(e^x+xe^x)|_{x=\ln 2}=2+\ln 2 e^{\ln 2}=2+2\ln 2.$

10. Let $y=f(x)$ be the solution to the differential equation $\dfrac{dy}{dx}=x-y$ with initial condition $f(2)=8$. What is the approximation for $f(3)$ obtained by using Euler's method with two steps of equal length, starting at $x=2$?

    (A) 2 (B) $\dfrac{5}{2}$ (C) $\dfrac{15}{4}$ (D) $\dfrac{61}{4}$

【答案】 C

题目大意：令 $y=f(x)$ 是微分方程 $\dfrac{dy}{dx}=x-y$ 满足初始条件 $f(2)=8$ 的特解.用等步长欧拉方法从 $x=2$ 开始近似计算 $f(3)$.

本题的考点是欧拉方法求微分方程的近似解.

$f(2)=8$,步长为：$\Delta x=\dfrac{3-2}{2}=0.5$

$f(2.5)\approx f(2)+f'(2)\cdot 0.5=8+(2-8)\times 0.5=5$

$f(3)\approx f(2.5)+f'(2.5)\cdot 0.5=5+(2.5-5)\times 0.5=\dfrac{15}{4}$

11. If $x^2+xy-3y=3$, then at the point $(2,1)$, $\dfrac{dy}{dx}=$

    (A) 5 (B) 4 (C) $\dfrac{7}{3}$ (D) 2

【答案】 A

本题的考点是隐函数求导.

$x^2+xy-3y=3$

方程两边同时对 x 求导,得：$2x+y+xy'-3y'=0$

当 $x=2,y=1$ 时,有：$4+1+2y'(2)-3y'(2)=0$

$y'(2)=5$

12. $\displaystyle\int\dfrac{3x+1}{x^2-4x+3}dx=$

    (A) $-2\ln|x-3|+5\ln|x-1|+C$

    (B) $-\dfrac{1}{5}\ln|x-3|-\dfrac{1}{2}\ln|x-1|+C$

    (C) $-\dfrac{1}{2}\ln|x-3|-\dfrac{1}{2}\ln|x-1|+C$

    (D) $5\ln|x-3|-2\ln|x-1|+C$

【答案】 D

本题的考点是部分分式法求不定积分.

设 $\dfrac{3x+1}{x^2-4x+3} = \dfrac{3x+1}{(x-1)(x-3)} = \dfrac{A}{x-1} + \dfrac{B}{x-3} = \dfrac{A(x-3)+B(x-1)}{(x-1)(x-3)}$

$\Rightarrow 3x+1 = A(x-3) + B(x-1)$

当 $x=3$ 时,$10=2B \Rightarrow B=5$

当 $x=1$ 时,$4=-2A \Rightarrow A=-2$

$\int \dfrac{3x+1}{x^2-4x+3} dx = -2\int \dfrac{1}{x-1} dx + 5\int \dfrac{1}{x-3} dx = 5\ln|x-3| - 2\ln|x-1| + C$

一般地,$\int \dfrac{1}{ax-b} dx = \dfrac{1}{a}\int \dfrac{1}{ax-b} d(ax-b) = \dfrac{1}{a}\ln|ax-b| + C$

13. Which of the following is a slope field for the differential equation $\dfrac{dy}{dx} = x^2 + y^2$?

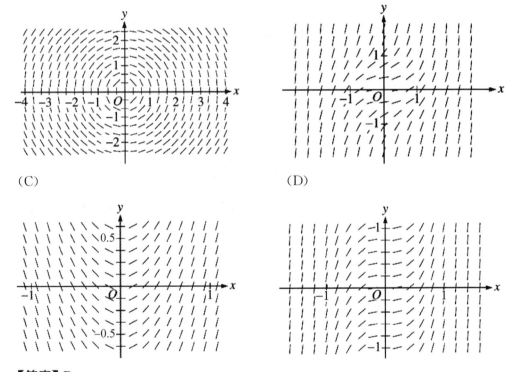

【答案】B

本题的考点是由微分方程判断斜率场.

先看斜率场的坐标轴上的小线段,当 $x=0$ 时,$\dfrac{dy}{dx} = y^2$,说明 y 轴上的小线段的斜率和 y 有关,选项 A、C、D 在 y 轴上的小线段全都平行,说明斜率为 0,即 $\dfrac{dy}{dx}=0$,所以可以排除,故选 B.

14. If $f(x) = 3x^2 + 2x$, then $f'(x) = $

(A) $\lim\limits_{h \to \infty} \dfrac{(3x^2+2x+h)-(3x^2+2x)}{h}$

(B)$\lim_{x\to 0}\frac{(3x^2+2x+h)-(3x^2+2x)}{h}$

(C)$\lim_{h\to 0}\frac{(3(x+h)^2+2(x+h))-(3x^2+2x)}{h}$

(D)$\lim_{x\to 0}\frac{(3(x+h)^2+2(x+h))-(3x^2+2x)}{h}$

【答案】C

本题的考点是导数的定义 $f'(x)=\lim_{h\to 0}\frac{f(x+h)-f(x)}{h}$.

$f(x)=3x^2+2x$, then $f'(x)=\lim_{h\to 0}\frac{f(x+h)-f(x)}{h}=\lim_{h\to 0}\frac{(3(x+h)^2+2(x+h))-(3x^2+2x)}{h}$.

15.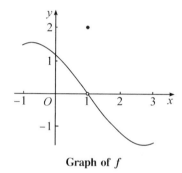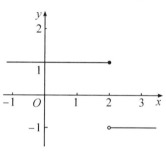

Graph of *f* 　　　　　　　　　**Graph of** *g*

The graphs of the functions f and g are shown in the figures above. Which of the following statements is false?

(A)$\lim_{x\to 1}f(x)=0$ 　　　　　　　　(B)$\lim_{x\to 2}g(x)$ does not exist

(C)$\lim_{x\to 1}(f(x)g(x+1))$ does not exist 　(D)$\lim_{x\to 1}(f(x+1)g(x))$ exist

【答案】D

本题的考点是根据图像判断函数的性质.

从 $f(x)$ 的图像可以看出 $\lim_{x\to 1^-}f(x)=\lim_{x\to 1^+}f(x)=0$,所以$\lim_{x\to 1}f(x)=0$,A 选项正确.

从 $g(x)$ 的图像可以看出 $\lim_{x\to 2^-}g(x)=1\neq -1=\lim_{x\to 2^+}g(x)$,所以$\lim_{x\to 2}g(x)$ does not exist,B 选项正确.

因为 $\lim_{x\to 1^-}(f(x)g(x+1))=\lim_{x\to 1^-}f(x)\lim_{x\to 1^-}g(x+1)=0 \cdot \lim_{x\to 2^-}g(x)=0\times 1=0$

$\lim_{x\to 1^+}(f(x)g(x+1))=\lim_{x\to 1^+}f(x)\lim_{x\to 1^+}g(x+1)=0 \cdot \lim_{x\to 2^+}g(x)=0\times (-1)=0$

所以 $\lim_{x\to 1^-}(f(x)g(x+1))=\lim_{x\to 1^+}(f(x)g(x+1))=0 \Rightarrow \lim_{x\to 1}(f(x)g(x+1))=0$,所以 D 正确,C 错误.

16. Which of the following is the interval of convergence for the series $\sum_{n=0}^{\infty}\frac{(x+2)^n}{2^n}$?

(A)$-4<x<0$　　(B)$-4\leqslant x<0$　　(C)$-2<x<2$　　(D)$-2\leqslant x<2$

【答案】A

本题的考点是函数项级数的收敛区间.

$$\lim_{n\to\infty}\left|\frac{u_{n+1}}{u_n}\right|=\lim_{n\to\infty}\left|\frac{(x+2)^{n+1}/2^{n+1}}{(x+2)^n/2^n}\right|=\left|\frac{x+2}{2}\right|<1 \Rightarrow |x+2|<2 \Rightarrow -4<x<0$$

当 $x=-4$ 时,$\sum_{n=0}^{\infty}\frac{(x+2)^n}{2^n}=\sum_{n=0}^{\infty}(-1)^n$ by nth term test 发散.

当 $x=0$ 时,$\sum_{n=0}^{\infty}\frac{(x+2)^n}{2^n}=\sum_{n=0}^{\infty}1$ by nth term test 发散.

所以,收敛区间为 $-4<x<0$.

17. $\int_0^5 \sqrt{\frac{5-x}{5}}\,dx=$

(A) $\frac{2}{3}$ (B) $\frac{10}{3}$ (C) 5   (D) $\frac{50\sqrt{5}}{3}$

【答案】B

本题的考点是凑微分法计算定积分.

$$\int_0^5 \sqrt{\frac{5-x}{5}}\,dx = -\frac{1}{\sqrt{5}}\int_0^5 \sqrt{5-x}\,d(5-x) = -\frac{1}{\sqrt{5}}\times\frac{2}{3}\cdot(5-x)^{\frac{3}{2}}\Big|_0^5$$

$$= -\frac{2}{3\sqrt{5}}\times(0-5\sqrt{5}) = \frac{10}{3}.$$

18. Which of the following limits are equal to -1?

    I. $\lim_{x\to 0^-}\frac{|x|}{x}$

    II. $\lim_{x\to 3}\frac{x^2-7x+12}{3-x}$

    III. $\lim_{x\to\infty}\frac{1-x}{1+x}$

(A) I only   (B) I and III only   (C) II and III   (D) I, II and III

【答案】B

$\lim_{x\to 0^-}\frac{|x|}{x}=\lim_{x\to 0^-}\frac{-x}{x}=-1$(左极限)

$\lim_{x\to 3}\frac{x^2-7x+12}{3-x}=\lim_{x\to 3}\frac{2x-7}{-1}=\frac{6-7}{-1}=1$(洛必达法则)

$\lim_{x\to\infty}\frac{1-x}{1+x}=\lim_{x\to\infty}\frac{-1}{1}=-1$(抓大头准则)

19. Let f be the function given by $f(x)=2\cos x+1$. What is the approximation for $f(1.5)$ found by using the line tangent to the graph of f at $x=\frac{\pi}{2}$?

(A) -2 (B) 1   (C) $\pi-2$ (D) $4-\pi$

【答案】C

题目大意:已知 $f(x)=2\cos x+1$,用过 $x=\frac{\pi}{2}$ 的切线近似计算 $f(1.5)$.

本题的考点是切线近似.

$$f\left(\frac{\pi}{2}\right)=1 \qquad f'\left(\frac{\pi}{2}\right)=-2\sin x\big|_{x=\frac{\pi}{2}}=-2$$

$$f(1.5)\approx f\left(\frac{\pi}{2}\right)+f'\left(\frac{\pi}{2}\right)\left(1.5-\frac{\pi}{2}\right)=1+(-2)\times\left(\frac{3}{2}-\frac{\pi}{2}\right)=1-3+\pi=\pi-2$$

20. A particle moves in the xy-plane, so that its position for $t\geq 0$ is given by the parametric equations $x=\ln(t+1)$ and $y=kt^2$, where k is a positive constant. The line tangent to the particle's path at the point where $t=3$ has slope 8. What is the value of k?

(A) $\dfrac{1}{192}$ (B) $\dfrac{1}{3}$ (C) $\dfrac{4}{3}$ (D) $\dfrac{16}{3}$

【答案】B

题目大意：质点在 $xy-$平面运动，$t\geq 0$ 时，其位置函数为参数方程 $\begin{cases} x=\ln(t+1) \\ y=kt^2(k>0) \end{cases}$，当 $t=3$ 时，质点运动路径的斜率为 8，求 k 的值。

本题的考点是参数方程求导数 $\dfrac{dy}{dx}=\dfrac{\frac{dy}{dt}}{\frac{dx}{dt}}$。

$$\frac{dy}{dx}=\frac{\frac{dy}{dt}}{\frac{dx}{dt}}=\frac{2kt}{\frac{1}{t+1}}=2kt(t+1)$$

$$\frac{dy}{dx}\Big|_{t=3}=24k=8\Rightarrow k=\frac{1}{3}$$

21.
Time(weeks)	0	2	6	10
Level	210	200	190	180

The table above gives the level of a person's cholesterol at different times during a 10-week treatment period. What is the average level over this 10-week period obtained by using a trapezoidal approximation with the subintervals $[0,2]$, $[2,6]$ and $[6,10]$?

(A) 188   (B) 193   (C) 195   (D) 198

【答案】B

题目大意：上表是某人在不同时间的胆固醇水平，将区间分割成小区间 $[0,2]$, $[2,6]$ 和 $[6,10]$，用梯形法近似计算 10 周内胆固醇水平的平均值。

本题的考点是 Average Value of a Continuous Function(连续函数的平均值)与定积分近似计算的梯形法。

The average value of $f(x)$ on $[a,b]=\dfrac{1}{b-a}\int_a^b f(x)dx$

$$\int_a^b f(x)dx\approx T(n)$$

$$=\frac{f(x_0)+f(x_1)}{2}\Delta x_1+\frac{f(x_1)+f(x_2)}{2}\Delta x_2+\cdots+\frac{f(x_{n-1})+f(x_n)}{2}\Delta x_n$$

$$\frac{1}{10}\int_0^{10}f(x)\mathrm{d}x=\frac{1}{10}\times\left(\frac{210+200}{2}\times 2+\frac{200+190}{2}\times 4+\frac{190+180}{2}\times 4\right)$$

$$=41+40+76+36$$

$$=193$$

22. $\int \dfrac{x}{2}\mathrm{e}^{-3x/4}\mathrm{d}x=$

(A) $-\dfrac{3x}{4}\mathrm{e}^{-3x/4}+\dfrac{3}{4}\mathrm{e}^{-3x/4}+C$ (B) $-\dfrac{2x}{3}\mathrm{e}^{-3x/4}-\dfrac{8}{9}\mathrm{e}^{-3x/4}+C$

(C) $-\dfrac{x}{2}\mathrm{e}^{-3x/4}+\dfrac{3}{8}\mathrm{e}^{-3x/4}+C$ (D) $\dfrac{x}{2}\mathrm{e}^{-3x/4}-\dfrac{1}{2}\mathrm{e}^{-3x/4}+C$

【答案】B

本题的考点是分部积分法计算不定积分.

$$\int\frac{x}{2}\mathrm{e}^{-3x/4}\mathrm{d}x=-\frac{4}{3}\int\frac{x}{2}\mathrm{d}(\mathrm{e}^{-3x/4})$$

$$=-\frac{4}{3}\left[\frac{x}{2}\mathrm{e}^{-3x/4}-\int\mathrm{e}^{-3x/4}\mathrm{d}\left(\frac{x}{2}\right)\right]$$

$$=-\frac{4}{3}\left[\frac{x}{2}\mathrm{e}^{-3x/4}-\frac{1}{2}\left(-\frac{4}{3}\right)\int\mathrm{e}^{-3x/4}\mathrm{d}\left(-\frac{4}{3}x\right)\right]$$

$$=-\frac{2x}{3}\mathrm{e}^{-3x/4}-\frac{8}{9}\mathrm{e}^{-3x/4}+C$$

23. If $f(x)=\sum\limits_{n=1}^{\infty}\dfrac{x^{2n}}{n!}$, then $f'(x)=$

(A) $\dfrac{x^3}{3}+\dfrac{x^5}{5\times 2!}+\dfrac{x^7}{7\times 3!}+\dfrac{x^9}{9\times 4!}+\cdots+\dfrac{x^{(2n-1)}}{(2n+1)n!}+\cdots$

(B) $x+\dfrac{3x^3}{2!}+\dfrac{5x^5}{3!}+\dfrac{7x^7}{4!}+\cdots+\dfrac{(2n-1)x^{2n-1}}{n!}+\cdots$

(C) $2+2x^2+x^4+\dfrac{x^6}{3}+\cdots+\dfrac{2x^{2n-1}}{(n-1)!}+\cdots$

(D) $2x+2x^3+x^5+\dfrac{x^7}{3}+\cdots+\dfrac{2nx^{2n-1}}{(n-1)!}+\cdots$

【答案】D

本题的考点是幂级数逐项求导.

$$f(x)=\sum_{n=1}^{\infty}\frac{x^{2n}}{n!}=x^2+\frac{x^4}{2!}+\frac{x^6}{3!}+\frac{x^8}{4!}\cdots+\frac{x^{2n}}{n!}+\cdots$$

逐项求导得：

$$f'(x)=2x+2x^3+x^5+\frac{x^7}{3}+\cdots+\frac{2nx^{2n-1}}{(n-1)!}+\cdots$$

24. If the average value of a continuous function f on the interval $[-2,4]$ is 12, what is $\int_{-2}^{4}\dfrac{f(x)}{8}\mathrm{d}x$?

(A)$\frac{3}{2}$ (B)3   (C)9   (D)72

【答案】C

本题的考点是连续函数的平均值与定积分的性质.

由题意知：$\frac{1}{6}\int_{-2}^{4}f(x)\mathrm{d}x=12$

所以 $\int_{-2}^{4}\frac{f(x)}{8}\mathrm{d}x=\frac{12\times 6}{8}=9$

25. What is the radius of convergence of the Maclaurin series for $\frac{2x}{1+x^2}$?

(A)$\frac{1}{2}$ (B)1   (C)2   (D)infinite

【答案】B

本题的考点是麦克劳林级数的展开式和收敛半径.

因为 $\frac{1}{1-x}=1+x+x^2+\cdots+x^n+\cdots(-1<x<1)$

$\frac{1}{1+x^2}=1-x^2+x^4-x^6+\cdots+(-1)^{n-1}x^{2n}+\cdots(-1<x<1)$

所以 $\frac{2x}{1+x^2}=2x-2x^3+2x^5-2x^7+\cdots+(-1)^{n-1}2x^{2n+1}+\cdots(-1<x<1)$

26. Let f be the function with $f(0)=\frac{1}{\pi^2}$, $f(2)=\frac{1}{\pi^2}$, and derivative given by $f'(x)=(x+1)\cos(\pi x)$. How many values of x in the open interval $(0,2)$ satisfy the conclusion of the Mean Value Theorem for the function f on the closed interval $[0,2]$?

(A)none   (B)one   (C)two   (D)more than two

【答案】C

题目大意：函数 $f(x)$ 在区间 $[0,2]$ 满足 $f(0)=\frac{1}{\pi^2}$, $f(2)=\frac{1}{\pi^2}$, 且 $f'(x)=(x+1)\cos(\pi x)$, 则在 $(0,2)$ 有多少点满足微分中值定理？

本题的考点是求满足微分中值定理 $f'(c)=\frac{f(b)-f(a)}{b-a}$ 的点的个数.

$f'(c)=\frac{f(b)-f(a)}{b-a}=\frac{f(2)-f(0)}{2-0}=0$

因为 $f'(x)=(x+1)\cos(\pi x)$

所以 $(c+1)\cos(\pi c)=0, c\in(0,2) \Rightarrow c=-1[-1\notin(0,2)舍去]$ 或 $\cos(\pi c)=0$

故 $c=\frac{1}{2}$ 或 $\frac{3}{2}$.

27. The number of students in a cafeteria is modeled by the function P that satisfies the logistic differential equation $\frac{\mathrm{d}P}{\mathrm{d}t}=\frac{1}{2000}P(200-P)$, where t is the time in

seconds and $P(0)=25$. What is the greatest rate of change, in students per second, of the number of students in the cafeteria?

(A) 5 (B) 25 (C) 100 (D) 200

【答案】A

题目大意：自助餐厅的学生人数的数学模型为函数 P，假设 P 满足 logistic 微分方程 $\dfrac{dP}{dt}=\dfrac{1}{2000}P(200-P)$，且 $P(0)=25$，求人数 P 的变化率的最大值.

本题的考点是 logistic 微分方程的性质.

对 logistic 微分方程 $\dfrac{dy}{dt}=ky(A-y)$，当 $y=\dfrac{A}{2}$ 时增长速度达到最大值.

$\dfrac{dP}{dt}=\dfrac{1}{2000}P(200-P) \Rightarrow A=200$

$\Rightarrow$ 当 $P=\dfrac{200}{2}=100$ 时，P 的变化率达到最大值.

$\dfrac{dP}{dt}\Big|_{P=100}=\dfrac{1}{2000}\times 100\times(200-100)=5$

28. A cube with edges of length x centimeters has volume $V(x)=x^3$ cubic centimeters. The volume is increasing at a constant rate of 40 cubic centimeters per minute. At the instant when $x=2$, what is the rate of change of x, in centimeters per minute, with respect to time?

(A) $\dfrac{10}{3}$ (B) $\sqrt{\dfrac{40}{3}}$ (C) 5   (D) 10

【答案】A

题目大意：一个边长为 x 厘米的正方体体积为 $V(x)=x^3$ 立方厘米，体积的增长速度为 40 立方厘米/分钟. 求 $x=2$ 时，边长 x 的增长速度.

本题的考点是相关变化率：

$\dfrac{dV}{dt}=40$

$\dfrac{dV}{dt}=\dfrac{dV}{dx}\cdot\dfrac{dx}{dt}=3x^2\cdot\dfrac{dx}{dt}=40$

当 $x=2$ 时，$12\dfrac{dx}{dt}\Big|_{x=2}=40 \Rightarrow \dfrac{dx}{dt}\Big|_{x=2}=\dfrac{10}{3}$

29. Which of the following is a power series expansion of $\dfrac{e^x+e^{-x}}{2}$?

(A) $1+\dfrac{x^2}{2!}+\dfrac{x^4}{4!}+\dfrac{x^6}{6!}+\cdots+\dfrac{x^{2n}}{(2n)!}+\cdots$

(B) $1-\dfrac{x^2}{2!}+\dfrac{x^4}{4!}-\dfrac{x^6}{6!}+\cdots+(-1)^n\dfrac{x^{2n}}{(2n)!}+\cdots$

(C) $x+\dfrac{x^3}{3!}+\dfrac{x^5}{5!}+\dfrac{x^7}{7!}+\cdots+\dfrac{x^{2n+1}}{(2n+1)!}+\cdots$

(D) $x - \dfrac{x^3}{3!} + \dfrac{x^5}{5!} - \dfrac{x^7}{7!} + \cdots + (-1)^n \dfrac{x^{2n+1}}{(2n+1)!} + \cdots$

【答案】 A

本题的考点是常见的麦克劳林级数.

因为 $\mathrm{e}^x = 1 + x + \dfrac{1}{2!}x^2 + \cdots + \dfrac{1}{n!}x^n + \cdots$

$\mathrm{e}^{-x} = 1 - x + \dfrac{1}{2!}x^2 + \cdots + (-1)^{n-1}\dfrac{1}{n!}x^n + \cdots$

所以 $\dfrac{\mathrm{e}^x + \mathrm{e}^{-x}}{2} = 1 + \dfrac{x^2}{2!} + \dfrac{x^4}{4!} + \dfrac{x^6}{6!} + \cdots + \dfrac{x^{2n}}{(2n)!} + \cdots$

30. Which of the following statements about the series $\sum\limits_{n=1}^{\infty} \dfrac{1}{2^n - n}$ is true?

(A) The series diverges by the nth term test.

(B) The series diverges by limit comparison to the harmonic series $\sum\limits_{n=1}^{\infty} \dfrac{1}{n}$.

(C) The series converges by the nth term test.

(D) The series converges by limit comparison to the geometric series $\sum\limits_{n=1}^{\infty} \dfrac{1}{2^n}$.

【答案】 D

本题的考点是级数的收敛判别法.

因为 $\lim\limits_{n \to \infty} \dfrac{1}{2^n - n} = 0$,所以无法根据 nth term test 判断其发散,故 A、C 错误.

因为 $\dfrac{1}{2^n - n} < \dfrac{1}{n}(n > 3)$,调和级数 $\sum\limits_{i=1}^{n} \dfrac{1}{n}$ 发散,根据比较判别法无法判断 $\sum\limits_{n=1}^{\infty} \dfrac{1}{2^n - n}$ 发散,故 B 错误.

因为 $\lim\limits_{n \to \infty} \dfrac{\dfrac{1}{2^n - n}}{\dfrac{1}{2^n}} = 1$,所以级数 $\sum\limits_{n=1}^{\infty} \dfrac{1}{2^n - n}$ 与 $\sum\limits_{n=1}^{\infty} \dfrac{1}{2^n}$ 同敛散性.

又 $\sum\limits_{n=1}^{\infty} \dfrac{1}{2^n}$ 是公比为 $\dfrac{1}{2} < 1$ 的几何级数,故 $\sum\limits_{n=1}^{\infty} \dfrac{1}{2^n}$ 收敛,即 $\sum\limits_{n=1}^{\infty} \dfrac{1}{2^n - n}$ 收敛,故 D 正确.

CALCULUS BC

SECTION I, Part B
Time—45 minutes
Number of questions—15

(45分钟,共15道题,允许使用图形计算器)

A GRAPHING CALCULATOR IS REQUIRED FOR SOME QUESTIONS ON THIS PART OF THE EXAM.

Directions: Solve each of the following problems, using the available space for scratch work. After examining the form of the choices, decide which is the best of the choices given and fill in the corresponding circle on the answer sheet. No credit will be given for anything written in this exam booklet. Do not spend too much time on any one problem.

BE SURE YOU ARE USING PAGE 3 OF THE ANSWER SHEET TO RECORD YOUR ANSWERS TO QUESTIONS NUMBERED 76~90.

YOU MAY NOT RETURN TO PAGE 2 OF THE ANSWER SHEET.

76. Let f be a twice-differentiable function for all real numbers x. Which of the following additional properties guarantees that f has a relative minimum at $x=c$?

(A) $f'(c)=0$　　　　　　　　　　　　(B) $f'(c)=0$ and $f''(c)<0$
(C) $f'(c)=0$ and $f''(c)>0$　　　　　(D) $f'(x)>0$ for $x<c$ and $f'(x)<0$ for $x>c$

【答案】C

题目大意:令 f 对于一切实数 x 均二阶可导,还需以下哪些条件可使得 f 在 $x=c$ 处取得极小值?

本题的考点是极值的二阶导数判别法.

Suppose $f''(x)$ is continuous on an open interval that contain $x=c$.
(1) If $f'(c)=0$ and $f''(c)<0$, then $f(x)$ has a local maximum at $x=c$.
(2) If $f'(c)=0$ and $f''(c)>0$, then $f(x)$ has a local minimum at $x=c$.

77. Let $H(x)$ be an antiderivative of $\dfrac{x^3+\sin x}{x^2+2}$. If $H(5)=\pi$, then $H(2)=$

　　(A) -9.008　　　　(B) -5.867　　　　(C) 4.626　　　　(D) 12.150

【答案】 B

题目大意: 令 $H(x)$ 为 $\dfrac{x^3+\sin x}{x^2+2}$ 的一个原函数,若 $H(5)=\pi$,则 $H(2)$ 的值是多少?

本题的考点是 FTC 的应用:$F(b)=F(a)+\displaystyle\int_a^b f(x)\mathrm{d}x$.

$H(x)$ 为 $\dfrac{x^3+\sin x}{x^2+2}$ 的一个原函数,即 $H'(x)=\dfrac{x^3+\sin x}{x^2+2}$.

由 FTC 可知:$H(2)=H(5)+\displaystyle\int_5^2 \dfrac{x^3+\sin x}{x^2+2}\mathrm{d}x=\pi+\displaystyle\int_5^2 \dfrac{x^3+\sin x}{x^2+2}\mathrm{d}x=-5.86666435$ (用计算器计算定积分).

78. The continuous function f is positive and has domain $x>0$. If the asymptotes of the graph of f are $x=0$ and $y=2$, which of the following statements must be true?

(A) $\lim\limits_{x\to 0^+}f(x)=\infty$ and $\lim\limits_{x\to 2}f(x)=\infty$
(B) $\lim\limits_{x\to 0^+}f(x)=2$ and $\lim\limits_{x\to\infty}f(x)=0$
(C) $\lim\limits_{x\to 0^+}f(x)=\infty$ and $\lim\limits_{x\to\infty}f(x)=2$
(D) $\lim\limits_{x\to 2}f(x)=\infty$ and $\lim\limits_{x\to\infty}f(x)=2$

【答案】 C

题目大意: 连续函数 $f>0$,$\{x\mid x>0\}$ 有渐近线 $x=0$ 和 $y=2$,下列说法正确的是哪一个?

本题的考点是渐近线的定义,由于渐近线为 $x=0$ 和 $y=2$,所以 C 为正确答案.

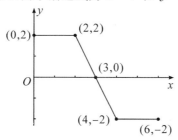

79. The graph of a function f, consisting of three line segments, is shown above. The function f is defined on the x closed interval $[0,6]$. Let $g(x)=\displaystyle\int_2^x f(t)\mathrm{d}t$. What is the maximum value of $g(x)$ for $0\leqslant x\leqslant 6$?

(A) 0   (B) 1   (C) 5   (D) 10

【答案】 B

题目大意: 函数 f 的图像由上图中的三段线段组成.函数 f 的定义域为闭区间 $[0,6]$,令 $g(x)=\displaystyle\int_2^x f(t)\mathrm{d}t$,求 $g(x)$ 在闭区间 $[0,6]$ 上的最大值.

本题的考点是函数求最值、变上限积分的导数和定积分的几何意义综合.

由 $g'(x)=f(x)=0\Rightarrow$ critical value(驻点)为 $x=3$.

比较驻点和区间端点的函数值:

$g(0)=\displaystyle\int_2^0 f(t)\mathrm{d}t=-4$

$g(3)=\displaystyle\int_2^3 f(t)\mathrm{d}t=\dfrac{1}{2}\times 1\times 2=1$

$g(6)=\displaystyle\int_2^6 f(t)\mathrm{d}t=-4$

所以，$g(x)$在闭区间$[0,6]$上的最大值为1.

80. The position of an object moving along a path in the xy-plane is given by the parametric equations $x(t)=5\sin(\pi t)$ and $y(t)=(2t-1)^2$. The speed of the particle at time $t=0$ is

(A) 3.422　　　　(B) 11.708　　　　(C) 15.580　　　　(D) 16.209

【答案】D

题目大意：质点在xy-平面上运动，其位置函数为参数方程$x(t)=5\sin(\pi t)$和$y(t)=(2t-1)^2$. 求$t=0$时质点的速率.

根据速率公式：$\text{speed}=\sqrt{\left(\dfrac{\mathrm{d}x}{\mathrm{d}t}\right)^2+\left(\dfrac{\mathrm{d}y}{\mathrm{d}t}\right)^2}\bigg|_{t=0}=16.20926001$（使用图形计算器直接计算）.

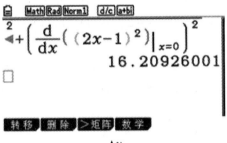

81.

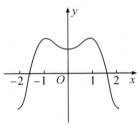

Graph of f

The graph of the function f is shown above for $-2\leqslant x\leqslant 2$. Which of the following could be the graph of an antiderivative of f?

(A)
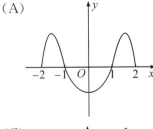

(B)
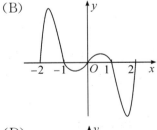

(C)
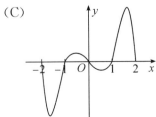

(D)
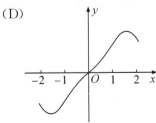

【答案】D

题目大意： 上图是函数 f 在 $-2 \leqslant x \leqslant 2$ 的图像，则下图中哪一个是其原函数的图像？

本题的考点是由导函数的图像判断原函数的图像，令 $F'(x)=f(x)$.

从函数 f 在 $-2 \leqslant x \leqslant 2$ 的图像可知：

$f(x)$ 在 $x=-1,x=0,x=1$ 处取得极值 $\Rightarrow$ 原函数 $F(x)$ 的拐点为 $x=-1,x=0,x=1$.

$f(-1.5)=0$ 且 $f(x)$ 在 $x=-1.5$ 两侧的符号由负变正 $\Rightarrow F(x)$ 在 $x=-1.5$ 取得极小值.

$f(1.5)=0$ 且 $f(x)$ 在 $x=1.5$ 两侧的符号由正变负 $\Rightarrow F(x)$ 在 $x=1.5$ 取得极大值，故选 D.

82. The derivative of the function f is given by $f'(x)=e^{-x}\cos(x^2)$, for all real numbers x. What is the minimum value of $f(x)$ for $-1 \leqslant x \leqslant 1$?

    (A) $f(-1)$

    (B) $f(-0.762)$

    (C) $f(1)$

    (D) There is no minimum value of $f(x)$ for $-1 \leqslant x \leqslant 1$

【答案】A

因为当 $-1 \leqslant x \leqslant 1$ 时，$f'(x)=e^{-x}\cos(x^2)>0$

所以 $f(x)$ 在 $-1 \leqslant x \leqslant 1$ 单调递增，故 $f(-1)$ 为最小值.

83.

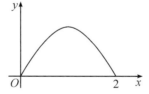

The base of a solid is the region bounded by a portion of the graph of $y=\sin\left(\dfrac{\pi}{2}x\right)$ and the x-axis, as shown in the figure above. For the solid, each cross section perpendicular to the x-axis is a rectangle of height 3. Which of the following expressions gives the volume of the solid?

(A) $\displaystyle\int_0^2 3\sin\left(\dfrac{\pi}{2}x\right)dx$

(B) $\displaystyle\int_0^2 3\sin^2\left(\dfrac{\pi}{2}x\right)dx$

(C) $\displaystyle\int_0^2 3\pi\sin\left(\dfrac{\pi}{2}x\right)dx$

(D) $\displaystyle\int_0^2 3\pi\sin^2\left(\dfrac{\pi}{2}x\right)dx$

【答案】A

题目大意： 如图所示，一个立体的底面是由 $y=\sin\left(\dfrac{\pi}{2}x\right)$ 和 x 轴围成的部分，立体与 x 轴垂直的截面是高为 3 的长方形，下列哪个表达式为立体的体积？

本题的考点是截面面积已知的物体的体积，体积元素为：$dV=3\sin\left(\dfrac{\pi}{2}x\right)dx$

所以 $V=\displaystyle\int_0^2 3\sin\left(\dfrac{\pi}{2}x\right)dx$.

84. If g is a twice-differentiable function, where $g(1)=0.5$ and $\lim_{x\to\infty}g(x)=4$, then $\int_1^\infty g'(x)\mathrm{d}x$ is

(A) -3.5 (B) 3.5 (C) 4.5 (D) nonexistent

【答案】B

题目大意：若 $g(x)$ 为二阶可导的函数且 $g(1)=0.5, \lim_{x\to\infty}g(x)=4$，则 $\int_1^\infty g'(x)\mathrm{d}x$ 为多少？

本题的考点是无穷限反常积分.

$\int_1^\infty g'(x)\mathrm{d}x = \lim_{b\to\infty}\int_1^b g'(x)\mathrm{d}x = \lim_{b\to\infty}g(x)\Big|_1^b = \lim_{b\to\infty}g(b)-g(1) = 4-0.5 = 3.5$

85.

Graph of f

The graph of the function f is shown above. If g is the function defined by $g(x)=\int_2^x f(t)\mathrm{d}t$, what is the value of $g(10)\cdot g'(10)$?

(A) $\dfrac{25}{4}$ (B) $\dfrac{5}{4}$ (C) $-\dfrac{5}{2}$ (D) $-\dfrac{25}{2}$

【答案】C

本题的考点是定积分的几何意义以及变上限积分的导数.

$\because g(x)=\int_2^x f(t)\mathrm{d}t$

$\therefore g'(x)=f(x)$

故 $g(10)\cdot g'(10)=\int_2^{10}f(x)\mathrm{d}x\cdot f(10)=(7.5-5)\times(-1)=-\dfrac{5}{2}$.

86. $f''(x)=x\ (x-1)^2\ (x+2)^3$

$g''(x)=x\ (x-1)^2\ (x+2)^3+1$

$h''(x)=x\ (x-1)^2\ (x+2)^3-1$

The twice-differentiable functions f, g and h have second derivatives given above. Which of the functions f, g and h have a graph with exactly two points of inflection?

(A) g only   (B) h only

(C) f and g only   (D) f, g and h

【答案】C

题目大意：$f(x), g(x)$ 和 $h(x)$ 的二阶导数为：
$$f''(x) = x\,(x-1)^2\,(x+2)^3$$
$$g''(x) = x\,(x-1)^2\,(x+2)^3 + 1$$
$$h''(x) = x\,(x-1)^2\,(x+2)^3 - 1$$

其中哪些(个)函数正好有两个拐点？

本题的考点是拐点的判定和函数图像的上、下平移．

由 $f''(x) = x\,(x-1)^2\,(x+2)^3$ 可知，$f''(x)$ 在 $x=0$ 和 $x=-2$ 两侧异号，所以 $f(x)$ 正好有两个拐点．

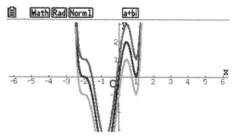

用图形计算器作图可知，$y = g''(x)$ 的图像是 $y = f''(x)$ 向上平移 1 个单位，依然有两个点的两侧 $g''(x)$ 异号，所以 $g(x)$ 正好有两个拐点．$y = h''(x)$ 的图像是 $y = f''(x)$ 向下平移 1 个单位，从图中可见 4 个点处 $y = h''(x)$ 异号，所以 $h(x)$ 有 4 个拐点．

87. The velocity vector of a particle moving in the xy-plane has components given by $\dfrac{\mathrm{d}x}{\mathrm{d}t} = \sin(t^2)$ and $\dfrac{\mathrm{d}y}{\mathrm{d}t} = \mathrm{e}^{\cos t}$. At time $t=4$, the position of the particle is $(2,1)$. What is the y-coordinate of the position vector at time $t=3$?

   (A) 0.410 (B) 0.590 (C) 0.851 (D) 1.410

【答案】B

题目大意：质点在 xy-平面移动的速度向量为：$\dfrac{\mathrm{d}x}{\mathrm{d}t} = \sin(t^2)$，$\dfrac{\mathrm{d}y}{\mathrm{d}t} = \mathrm{e}^{\cos t}$．在 $t=4$ 时，质点的位置为 $(2,1)$，则 $t=3$ 时位置的纵坐标为多少？

$$y(3) = y(4) + \int_4^3 \dfrac{\mathrm{d}y}{\mathrm{d}t}\mathrm{d}t = 1 + \int_4^3 \mathrm{e}^{\cos t}\mathrm{d}t = 0.590292404$$

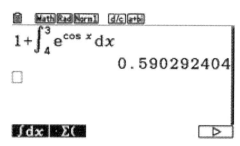

88. The function f is increasing on the interval $[1,3]$ and nowhere else. The first derivative of f, f', is continuous for all real numbers. Which of the following could be a table of values for $f'(x)$?

(A)		(B)		(C)		(D)	
x	$f'(x)$	x	$f'(x)$	x	$f'(x)$	x	$f'(x)$
0	−1	0	−1	0	1	0	1
1	0	1	1	1	0	1	0
2	2	2	2	2	1	2	2
3	0	3	1	3	2	3	0
4	−2	4	−2	4	0	4	−2

【答案】 A

题目大意：函数 $f(x)$ 仅在区间 $[1,3]$ 单调递增，$f'(x)$ 对 x 取一切实数连续，下表中哪一个可能是 $f'(x)$ 的值？

因为 $f'(x)$ 对 x 取一切实数连续，所以函数 $f(x)$ 处处可导．由可导必然连续知，$f(x)$ 处处连续．由于函数 $f(x)$ 仅在区间 $[1,3]$ 单调递增，所以仅在区间 $[1,3]$ 有 $f'(x)\geqslant 0$，故可以排除选项 C、D．

B 选项中 $f'(0)=-1$，$f'(1)=1$，则由 MVT 至少存在一点 $c\in(0,1)$ 使得 $f'(c)=0$，即 $(c,1)$ 也是单调递增区间，与函数 $f(x)$ 仅在区间 $[1,3]$ 单调递增矛盾，所以 B 可以排除．

故选项 A 正确．

89.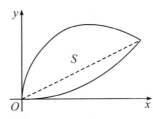

Let S be the region in the first quadrant bounded above by the graph of the polar curve $r=\cos\theta$ and bounded below by the graph of the polar curve $r=2\theta$, as shown in the figure above. The two curves intersect when $\theta=0.450$. What is the area of S?

(A) 0.232　　(B) 0.243　　(C) 0.271　　(D) 0.384

【答案】 B

题目大意：令 S 表示由极坐标曲线 $r=\cos\theta$ 和 $r=2\theta$ 围成的第一象限的图形的面积，已知两条曲线的交点为 $\theta=0.450$，求 S．

本题的考点是极坐标曲线求面积．

$$S=\frac{1}{2}\int_{0.450}^{\frac{\pi}{2}}\cos^2\theta\,d\theta+\frac{1}{2}\int_{0}^{0.450}4\theta^2\,d\theta=0.243033218$$

90. If the infinite series $S=\sum_{n=1}^{\infty}(-1)^{n+1}\dfrac{2}{n}$ is approximated by $P_k=\sum_{n=1}^{k}(-1)^{n+1}\dfrac{2}{n}$, what is the least value of k for which the alternating series error bound guarantees that $|S-P_k|<\dfrac{3}{100}$?

(A) 64　　(B) 66　　(C) 68　　(D) 70

【答案】 B

题目大意: 用 $P_k = \sum\limits_{n=1}^{k} (-1)^{n+1} \dfrac{2}{n}$ 近似计算 $S = \sum\limits_{n=1}^{\infty} (-1)^{n+1} \dfrac{2}{n}$,求 k 的最小值使得误差 $|S - P_k| < \dfrac{3}{100}$.

本题的考点是交错级数估值定理. $|S - P_k| < \dfrac{2}{k+1} < \dfrac{3}{100}$, k 最小值取 66.

CALCULUS BC

SECTION II, Part A

Time—30 minutes

Number of questions—2

A GRAPHING CALCULATOR IS REQUIRED FOR THESE QUESTIONS.

（自由问答 A 部分，共 2 道题，30 分钟时间作答，可以使用计算器）

1. For $0 \leqslant t \leqslant 8$, a particle moving in the xy-plane has position vector $\langle x(t), y(t) \rangle = \langle \sin(2t), t^2 - t \rangle$, where $x(t)$ and $y(t)$ are measured in meters and t is measured in seconds.

   (a) Find the speed of the particle at time $t = 2$ seconds. Indicate units of measure.

   (b) At time $t = 4$ seconds, is the speed of the particle increasing or decreasing? Explain your answer.

   (c) Find the total distance the particle travels over the time interval $0 \leqslant t \leqslant 5$ seconds.

   (d) At time $t = 8$ seconds, the particle begins moving in a straight line. For $t \geqslant 8$, the particle travels with the same velocity vector that it had at time $t = 8$ seconds. Find the position of the particle at time $t = 10$ seconds.

   题目大意：质点在 $x-y$ 平面运动的位置向量为 $\langle x(t), y(t) \rangle = \langle \sin(2t), t^2 - t \rangle$（$0 \leqslant t \leqslant 8$），$x(t), y(t)$ 的单位是米，时间 t 的单位是秒.

   (a) 求质点在 $t = 2$ 秒时的速度，并标明速度单位.

   (b) 当 $t = 4$ 秒时，质点的速率是单调递增还是单调递减？请说明理由.

   (c) 求质点在时间 $0 \leqslant t \leqslant 5$ 秒过程中运动的总距离.

   (d) 从 $t = 8$ 秒开始，质点以 $t = 8$ 秒时的速度做匀速直线运动，求质点在 $t = 10$ 秒时的位置.

   Solution

   (a) $\sqrt{(x'(2))^2 + (y'(2))^2} = 3.272461$（速度的表达式及答案正确得 1 分）

   The speed of the particle at time $t = 2$ seconds is 3.272 meters per second.（速度单位正确得 1 分）

   (b) $s(t) = \sqrt{(x'(2))^2 + (y'(2))^2} = \sqrt{(2\cos(2t))^2 + (2t-1)^2}$

   $s'(4) = 2.16265$（$s'(4)$ 结果正确得 1 分）

   Since $s'(4) > 0$, the speed of the particle is increasing at time $t = 4$.（说理正确得 1 分）

(c) $\int_0^5 \sqrt{(x'(t))^2+(y'(t))^2}\,dt = 22.381767$（积分表达式正确得 1 分，结果计算正确得 1 分）

The total distance the particle travels over the time interval $0 \leqslant t \leqslant 5$ seconds is 22.382 (or 22.381) meters.

(d) $x(10) = x(8) + x'(8) \cdot 2 = \sin 16 + x'(8) \cdot 2 = -4.118541$（结果中有 $t=8$ 处的位置得 1 分）

$y(10) = y(8) + y'(8) \cdot 2 = (8^2-8) + y'(8) \cdot 2 = 86$（结果中有 $t=8$ 处的速度得 1 分）

The position of the particle at time $t=10$ seconds is $(-4.119, 86)$ [or $(-4.118, 86)$]. (有 $t=8$ 处的位置得 1 分)

2.

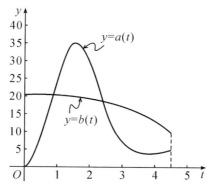

During the time interval $0 \leqslant t \leqslant 4.5$ hours, water flows into tank A at a rate of $a(t) = (2t-5) + 5e^{2\sin t}$ liters per hour. During the same time interval, water flows into tank B at a rate of $b(t)$ liters per hour. Both tanks are empty at time $t=0$. The graphs of $y=a(t)$ and $y=b(t)$, shown in the figure above, intersect at $t=k$ and $t=2.416$.

(a) How much water will be in tank A at time $t=4.5$?

(b) During the time interval $0 \leqslant t \leqslant k$ hours, water flows into tank B at a constant rate of 20.5 liters per hour. What is the difference between the amount of water in tank A and the amount of water in tank B at time $t=k$?

(c) The area of the region bounded by the graphs of $y=a(t)$ and $y=b(t)$ for $k \leqslant t \leqslant 2.416$ is 14.470. How much water is in tank B at time $t=2.416$?

(d) During the time interval $2.7 \leqslant t \leqslant 4.5$ hours, the rate at which water flows into tank B is modeled by $w(t) = 21 - \dfrac{30t}{(t-8)^2}$ liters per hour. Is the difference $w(t) - a(t)$ increasing or decreasing at time $t=3.5$? Show the work that leads to your answer.

题目大意： 当 $0 \leqslant t \leqslant 4.5$ 时，$a(t) = (2t-5) + 5e^{2\sin t}$ 表示水流入水槽 A 的速度（单位：升/小时）.在同一时间间隔，水流入水槽 B 的速度是 $b(t)$ 升/小时.两个水槽在 $t=0$ 时都是空的，如上图所示，$y=a(t)$ 和 $y=b(t)$ 相交于 $t=k$ 和 $t=2.416$.

(a) 在 $t=4.5$ 时，水槽 A 中有多少水？

(b)当 $0 \leqslant t \leqslant k$ 时,水流入水槽 B 的速度是 20.5 升/小时,求 $t=k$ 时两个水槽中水量的差.

(c)如图所示,$y=a(t)$ 与 $y=b(t)$ 在 $k \leqslant t \leqslant 2.416$ 时围成的面积为 14.470.求水槽 B 在 $t=2.416$ 时所装的水的体积.

(d)当 $2.7 \leqslant t \leqslant 4.5$ 时,水流入水槽 B 的速度为 $w(t)=21-\dfrac{30t}{(t-8)^2}$ 升/小时.当 $t=3.5$ 时,$w(t)-a(t)$ 是单调递增还是单调递减? 请写明判断过程.

Solution

(a) $\int_0^{4.5} a(t)\mathrm{d}t = 66.532128$（积分表达式正确得 1 分,计算结果正确得 1 分）

At time $t=4.5$, tank A contains 66.532 liters of water.

(b) $a(k)=20.5 \Rightarrow k=0.892040$（1 分）

$\int_0^k (20.5-a(t))\mathrm{d}t = 10.599191$（积分表达式正确得 1 分,结果正确得 1 分）

At time $t=k$, the difference in the amounts of water in the tanks is 10.599 liters.

(c) $\int_0^{2.416} b(t)\mathrm{d}t = \int_0^k b(t)\mathrm{d}t + \int_k^{2.416} b(t)\mathrm{d}t$

$\int_0^k b(t)\mathrm{d}t = 20.5 \cdot k = 18.286826.$

On $k < t < 2.416$, tank A receives $\int_k^{2.416} a(t)\mathrm{d}t = 44.497051$ liters of water, which is 14.470 more liters of water than tank B.

Therefore, $\int_k^{2.416} b(t)\mathrm{d}t = \int_k^{2.416} a(t)\mathrm{d}t - 14.470 = 30.027051.$（有 $\int_k^{2.416} a(t)\mathrm{d}t$ 得 1 分）

$\int_0^k b(t)\mathrm{d}t + \int_k^{2.416} b(t)\mathrm{d}t = 48.313876$（结果正确得 1 分）.

At time $t=2.416$, tank B contains 48.314 (or 48.313) liters of water.

(d) $w'(3.5) - a'(3.5) = -1.14298 < 0.$（有 $w'(3.5)-a'(3.5)<0$ 得 1 分）

The difference $w(t)-a(t)$ is decreasing at $t=3.5$.（结论正确得 1 分）

CALCULUS BC

SECTION Ⅱ, Part B

Time—1 hour
Number of questions—4
（自由问答 B 部分,共 4 道题,1 小时作答,可以使用计算器）

NO CALCULATOR IS ALLOWED FOR THESE QUESTIONS. DO NOT BREAK THE SEALS UNTIL YOU ARE TOLD TO DO SO.

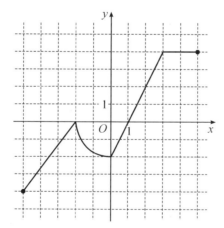

3. The graph of the function f, consisting of three line segments and a quarter of a circle, is shown above. Let g be the function defined by $g(x)=\int_1^x f(t)dt$.

(a) Find the average rate of change of g from $x=-5$ to $x=5$.

(b) Find the instantaneous rate of change of g with respect to x at $x=3$, or state that it does not exist.

(c) On what open intervals, if any, is the graph of g concave up? Justify your answer.

(d) Find all x-values in the interval $-5<x<5$ at which g has a critical point. Classify each critical point as the location of a local minimum, a local maximum, or neither. Justify your answers.

题目大意：函数 f 的图像由三条线段和一个 1/4 圆构成，令 $g(x)=\int_1^x f(t)dt$.

(a) 求 $g(x)$ 从 $x=-5$ 到 $x=5$ 的平均值.

(b) 求 $g(x)$ 在 $x=3$ 时的瞬时变化率,若不存在请加以说明.

(c) $g(x)$ 是否有 concave up 的开区间? 请说明理由.

(d) 求 $g(x)$ 在 $-5<x<5$ 的驻点,并判定在该点处取得极小值还是极大值或没有极值,请说明理由.

Solution

(a) $\dfrac{g(5)-g(-5)}{5-(-5)}$ (差商正确得 1 分)

$=\dfrac{12-(\pi+7)}{10}$

$=\dfrac{5-\pi}{10}$ (计算结果正确得 2 分)

(b) $g'(x)=f(x)$

$g'(3)=f(3)=4$

The instantaneous rate of change of g at $x=3$ is 4 (结果正确得 1 分).

(c) The graph of g is concave up on $-5<x<-2$ and $0<x<3$, because $g'(x)=f(x)$ is increasing on these intervals. (凹凸区间正确得 2 分)

(d) $g'(x)=f(x)$ is defined at all x with $-5<x<5$.

$g'(x)=f(x)=0$ at $x=-2$ and $x=1$ ($f(x)=0$ 得 1 分, critical points 正确得 1 分)

Therefore, g has critical points at $x=-2$ and $x=1$.

g has neither a local maximum nor a local minimum at $x=-2$, because g' does not change sign there.

g has a local minimum at $x=1$, because g' changes from negative to positive there (结论正确得 1 分).

4.	x	0	1	2	3	4	5	6
	$f'(x)$	4	3.5	2	0.8	1.7	5.8	7

The function f satisfies $f(0)=20$. The first derivative of f satisfies the inequality $0 \leqslant f'(x) \leqslant 7$ for all x in the closed interval $[0,6]$. Selected values of f' are shown in the table above. The function f has a continuous second derivative for all real numbers.

(a) Use a midpoint Riemann sum with three subintervals of equal length indicated by the data in the table to approximate the value of $f(6)$.

(b) Determine whether the actual value of $f(6)$ could be 70. Explain your reasoning.

(c) Evaluate $\int_2^4 f''(x)\,dx$.

(d) Find $\lim\limits_{x\to\infty}\dfrac{f(x)-20e^x}{0.5f(x)-10}$.

题目大意： 函数 $f(x)$ 满足 $f(0)=20$，且对所有 $x\in[0,6]$ 有 $0\leqslant f'(x)\leqslant 7$. 上表中列出了部分 $f'(x)$ 的值，$f(x)$ 在实数范围内有连续的二阶导数.

(a) 根据上表中的数据，基于 3 个等长的小区间，用中点黎曼和方法近似计算 $f(6)$.

(b) 请问 $f(6)$ 是否能为 70？请说明理由.

(c) 计算 $\int_2^4 f''(x)\,\mathrm{d}x$.

(d) 求 $\lim\limits_{x\to 0}\dfrac{f(x)-20\mathrm{e}^x}{0.5f(x)-10}$.

Solution

(a) $\int_0^6 f'(x)\,\mathrm{d}x \approx 2\times 3.5+2\times 0.8+2\times 5.8=20.2$ (1 分)

$$f(6)-f(0)=\int_0^6 f'(x)\,\mathrm{d}x \quad (1\text{ 分})$$

$$f(6)=f(0)+\int_0^6 f'(x)\,\mathrm{d}x \approx 20+20.2=40.2 \text{（答案正确得 1 分）}$$

(b) Since $f'(x)\leqslant 7$, $\int_0^6 f'(x)\,\mathrm{d}x \leqslant 6\times 7=42$ (1 分)

$$f(6)-f(0)\leqslant 42 \quad\Rightarrow\quad f(6)\leqslant 20+42=62$$

Therefore, the actual value of $f(6)$ could not be 70 (1 分).

(c) $\int_2^4 f''(x)\,\mathrm{d}x=f'(4)-f'(2)$ (1 分)

$$=1.7-2=-0.3 \text{ (1 分)}$$

(d) $\lim\limits_{x\to 0}(f(x)-20\mathrm{e}^x)=0$

$\lim\limits_{x\to 0}(0.5f(x)-10)=0$

Using L'Hôpital's Rule,

$$\lim_{x\to 0}\frac{f(x)-20\mathrm{e}^x}{0.5f(x)-10}=\lim_{x\to 0}\frac{f'(x)-20\mathrm{e}^x}{0.5f'(x)} \text{（应用洛必达法则得 1 分）}$$

$$=\frac{4-20}{0.5\times 4}=-8 \text{（结果正确得 1 分）}$$

5. Consider the differential equation $\dfrac{\mathrm{d}y}{\mathrm{d}x}=-1+\dfrac{y^2}{x}$.

(a) Show that $\dfrac{\mathrm{d}^2 y}{\mathrm{d}x^2}=\dfrac{2y^3-y^2-2xy}{x^2}$.

(b) Let $y=g(x)$ be the particular solution to the differential equation $\dfrac{\mathrm{d}y}{\mathrm{d}x}=-1+\dfrac{y^2}{x}$ with initial condition $g(4)=2$. Does g have a relative minimum, a relative maximum, or neither at $x=4$? Justify your answer.

(c) Let $y=h(x)$ be the particular solution to the differential equation $\dfrac{\mathrm{d}y}{\mathrm{d}x}=-1+\dfrac{y^2}{x}$ with initial condition $h(1)=2$. Write the second-degree Taylor polynomial for $h(x)$ about $x=1$.

(d) For the function $h(x)$ given in part (c), it is known that $|h'''(x)| \leqslant 60$ for all x in the interval $0.9 \leqslant x \leqslant 1.1$. Let A represent the approximation of $h(1.1)$ found by using the second-degree Taylor polynomial for $h(x)$ about $x=1$ from part (c). Use the Lagrange error bound to show that A differs from $h(1.1)$ by at most 0.01.

题目大意：微分方程：$\dfrac{dy}{dx} = -1 + \dfrac{y^2}{x}$.

(a) 证明：$\dfrac{d^2 y}{dx^2} = \dfrac{2y^3 - y^2 - 2xy}{x^2}$.

(b) 令 $y = g(x)$ 为微分方程 $\dfrac{dy}{dx} = -1 + \dfrac{y^2}{x}$ 满足初始条件 $g(4) = 2$ 的特解，则 $g(x)$ 在 $x = 4$ 有极大值还是极小值或没有极值？请说明理由.

(c) 令 $y = h(x)$ 为微分方程 $\dfrac{dy}{dx} = -1 + \dfrac{y^2}{x}$ 满足初始条件 $h(1) = 2$ 的特解，写出 $h(x)$ 在 $x = 1$ 处的二次泰勒多项式.

(d) 对于 (c) 中的 $y = h(x)$，若 $0.9 \leqslant x \leqslant 1.1$ 时，有 $|h'''(x)| \leqslant 60$，令 A 是用 $h(x)$ 在 $x = 1$ 处的二次泰勒多项式计算的 $h(1.1)$ 的近似值，用拉格朗日误差界证明误差最多为 0.01.

Solution

(a) $\dfrac{d^2 y}{dx^2} = \dfrac{x \cdot 2y \dfrac{dy}{dx} - y^2 \cdot 1}{x^2}$ （1 分）

$= \dfrac{2xy\left(-1 + \dfrac{y^2}{x}\right) - y^2}{x^2} = \dfrac{2y^3 - y^2 - 2xy}{x^2}$ （结果正确得 1 分）

(b) $\dfrac{dy}{dx}\bigg|_{(x,y)=(4,2)} = -1 + \dfrac{4}{4} = 0$ （1 分）

$\dfrac{d^2 y}{dx^2}\bigg|_{(x,y)=(4,2)} = \dfrac{2 \times 8 - 4 - 16}{16} = -\dfrac{1}{4} < 0$

By the Second Derivative Test, g has a relative maximum at $x = 4$（说明理由得 1 分）

(c) $\dfrac{dy}{dx}\bigg|_{(x,y)=(1,2)} = -1 + \dfrac{4}{1} = 3$

$\dfrac{d^2 y}{dx^2}\bigg|_{(x,y)=(1,2)} = \dfrac{2 \times 8 - 4 - 4}{1} = 8$ ($\dfrac{dy}{dx}, \dfrac{d^2 y}{dx^2}$ 结果正确得 1 分)

The second-degree Taylor polynomial for h about $x = 1$ is

$T_2(x) = 2 + 3(x-1) + \dfrac{8}{2!}(x-1)^2 = 2 + 3(x-1) + 4(x-1)^2$（泰勒多项式书写正确得 2 分）.

(d) $|h(1.1) - A| \leqslant \dfrac{\max\limits_{1.0 \leqslant x \leqslant 1.1} |h'''(x)| \, |1.1 - 1|^3}{3!}$ （写出误差界得 1 分）

$\leqslant \dfrac{60}{6} \times \dfrac{1}{1000} = \dfrac{10}{1000} = \dfrac{1}{100}$（推导出结论得 1 分）

6. Let f be the function defined by $f(x)=\dfrac{1}{x^2+9}$.

(a) Evaluate the improper integral $\displaystyle\int_3^\infty f(x)\,\mathrm{d}x$, or show that the integral diverges.

(b) Determine whether the series $\displaystyle\sum_{n=3}^\infty f(n)$ converges or diverges. State the conditions of the test used for determining convergence or divergence.

(c) Determine whether the series $\displaystyle\sum_{n=1}^\infty \dfrac{(-1)^n}{(e^n\cdot f(n))}=\sum_{n=1}^\infty\dfrac{(-1)^n(n^2+9)}{e^n}$ converges absolutely, converges conditionally, or diverges.

题目大意： $f(x)=\dfrac{1}{x^2+9}$.

(a) 计算反常积分 $\displaystyle\int_3^\infty f(x)\,\mathrm{d}x$.

(b) 判断级数 $\displaystyle\sum_{n=3}^\infty f(n)$ 的敛散性，并说明所用判定法满足的条件.

(c) 判断级数 $\displaystyle\sum_{n=1}^\infty \dfrac{(-1)^n}{(e^n\cdot f(n))}=\sum_{n=1}^\infty\dfrac{(-1)^n(n^2+9)}{e^n}$ 是绝对收敛、条件收敛还是发散.

Solution

(a) $\displaystyle\int_3^\infty\dfrac{1}{x^2+9}\mathrm{d}x=\lim_{b\to\infty}\int_3^b\dfrac{1}{x^2+9}\,\mathrm{d}x=\lim_{b\to\infty}\left(\dfrac{1}{3}\tan^{-1}\left(\dfrac{x}{3}\right)\Big|_3^b\right)$ （1 分）

$=\displaystyle\lim_{b\to\infty}\left[\dfrac{1}{3}\tan^{-1}\left(\dfrac{b}{3}\right)-\dfrac{1}{3}\tan^{-1}(1)\right]$（1 分）

$=\dfrac{\pi}{6}-\dfrac{\pi}{12}=\dfrac{\pi}{12}$（1 分）

(b) The function f is continuous, positive, and decreasing on $[3,\infty)$. By the integral test, since $\displaystyle\int_3^\infty f(x)\,\mathrm{d}x$ converges, $\displaystyle\sum_{n=3}^\infty f(n)$ converges. (条件和结论共 2 分)

—OR—

$0<\dfrac{1}{n^2+9}<\dfrac{1}{n^2}$ for $n\geqslant 3$.

Since the series $\displaystyle\sum_{n=3}^\infty\dfrac{1}{n^2}$ converges, the series $\displaystyle\sum_{n=3}^\infty f(n)=\sum_{n=3}^\infty\dfrac{1}{n^2+9}$ converges by the comparison test.

(c) Consider the series $\displaystyle\sum_{n=1}^\infty\dfrac{1}{(e^n\cdot f(n))}=\sum_{n=1}^\infty\dfrac{n^2+9}{e^n}$. (1 分)

$\displaystyle\lim_{n\to\infty}\left|\dfrac{\dfrac{(n+1)^2+9}{e^{n+1}}}{\dfrac{n^2+9}{e^n}}\right|=\lim_{n\to\infty}\left|\dfrac{(n+1)^2+9}{n^2+9}\cdot\dfrac{1}{e}\right|=\dfrac{1}{e}<1$ (1 分)

By the ratio test, $\sum_{n=1}^{\infty} \dfrac{1}{(e^n \cdot f(n))}$ converges.(1 分)

Therefore, $\sum_{n=1}^{\infty} \dfrac{(-1)^n}{(e^n \cdot f(n))}$ converges absolutely.(1 分)

本套模拟试题评分标准

Section I : Multiple Choice

$$\frac{}{\text{Number Correct}} \times 12000 = \frac{}{\text{Weighted Section I Score}}$$
$$\text{(out of 45)} \qquad\qquad \text{(do not round)}$$

Section II : Free Response

Question 1 $\dfrac{}{\text{(out of 9)}} \times 1000 = \dfrac{}{\text{(do not round)}}$

Question 2 $\dfrac{}{\text{(out of 9)}} \times 1000 = \dfrac{}{\text{(do not round)}}$

Question 3 $\dfrac{}{\text{(out of 9)}} \times 1000 = \dfrac{}{\text{(do not round)}}$

Question 4 $\dfrac{}{\text{(out of 9)}} \times 1000 = \dfrac{}{\text{(do not round)}}$

Question 5 $\dfrac{}{\text{(out of 9)}} \times 1000 = \dfrac{}{\text{(do not round)}}$

Question 6 $\dfrac{}{\text{(out of 9)}} \times 1000 = \dfrac{}{\text{(do not round)}}$

$$\text{Sum} = \frac{}{\text{Weighted Section II Score}}$$
$$\text{(do not round)}$$

Composite Score

$$\frac{}{\text{Weighted Section I Score}} + \frac{}{\text{Weighted Section II Score}} = \frac{}{\text{Composite Score (round to nearest whole number)}}$$

AP Score Conversion Chart
Calculus BC

Composite Score Range	AP Score
65~108	5
54~64	4
39~53	3
24~38	2
0~23	1

本套试题转化为微积分 AB 成绩的方法：

Section Ⅰ: Multiple Choice

Questions(1～3, 6～7, 9, 11, 13～15, 17～19, 21, 24, 26, 28, 76～79, 81～83, 85～86, 88)

$$\frac{}{\text{Number Correct (out of 27)}} \times 1000 = \frac{}{\text{Weighted Section I Score (do not round)}}$$

Section Ⅱ: Free Response

Question 2 $\dfrac{}{(\text{out of 9})} \times 1000 = \dfrac{}{(\text{do not round})}$

Question 3 $\dfrac{}{(\text{out of 9})} \times 1000 = \dfrac{}{(\text{do not round})}$

Question 4 $\dfrac{}{(\text{out of 9})} \times 1000 = \dfrac{}{(\text{do not round})}$

$$\text{Sum} = \frac{}{\text{Weighted Section Ⅱ Score (do not round)}}$$

Composite Score

$$\frac{}{\text{Weighted Section Ⅰ Score}} + \frac{}{\text{Weighted Section Ⅱ Score}} = \frac{}{\text{Composite Score (round to nearest whole number}}$$

AP Score Conversion Chart
Calculus AB Subscore

Composite Score Range	AP Score
34～54	5
28～33	4
21～27	3
13～20	2
0～12	1

Chapter 14 主要公式及定理

14.1 Functions(函数)

The Vertical Line Test(垂线测试法)

A curve in the xy-plane is the graph of some function f if and only if no vertical line intersects the curve more than once.

常见的函数定义域：

(1) $y = \sqrt[n]{x}$ (n 为偶数) Domain: $\{x \mid x \geqslant 0\}$

(2) $y = \dfrac{1}{x}$ Domain: $\{x \mid x \neq 0\}$

(3) $y = \ln x$ Domain: $\{x \mid x > 0\}$

$y = \log_a x\,(a > 0, a \neq 1)$ Domain: $\{x \mid x > 0\}$

(4) $y = \tan x$ Domain: $\left\{x \mid x \neq k\pi + \dfrac{\pi}{2}, k \in \mathbf{Z}\right\}$

$y = \cot x$ Domain: $\{x \mid x \neq k\pi, k \in \mathbf{Z}\}$

$y = \arcsin x = \sin^{-1} x$ Domain: $[-1, 1]$

$y = \arccos x = \cos^{-1} x$ Domain: $[-1, 1]$

Monotony of Functions(函数的单调性)

Let f be a function defined on an interval I and let x_1 and x_2 be any two points in I.

(1) If $f(x_1) < f(x_2)$ whenever $x_1 < x_2$, then f is said to be **increasing** on I.

(2) If $f(x_2) < f(x_1)$ whenever $x_1 < x_2$, then f is said to be **decreasing** on I.

A function that is increasing or decreasing on I is called **monotonic** on I.

Boundedness of Functions(函数的有界性)

A function f is **bounded below** if there is some number b that is less than or equal to every number in the range of f. Any such number b is called a **lower bound** of f.

A function f is **bounded above** if there is some number B that is greater than or equal to every number in the range of f. Any such number B is called an **upper bound** of f.

A function f is **bounded**(有界) if it is bounded both above and below.

Symmetry of Functions(函数的对称性)

A function f is $\begin{cases}\text{odd}\\ \text{even}\end{cases}$ if, for all x in the domain of f, $\begin{cases}f(-x)=-f(x)\\ f(-x)=f(x)\end{cases}$.

The graph of an odd function is symmetric about the origin, the graph of an even function is symmetric about the y-axis.

Periodicity of Functions(周期性)

A function is **periodic** if there is a positive number p such that $f(x+p)=f(x)$ for every value of x. The smallest such value of p is the **period** of f.

Inverse Functions(反函数)

A function $f(x)$ is one-to-one on a domain D if $f(a)\neq f(b)$ whenever $a\neq b$.

The function defined by reversing a one-to-one function f is the inverse of f. The symbol for the inverse of f is f^{-1}.

$$\text{domain of } f^{-1}=\text{range of } f \qquad \text{range of } f^{-1}=\text{domain of } f$$

Inverse Properties(反函数的性质)

(1) $(f\circ f^{-1})(x)=(f^{-1}\circ f)(x)=x$.

(2) $f(x)=e^x$ and $f^{-1}(x)=\ln x$ are inverses of each other.

(3) $\ln e^x=e^{\ln x}=x\,(x>0)$.

Power Functions(幂函数)

The Identity Function: $y=x$

The Squaring Function: $y=x^2$

The Cubic Function: $y=x^3$

The Square Root Function: $y=\sqrt{x}=x^{1/2}\,(x\geqslant 0)$

The Reciprocal Function: $y=\dfrac{1}{x}\,(x\neq 0)$

Linear Functions(线性函数)

Point-Slope Form(点斜式)

Through $P_1(x_1,y_1)$ and with slope m: $y=m(x-x_1)+y_1$

Slope-Intercept Form(斜截式)

With slope m and y-intercept b: $y=mx+b$

Two-Point Form(两点式)

Through $P_1(x_1,y_1)$ and $P_2(x_2,y_2)$: $y=\dfrac{y_2-y_1}{x_2-x_1}(x-x_1)+y_1$

Intercept Form(截距式)

With x and y-intercept a and b, respectively: $\dfrac{x}{a}+\dfrac{y}{b}=1$

General Form(一般式)

$Ax+By+C=0$, where A and B are not both zero.

If $B\neq 0$, the slope is $-\dfrac{A}{B}$; the y-intercept is $-\dfrac{C}{B}$; the x-intercept is $-\dfrac{C}{A}$.

Quadratic Functions(二次函数)

Standard Form(标准式): $y=ax^2+bx+c\,(a\neq 0)$

Vertex Form(顶点式): $y=a(x-h)^2+k$

Laws of Exponents(指数的性质)

If a and b are positive numbers and m and n any real numbers, then

$a^m \cdot a^n = a^{m+n}$; $a^m \div a^n = a^{m-n}$; $(a^m)^n = a^{m \cdot n}$; $(ab)^n = a^n b^n$

Properties of Logarithms(对数的性质)

The logarithmic function $y=\log_a x\,(a>0, a\neq 1)$ has the following properties:

$\log_a 1 = 0$; $\log_a a = 1$; $\log_a mn = \log_a m + \log_a n$

$\log_a \dfrac{m}{n} = \log_a m - \log_a n$; $\log_a x^m = m\log_a x$

Formulas from Trigonometry(三角公式)

Reciprocal Identities:

$\sec x = \dfrac{1}{\cos x}$; $\csc x = \dfrac{1}{\sin x}$; $\cot x = \dfrac{1}{\tan x}$

Pythagorean Identities:

$\sin^2 x + \cos^2 x = 1$; $1+\tan^2 x = \sec^2 x$; $1+\cot^2 x = \csc^2 x$

Odd-Even Identities:

$\sin(-x) = -\sin x$; $\cos(-x) = \cos x$; $\tan(-x) = -\tan x$

$\csc(-x) = -\csc x$; $\sec(-x) = \sec x$; $\cot(-x) = -\cot x$

Sum and Difference Identities:

$\sin(x+y) = \sin x \cos y + \cos x \sin y$; $\sin(x-y) = \sin x \cos y - \cos x \sin y$

$\cos(x+y) = \cos x \cos y - \sin x \sin y$; $\cos(x-y) = \cos x \cos y + \sin x \sin y$

$\tan(x+y) = \dfrac{\tan x + \tan y}{1 - \tan x \tan y}$; $\tan(x-y) = \dfrac{\tan x - \tan y}{1 + \tan x \tan y}$

和差化积:

$\sin x + \sin y = 2\sin\dfrac{x+y}{2}\cos\dfrac{x-y}{2}$

$\sin x - \sin y = 2\cos\dfrac{x+y}{2}\sin\dfrac{x-y}{2}$

$\cos x + \cos y = 2\cos\dfrac{x+y}{2}\cos\dfrac{x-y}{2}$

$\cos x - \cos y = -2\sin\dfrac{x+y}{2}\sin\dfrac{x-y}{2}$

积化和差：

$$\sin x \cos y = \frac{1}{2}[\sin(x+y) + \sin(x-y)]$$

$$\cos x \sin y = \frac{1}{2}[\sin(x+y) - \sin(x-y)]$$

$$\cos x \cos y = \frac{1}{2}[\cos(x+y) + \cos(x-y)]$$

$$\sin x \sin y = -\frac{1}{2}[\cos(x+y) - \cos(x-y)]$$

Cofunction Identities：

$$\cos\left(\frac{\pi}{2} - x\right) = \sin x \,;\, \sin\left(\frac{\pi}{2} - x\right) = \cos x \,;\, \tan\left(\frac{\pi}{2} - x\right) = \cot x$$

$$\sec\left(\frac{\pi}{2} - x\right) = \csc x \,;\, \csc\left(\frac{\pi}{2} - x\right) = \sec x \,;\, \cot\left(\frac{\pi}{2} - x\right) = \tan x$$

Double-Angle Identities：

$$\sin 2x = 2\sin x \cos x$$

$$\cos 2x = \cos^2 x - \sin^2 x = 2\cos^2 x - 1 = 1 - 2\sin^2 x$$

$$\tan 2x = \frac{2\tan x}{1 - \tan^2 x}$$

Power-Reducing Identities：

$$\sin^2 x = \frac{1 - \cos 2x}{2} \,;\, \cos^2 x = \frac{1 + \cos 2x}{2} \,;\, \tan^2 x = \frac{1 - \cos 2x}{1 + \cos 2x}$$

Trigonometric Functions of Important Angels（特殊角的三角函数）

θ	radians	$\sin\theta$	$\cos\theta$	$\tan\theta$
0	0	0	1	0
30°	$\pi/6$	$1/2$	$\sqrt{3}/2$	$\sqrt{3}/3$
45°	$\pi/4$	$\sqrt{2}/2$	$\sqrt{2}/2$	1
60°	$\pi/3$	$\sqrt{3}/2$	$1/2$	$\sqrt{3}$
90°	$\pi/2$	1	0	—

14.2 Limits and Continuity（极限与连续）

Theorem 2.1

$$\lim_{x \to c} f(x) = L \Leftrightarrow \lim_{x \to c^-} f(x) = L \text{ and } \lim_{x \to c^+} f(x) = L.$$

Theorem 2.2

$$\lim_{x \to \infty} f(x) = L \Leftrightarrow \lim_{x \to -\infty} f(x) = L \text{ and } \lim_{x \to +\infty} f(x) = L$$

The squeeze or sandwich theorem(夹逼定理又称三明治定理)

Suppose that $g(x) \leqslant f(x) \leqslant h(x)$ for all x in some open interval containing c, except possibly at itself. Suppose that
$$\lim_{x \to c} g(x) = \lim_{x \to c} h(x) = L$$
Then
$$\lim_{x \to c} f(x) = L.$$

14.3 Definition of Derivative(导数的定义)

Derivative at a Point(点导数)

The derivative of $y = f(x)$ at $x = x_0$, denoted by $f'(x_0)$:

$$f'(x_0) = \lim_{\Delta x \to 0} \frac{\Delta y}{\Delta x} = \lim_{\Delta x \to 0} = \frac{f(x_0 + \Delta x) - f(x_0)}{\Delta x}.$$

$$f'(x_0) = \lim_{h \to 0} \frac{f(x_0 + h) - f(x_0)}{h}.$$

$$f'(x_0) = \lim_{x \to x_0} \frac{f(x) - f(x_0)}{x - x_0}.$$

Theorem Differentiability Implies Continuity

If a function $f(x)$ is differentiable at x_0, then $f(x)$ is continuous at x_0.

14.4 Computation of Derivative(导数的计算)

Arithmetic Operations on Derivative(导数的代数运算)

设 $u(x), v(x)$ 可导，c 为常数.

Constant Multiple Rule(数乘法则): $(cu)' = cu'$.

Sum and Difference Rule(和、差法则): $(u \pm v)' = u' \pm v'$.

Product Rule(乘法法则): $(uv)' = u'v + uv'$.

Quotient Rule(除法法则): $\left(\dfrac{u}{v}\right)' = \dfrac{u'v - uv'}{v^2}$.

Essential Formulas(基本公式)

(1) $(c)' = 0$; (2) $(x^M)' = M x^{M-1}$;

(3) $(\sin x)' = \cos x$; (4) $(\cos x)' = -\sin x$;

(5) $(\tan x)' = \sec^2 x$; (6) $(\cot x)' = -\csc^2 x$;

(7) $(\sec x)' = \sec x \tan x$; (8) $(\csc x)' = -\csc x \cot x$;

(9) $(a^x)' = a^x \ln a$; (10) $(e^x)' = e^x$;

(11) $(\log_a x)' = \dfrac{1}{x \ln a}$; (12) $(\ln x)' = \dfrac{1}{x}$;

(13) $(\arcsin x)' = \dfrac{1}{\sqrt{1-x^2}}$; (14) $(\arccos x)' = -\dfrac{1}{\sqrt{1-x^2}}$;

(15) $(\arctan x)' = \dfrac{1}{1+x^2}$; (16) $(\operatorname{arccot} x)' = -\dfrac{1}{1+x^2}$.

Derivative of Inverse Function(反函数的导数)

$$[f^{-1}(x)]' = \dfrac{1}{f'[f^{-1}(x)]} = \dfrac{1}{f'(y)} \text{ 或 } \dfrac{dy}{dx} = \dfrac{1}{\dfrac{dx}{dy}}$$

The Chain Rule(链式法则)

If $f(u)$ is differentiable at the point $u = g(x)$ and $g(x)$ is differentiable at x, then the composite function $(f \circ g)(x) = f(g(x))$ is differentiable at x, and

$$(f \circ g)'(x) = f'(g(x)) \cdot g'(x)$$

Parametric Function Derivative(参数方程的导数)

If $x = f(t)$ and $y = g(t)$ are differentiable functions of t, then

$$\dfrac{dy}{dx} = \dfrac{\dfrac{dy}{dt}}{\dfrac{dx}{dt}} \text{ and } \dfrac{d^2 y}{dx^2} = \dfrac{d}{dx}\left(\dfrac{dy}{dx}\right) = \dfrac{\dfrac{d}{dt}\left(\dfrac{dy}{dx}\right)}{\dfrac{dx}{dt}}$$

Polar Function Derivative(极坐标方程的导数)

$$\dfrac{dy}{dx} = \dfrac{\dfrac{dy}{d\theta}}{\dfrac{dx}{d\theta}} = \dfrac{\dfrac{dr}{d\theta}\sin\theta + r\cos\theta}{\dfrac{dr}{d\theta}\cos\theta - r\sin\theta}$$

14.5 Applications of Derivative(导数的应用)

Average and Instantaneous Rates of Change(平均变化率与瞬时变化率)

Average rate of change of f over the interval from a to $a+h = \dfrac{f(a+h) - f(a)}{h}$.

The (instantaneous) rate of change of f at $a = f'(a) = \lim\limits_{h \to 0} \dfrac{f(a+h) - f(a)}{h}$.

Tangents and Normals(切线和法线)

过曲线 $y = f(x)$ 的切点 (x_0, y_0) 的切线方程为: $y - y_0 = f'(x_0)(x - x_0)$.

过曲线 $y = f(x)$ 的切点 (x_0, y_0) 的法线方程为: $y - y_0 = -\dfrac{1}{f'(x_0)}(x - x_0)$.

The Mean Value Theorem for Derivatives(微分中值定理, 简记为 MVT)

Assume that $f(x)$ is continuous on the closed interval $[a, b]$ and differentiable on (a, b). Then there exists at least one value c in (a, b) such that

$$f'(c) = \frac{f(b)-f(a)}{b-a}.$$

Theorem:L'Hôpital's Rule(洛必达法则)

设(1)$\lim\limits_{x \to a} f(x) = \lim\limits_{x \to a} g(x) = 0$,或为$\infty$;

(2)$f(x), g(x)$可导,且$g(x) \neq 0$;

(3)$\lim\limits_{x \to a} \dfrac{f'(x)}{g'(x)}$存在,或为$\infty$;

则$\lim\limits_{x \to a} \dfrac{f(x)}{g(x)} = \lim\limits_{x \to a} \dfrac{f'(x)}{g'(x)}.$

Theorem:First Derivative Test for Monotonic Functions(函数单调性的一阶导数判别法)

Suppose that $f(x)$ is continuous on $[a,b]$ and differentiable on (a,b).

(1)If $f'(x) > 0$ at each point $x \in (a,b)$, then $f(x)$ is increasing on $[a,b]$.

(2)If $f'(x) < 0$ at each point $x \in (a,b)$, then $f(x)$ is decreasing on $[a,b]$.

(3)If $f'(x) = 0$ at each point $x \in (a,b)$, then $f(x)$ is a constant on $[a,b]$.

Theorem:First Derivative Test for Local Extrema(极值的一阶导数判别法)

Suppose that c is a critical point of a continuous function f, and that f is differentiable at every point in some interval containing c except possibly at c itself.

Moving across c from left to right.

(1)if f' changes from negative to positive at c, then f has a local minimum at c;

(2)if f' changes from positive to negative at c, then f has a local maximum at c;

(3)if f' does not change sign at c (that is, is positive on both sides of c or negative on both sides), then f has no local extrema at c.

Theorem:Second Derivative Test for Local Extrema(极值的二阶导数判别法)

Suppose $f''(x)$ is continuous on an open interval that contain $x = c$.

(1)If $f'(c) = 0$ and $f''(c) < 0$, then $f(x)$ has a local maximum at $x = c$.

(2)If $f'(c) = 0$ and $f''(c) > 0$, then $f(x)$ has a local minimum at $x = c$.

(3)If $f'(c) = 0$ and $f''(c) = 0$, then the test fails. The function $f(x)$ may have a local maximum, a local minimum, or neither at $x = c$.

Theorem:Test for Concavity(函数凹凸性的判断定理)

Assume that $f''(x)$ exists for all $x \in (a,b)$.

(1)If $f''(x) > 0$ for all $x \in (a,b)$, then f is concave up on (a,b).

(2)If $f''(x) < 0$ for all $x \in (a,b)$, then f is concave down on (a,b).

Move along a Line(直线运动)

令$x = s(t)$表示质点运动的position function(位置函数),则:

Displacement(位移):从时刻t_0到时刻t_1的位移为:

Displacement: $s(t_1) - s(t_0)$

Velocity(速度): $v(t) = \dfrac{\mathrm{d}s}{\mathrm{d}t} = \begin{cases} v > 0, \text{move to the right} \\ v < 0, \text{move to the left} \\ v = 0, \text{rest} \end{cases}$

转向点: $v = 0$ 且在其两侧 v 异号时,质点改变运动方向.

Acceleration(加速度): $a(t) = \dfrac{\mathrm{d}v}{\mathrm{d}t} = \dfrac{\mathrm{d}^2 s}{\mathrm{d}t^2} = \begin{cases} a > 0 \Rightarrow v \text{ is increasing} \\ a < 0 \Rightarrow v \text{ is decreasing} \end{cases}$

Speed(速率): $\text{speed} = |v(t)| \begin{cases} a \text{ 与 } v \text{ 同号} \Rightarrow \text{speed is increasing} \\ a \text{ 与 } v \text{ 异号} \Rightarrow \text{speed is decreasing} \end{cases}$

Move in a Plane(平面运动)

Position Vector(位置向量): $\vec{R}(t) = \langle x(t), y(t) \rangle$

Velocity Vector(速度向量): $\vec{v}(t) = \langle \dfrac{\mathrm{d}x}{\mathrm{d}t}, \dfrac{\mathrm{d}y}{\mathrm{d}t} \rangle$

The slope of $\vec{v}(t) = \dfrac{\frac{\mathrm{d}y}{\mathrm{d}t}}{\frac{\mathrm{d}x}{\mathrm{d}t}}$

$\dfrac{\mathrm{d}x}{\mathrm{d}t} > 0$ 表示质点向右运动, $\dfrac{\mathrm{d}x}{\mathrm{d}t} < 0$ 表示质点向左运动.

$\dfrac{\mathrm{d}y}{\mathrm{d}t} > 0$ 表示质点向上运动, $\dfrac{\mathrm{d}y}{\mathrm{d}t} < 0$ 表示质点向下运动.

Speed(速率): $\text{speed} = |\vec{v}(t)| = \sqrt{\left(\dfrac{\mathrm{d}x}{\mathrm{d}t}\right)^2 + \left(\dfrac{\mathrm{d}y}{\mathrm{d}t}\right)^2}$

Acceleration Vector(加速度向量): $\vec{a}(t) = \langle \dfrac{\mathrm{d}^2 x}{\mathrm{d}t^2}, \dfrac{\mathrm{d}^2 y}{\mathrm{d}t^2} \rangle$

加速度向量的模 $\vec{a}(t)$: $|\vec{a}(t)| = \sqrt{\left(\dfrac{\mathrm{d}^2 x}{\mathrm{d}t^2}\right)^2 + \left(\dfrac{\mathrm{d}^2 y}{\mathrm{d}t^2}\right)^2}$

The slope of $\vec{a}(t) = \dfrac{\frac{\mathrm{d}^2 y}{\mathrm{d}t^2}}{\frac{\mathrm{d}^2 x}{\mathrm{d}t^2}}$

14.6 Differential and Approximation(微分与近似计算)

基本初等函数的微分公式

$\mathrm{d}(x^M) = M x^{M-1} \mathrm{d}x$; $\mathrm{d}(\sin x) = \cos x \, \mathrm{d}x$; $\mathrm{d}(\cos x) = -\sin x \, \mathrm{d}x$;

$\mathrm{d}(\tan x) = \sec^2 x \, \mathrm{d}x$; $\mathrm{d}(\cot x) = -\csc^2 x \, \mathrm{d}x$; $\mathrm{d}(\sec x) = \sec x \tan x \, \mathrm{d}x$;

$\mathrm{d}(\csc x) = -\csc x \cot x \, \mathrm{d}x$; $\mathrm{d}(a^x) = a^x \ln a \, \mathrm{d}x$; $\mathrm{d}(e^x) = e^x \, \mathrm{d}x$;

$\mathrm{d}(\log_a x) = \dfrac{1}{x \ln a} \mathrm{d}x$; $\mathrm{d}(\ln x) = \dfrac{1}{x} \mathrm{d}x$; $\mathrm{d}(\arcsin x) = \dfrac{1}{\sqrt{1-x^2}} \mathrm{d}x$;

$$d(\arccos x) = -\frac{1}{\sqrt{1-x^2}}dx; \quad d(\arctan x) = \frac{1}{1+x^2}dx; \quad d(\text{arccot } x) = -\frac{1}{1+x^2}dx.$$

函数和、差、积、商的微分法则

$$d(u \pm v) = du \pm dv; \quad d(Cu) = Cdu;$$

$$d(uv) = vdu + udv; \quad d\left(\frac{u}{v}\right) = \frac{vdu - udv}{v^2}dx \, (v \neq 0).$$

Approximating a Derivative Value(导数的近似计算)

Difference Quotient(差商法): $f'(a) \approx \dfrac{f(a+\Delta x) - f(a)}{\Delta x}$（当 $\Delta x \to 0$ 时）

Symmetric Difference Quotient(对称差商法):

$$f'(a) \approx \frac{f(a+\Delta x) - f(a-\Delta x)}{2\Delta x}（当 \Delta x \to 0 时）$$

Local Linear Approximation(局部线性近似)

If f is differentiable at $x=a$ and x is close to a, then

$f(x) \approx f(a) + f'(a)(x-a)$ ——the local linear approximation of $f(x)$ at a

$f(x)$ is concave down at $x=a \Rightarrow$ approximates value $>$ actual value (over estimate)

$f(x)$ is concave up at $x=a \Rightarrow$ approximates value $<$ actual value (under estimate)

Newton's Method(牛顿法)

$$x_{n+1} = x_n - \frac{f(x_n)}{f'(x_n)}（牛顿法求方程 f(x) = 0 近似根的迭代公式）$$

14.7 Antidifferentiation(不定积分)

Definition of Indefinite Integral(不定积分的定义)

$f(x)$ 的全体"原函数"称为 $f(x)$ 的不定积分，记作：

$$\int f(x)dx = F(x) + C \, (C \text{ is an arbitrary constant})$$

基本积分表

(1) $\int k \, dx = kx + C$ (k 是常数).

(2) $\int x^\mu dx = \dfrac{1}{\mu+1} x^{\mu+1} + C$;

$\int \dfrac{1}{x^2} dx = -\dfrac{1}{x} + C; \int \dfrac{1}{\sqrt{x}} dx = 2\sqrt{x} + C; \int \sqrt{x} \, dx = \dfrac{2}{3} x^{\frac{3}{2}} + C.$

(3) $\int \dfrac{1}{x} dx = \ln |x| + C; \int \dfrac{1}{ax+b} dx = \dfrac{1}{a} \ln |ax+b| + C.$

(4) $\int e^x dx = e^x + C.$

(5) $\int a^x \, dx = \dfrac{a^x}{\ln a} + C.$

(6) $\int \cos x \, dx = \sin x + C.$

(7) $\int \sin x \, dx = -\cos x + C.$

(8) $\int \dfrac{1}{\cos^2 x} \, dx = \int \sec^2 x \, dx = \tan x + C.$

(9) $\int \dfrac{1}{\sin^2 x} \, dx = \int \csc^2 x \, dx = -\cot x + C.$

(10) $\int \dfrac{1}{1+x^2} \, dx = \arctan x + C; \int \dfrac{1}{a^2+x^2} \, dx = \dfrac{1}{a} \arctan \dfrac{x}{a} + C.$

(11) $\int \dfrac{1}{\sqrt{1-x^2}} \, dx = \arcsin x + C; \int \dfrac{1}{\sqrt{a^2-x^2}} \, dx = \arcsin \dfrac{x}{a} + C.$

(12) $\int \sec x \tan x \, dx = \sec x + C.$

(13) $\int \csc x \cot x \, dx = -\csc x + C.$

(14) $\int \tan x \, dx = \ln|\sec x| + C = \ln|\cos x| + C.$

(15) $\int \cot x \, dx = -\ln|\csc x| + C = \ln|\sin x| + C.$

(16) $\int \csc x \, dx = \ln|\csc x - \cot x| + C.$

(17) $\int \sec x \, dx = \ln|\sec x + \tan x| + C.$

不定积分的性质

$\int [f(x) + g(x)] \, dx = \int f(x) \, dx + \int g(x) \, dx; \int kf(x) \, dx = k \int f(x) \, dx \, (k \text{ 是常数}, k \neq 0).$

Parts Formula(分部积分公式)

设函数 $u = u(x), v = v(x)$ 具有连续导数.

$\int uv' \, dx = uv - \int u'v \, dx$, 或 $\int u \, dv = uv - \int v \, du.$

14.8　Definite Integrals(定积分)

Approximation of Definite Integral(定积分的近似计算)

Let $f(x)$ be defined on $[a, b]$ and x_i be points on $[a, b]$ such that $x_0 = a, x_n = b$, and $a < x_1 < x_2 < x_3 < \cdots < x_{n-1} < b$, Δx_i be the length of the ith interval $[x_{i-1}, x_i]$.

Left-Riemann Sum Rule(左和法):

$$\int_a^b f(x) \, dx \approx L(n) = f(x_0) \Delta x_1 + f(x_1) \Delta x_2 + \cdots + f(x_{n-1}) \Delta x_n$$

Midpoint Sum Rule(中点法):

$$\int_a^b f(x)\mathrm{d}x \approx M(n) = f(\bar{x}_1)\Delta x_1 + f(\bar{x}_2)\Delta x_2 + \cdots + f(\bar{x}_n)\Delta x_n$$

Where $\bar{x}_i = \dfrac{1}{2}(x_{i-1} + x_i) = $ midpoint of $[x_{i-1}, x_i]$.

Right-Riemann Sum Rule(右和法):

$$\int_a^b f(x)\mathrm{d}x \approx R(n) = f(x_1)\Delta x_1 + f(x_2)\Delta x_2 + \cdots + f(x_n)\Delta x_n$$

Trapezoid Sum Rule (梯形法):

$$\int_a^b f(x)\mathrm{d}x \approx T(n)$$
$$= \frac{f(x_0) + f(x_1)}{2}\Delta x_1 + \frac{f(x_1) + f(x_2)}{2}\Delta x_2 + \cdots + \frac{f(x_{n-1}) + f(x_n)}{2}\Delta x_n$$

Comparing Approximating Sums(近似和的比较)

利用单调性比较 $L(n)$ 与 $R(n)$:

$$f(x) \text{ is increasing} \Rightarrow L(n) < \int_a^b f(x)\mathrm{d}x < R(n)$$

$$f(x) \text{ is decreasing} \Rightarrow L(n) > \int_a^b f(x)\mathrm{d}x > R(n)$$

利用凹凸性比较 $M(n)$ 与 $T(n)$:

$$f(x) \text{ is concave up} \Rightarrow M(n) < \int_a^b f(x)\mathrm{d}x < T(n)$$

$$f(x) \text{ is concave down} \Rightarrow M(n) > \int_a^b f(x)\mathrm{d}x > T(n)$$

Properties of Definite Integrals(定积分的性质)

(1) 当 $a = b$ 时，$\int_a^b f(x)\mathrm{d}x = 0$.

(2) 当 $a > b$ 时，$\int_a^b f(x)\mathrm{d}x = -\int_b^a f(x)\mathrm{d}x$.

(3) $\int_a^b [f(x) \pm g(x)]\mathrm{d}x = \int_a^b f(x)\mathrm{d}x \pm \int_a^b g(x)\mathrm{d}x$.

(4) $\int_a^b kf(x)\mathrm{d}x = k\int_a^b f(x)\mathrm{d}x$.

(5) $\int_a^b f(x)\mathrm{d}x = \int_a^c f(x)\mathrm{d}x + \int_c^b f(x)\mathrm{d}x$.

(6) $\int_a^b 1\mathrm{d}x = \int_a^b \mathrm{d}x = b - a$.

(7) 如果在区间 $[a,b]$ 上 $f(x) \geqslant 0$，则 $\int_a^b f(x)\mathrm{d}x \geqslant 0\,(a < b)$.

(8) 如果在区间 $[a,b]$ 上 $f(x) \leqslant g(x)$，则 $\int_a^b f(x)\mathrm{d}x \leqslant \int_a^b g(x)\mathrm{d}x\,(a < b)$.

(9) 设 M 和 m 分别是函数 $f(x)$ 在区间 $[a,b]$ 上的最大值和最小值，则 $m(b-a) \leqslant$

$$\int_a^b f(x)dx \leq M(b-a) \quad (a<b).$$

MVT for Integrals(定积分中值定理)

If $f(x)$ is continuous on $[a,b]$, then at some point c in $[a,b]$,
$$\int_a^b f(x)dx = f(c)(b-a).$$

Average Value of a Continuous Function(连续函数的平均值)

If f is integrable on $[a,b]$, its average (mean) value on $[a,b]$ is $\dfrac{1}{b-a}\int_a^b f(x)dx$.

First Fundamental Theorem of Calculus(微积分基本定理,FTC)

If $f(x)$ is continuous on $[a,b]$ and $F(x)$ is an antiderivative of $f(x)$ on $[a,b]$, then
$$\int_a^b f(x)dx = F(x)\big|_a^b = F(b)-F(a).$$

Second Fundamental Theorem of Calculus

If $f(x)$ is continuous on $[a,b]$ and $F(x)=\int_a^x f(t)dt$, then
$$\frac{d}{dx}\int_a^x f(t)dt = f(x) \quad \text{or} \quad \left(\int_a^x f(t)dt\right)' = f(x)$$
at every point x in $[a,b]$.

一般地,变限积分的导数为:
$$\frac{d}{dx}\int_{a(x)}^{b(x)} f(t)dt = f[b(x)]\cdot b'(x) - f[a(x)]\cdot a'(x) \quad \text{or} \quad \left(\int_{a(x)}^{b(x)} f(t)dt\right)'$$
$$= f[b(x)]\cdot b'(x) - f[a(x)]\cdot a'(x)$$

Infinite Limits of Integration(无穷限反常积分)

(1) If $f(x)$ is continuous on $[a,\infty)$, then $\int_a^\infty f(x)dx = \lim\limits_{b\to\infty}\int_a^b f(x)dx$.

(2) If $f(x)$ is continuous on $(-\infty,b]$, then $\int_{-\infty}^b f(x)dx = \lim\limits_{a\to-\infty}\int_a^b f(x)dx$.

(3) If $f(x)$ is continuous on $(-\infty,\infty)$, then
$$\int_{-\infty}^\infty f(x)dx = \int_{-\infty}^c f(x)dx + \int_c^\infty f(x)dx = \lim_{a\to-\infty}\int_a^c f(x)dx + \lim_{b\to\infty}\int_c^b f(x)dx$$
where c is any real number.

Integrands with Infinite Discontinuities(无界函数的反常积分)

(1) If $f(x)$ is continuous on $(a,b]$ and is discontinuous at a, then $\int_a^b f(x)dx = \lim\limits_{c\to a^+}\int_c^b f(x)dx$.

(2) If $f(x)$ is continuous on $[a,b)$ and is discontinuous at b, then $\int_a^b f(x)dx = \lim\limits_{c\to b^-}\int_a^c f(x)dx$.

(3) If $f(x)$ is discontinuous at c, where $a<c<b$, and continuous on $[a,c)\cup(c,b]$, then
$$\int_a^b f(x)dx = \int_a^c f(x)dx + \int_c^b f(x)dx.$$

14.9 Applications of the Integral to Geometry(定积分的几何应用)

Area Between Two Curves(由两条曲线所围成的图形的面积)

The area A of the region bounded above by $y=f(x)$, below by $y=g(x)$, and on the sides by the lines $x=a$ and $x=b$.

$$A=\int_a^b [f(x)-g(x)]\mathrm{d}x.$$

If a region's bounding curves are described by functions of y, the approximating rectangles are horizontal instead of vertical and the basic formula has y in place of x.

$$A=\int_c^d \mathrm{d}A=\int_c^d [f(y)-g(y)]\mathrm{d}y.$$

Area in Polar Coordinates(极坐标方程求面积)

If $f(\theta)$ is a continuous function, then the area bounded by a curve in polar form $r=f(\theta)$ and the rays $\theta=\alpha$ and $\theta=\beta$ (with $\alpha<\beta$) is equal to

$$A=\frac{1}{2}\int_\alpha^\beta r^2 \mathrm{d}\theta=\frac{1}{2}\int_\alpha^\beta f(\theta)^2 \mathrm{d}\theta.$$

Solids with Known Cross Sections(截面面积已知的物体的体积)

The volume of a solid of known integrable cross-sectional area $A(x)$ from $x=a$ to $x=b$ is the integral of A from a to b,

$$V=\int_a^b A(x)\mathrm{d}x.$$

Let $A(y)$ be the area of the horizontal cross section at height y of a solid body extending from $y=a$ to $y=b$. Then

$$V=\int_a^b A(y)\mathrm{d}y.$$

Solids of Revolution: Disks and Washers(旋转体的体积:圆盘法和垫圈法)

Disk(圆盘法)

Let f be continuous and nonnegative on $[a,b]$, and let R be the region that is bounded above by $y=f(x)$, below by the x-axis, and on the sides by the lines $x=a$ and $x=b$. The volume of the solid of revolution that is generated by revolving the region R about the x-axis is

$$V=\int_a^b \pi [f(x)]^2 \mathrm{d}x.$$

The volume of the solid of revolution that is generated by revolving the region R about the y-axis is

$$V=\int_c^d \pi [u(y)]^2 \mathrm{d}y.$$

Washer(垫圈法)

Let f and g be continuous and nonnegative on $[a,b]$, and suppose that $f(x) \geqslant g(x)$ for all x in the interval $[a,b]$. Let R be the region that is bounded above by $y=f(x)$, below by $y=g(x)$, and on the sides by the lines $x=a$ and $x=b$. The volume of the solid of revolution that is generated by revolving the region R about the x-axis is

$$V = \int_a^b \pi [f(x)^2 - g(x)^2] dx.$$

The volume of the solid of revolution that is generated by revolving the region R about the y-axis is

$$V = \int_a^b \pi [w(y)^2 - v(y)^2] dy.$$

Length of a Plan Curve(平面曲线的弧长)

Arc Length Formula in Cartesian Coordinates(直角坐标方程的弧长公式)

If $y=f(x)$ is a smooth curve on the interval $[a,b]$, then the arc length L of this curve over $[a,b]$ is defined as

$$L = \int_a^b \sqrt{1+[f'(x)]^2} \, dx = \int_a^b \sqrt{1+\left(\frac{dy}{dx}\right)^2} \, dx.$$

More over, for a curve expressed in the form $x=g(y)$, where g is continuous on $[c,d]$, the arc length L from $y=c$ to $y=d$ can be expressed as

$$L = \int_c^d \sqrt{1+[g'(y)]^2} \, dy = \int_a^b \sqrt{1+\left(\frac{dx}{dy}\right)^2} \, dy.$$

Arc Length Formula for Parametric Curves(参数方程的弧长公式)

If no segment of the curve represented by the parametric equations

$$x=x(t), y=y(t) \quad (a \leqslant t \leqslant b)$$

is traced more than once as t increases from a to b, and if $\frac{dx}{dt}$ and $\frac{dy}{dt}$ are continuous functions for $a \leqslant t \leqslant b$, then the arc length L of the curve is given by

$$L = \int_a^b \sqrt{\left(\frac{dx}{dt}\right)^2 + \left(\frac{dy}{dt}\right)^2} \, dt$$

Arc Length of a Polar Curve(极坐标曲线求弧长)

If no segment of the polar curve $r=f(\theta)$ is traced more than once as θ increases from α to β, and if $\frac{dr}{d\theta}$ is continuous for $\alpha \leqslant \theta \leqslant \beta$, then the arc length L from $\theta=\alpha$ to $\theta=\beta$ is

$$L = \int_\alpha^\beta \sqrt{r^2 + \left(\frac{dr}{d\theta}\right)^2} \, d\theta = \int_\alpha^\beta \sqrt{[f(\theta)]^2 + [f'(\theta)]^2} \, d\theta.$$

14.10 Differential Equations(微分方程)

Euler's Method(欧拉方法)

To approximate the solution of the initial-value problem:
$$\frac{dy}{dx}=f(x,y), f(x_0)=y_0$$
$$f(x_1)\approx f(x_0)+f'(x_0,y_0)\Delta x$$
$$f(x_2)\approx f(x_1)+f'(x_1,y_1)\Delta x$$
$$\vdots$$
$$f(x_{n+1})\approx f(x_n)+f'(x_n,y_n)\Delta x$$

Δx— step size(步长), $x_1=x_0+\Delta x, x_2=x_1+\Delta x, x_3=x_2+\Delta x, \cdots$

Exponential Growth(指数增长)

A quantity $y=y(t)$ is said to have an exponential growth model if it increases at a rate that is proportional to the amount of the quantity present, the quantity $y(t)$ satisfies an equation of the form
$$\frac{dy}{dt}=ky \quad (k>0)$$

Initial condition: $y=y_0$ when $t=0, y=y_0 e^{kt}$ (指数增长的特解).

Restrict Growth and Decay(约束增长和衰减)

A quantity $y=y(t)$ whose rate of change is proportional to the difference $y-b$.

Differential equation:

$y'=k(y-b)$

General solution:

$y(t)=b+Ce^{kt}$

The Logistic Equation(逻辑斯蒂方程)

The rate of change of a quantity may be proportional both to the amount of the quantity and to the difference between a fixed constant A and its amount.

Logistic differential equation: $\frac{dy}{dt}=ky(A-y)$

The general solution: $y=\dfrac{A}{1+Ce^{-Akt}}$

Here $k>0$ is the growth constant, and $A>0$ is called the carrying capacity.

14.11 Sequences and Series(序列和级数)

Sequence Converges or Diverges(序列收敛或发散)

A sequence $\{a_n\}$ has the limit L and we write
$$\lim_{n\to\infty} a_n = L.$$

If $\lim_{n\to\infty} a_n$ exists, we say the sequence converges. Otherwise, we say the sequence diverges.

Geometric Series(几何级数)

A geometric series
$$\sum_{n=0}^{\infty} ar^n = a + ar + ar^2 + \cdots + ar^n + \cdots \ (r \neq 0 \text{ is the common ratio})$$
converges if $|r| < 1$ and **diverges** if $|r| \geq 1$.

If the series converges, then the sum is
$$\sum_{n=0}^{\infty} ar^n = \frac{a}{1-r}.$$

p-Series(p-级数)

$$\sum_{n=1}^{\infty} \frac{1}{n^p} = 1 + \frac{1}{2^p} + \frac{1}{3^p} + \cdots + \frac{1}{n^p} + \cdots \ (p > 0) \text{ converges if } p > 1 \text{ and diverges if } 0 < p \leq 1.$$

Harmonic Series(调和级数)

$$\sum_{n=1}^{\infty} \frac{1}{n} = 1 + \frac{1}{2} + \frac{1}{3} + \cdots + \frac{1}{n} + \cdots \text{ diverges}(调和级数即 p=1 \text{ 的 } p-级数,发散).$$

nth Term Test (The Divergence Test,发散判别法)

(a) If $\lim_{n\to\infty} a_n \neq 0$, then the series $\sum a_n$ diverges.

(b) If $\lim_{n\to\infty} a_n = 0$, then the series $\sum a_n$ may either converge or diverge.

The Integral Test(积分判别法)

Let $a_n = f(n)$, where $f(x)$ is positive, decreasing, and continuous for $x \geq 1$.

(a) If $\int_1^{\infty} f(x) dx$ converges, then $\sum_{n=1}^{\infty} a_n$ converges.

(b) If $\int_1^{\infty} f(x) dx$ diverges, then $\sum_{n=1}^{\infty} a_n$ diverges.

The Comparison Test(比较判别法)

Assume that there exists $M > 0$ such that $0 \leq a_n \leq b_n$ for $n \geq M$.

(a) If $\sum_{n=1}^{\infty} b_n$ converges, then $\sum_{n=1}^{\infty} a_n$ also converges.

(b) If $\sum\limits_{n=1}^{\infty} a_n$ diverges, then $\sum\limits_{n=1}^{\infty} b_n$ also diverges.

Comparison Test in Limit Form(比较判别法的极限形式)

Suppose $\sum a_n$ and $\sum b_n$ are series with positive terms. If
$$\lim_{n\to\infty} \frac{a_n}{b_n} = L$$
where L is a finite number and $L>0$, then either both series converge or both diverge.

The Ratio Test(比值判别法)

Let $\sum a_n$ be a series with positive terms and suppose that:
$$\rho = \lim_{n\to\infty} \frac{a_{n+1}}{a_n}$$

(a) If $\rho<1$, then $\sum a_n$ converges.

(b) If $\rho>1$ or $\rho=\infty$, then $\sum a_n$ diverges.

(c) If $\rho=1$, then $\sum a_n$ may converges or diverges, so that another test must be tried.

The nth Root Test(根值判别法)

Let $\sum a_n$ be a series with positive terms and suppose that:
$$\rho = \lim_{n\to\infty} \sqrt[n]{a_n}$$

(a) If $\rho<1$, then $\sum a_n$ converges.

(b) If $\rho>1$ or $\rho=\infty$, then $\sum a_n$ diverges.

(c) If $\rho=1$, then $\sum a_n$ may converges or diverges, so that another test must be tried.

The Alternating Series Test(交错级数判别法)

Assume that $\{a_n\}$ is a positive sequence that is decreasing and converges to 0:
$$a_1 > a_2 > a_3 > a_4 > \cdots > 0, \qquad \lim_{n\to\infty} a_n = 0$$

Then the following alternating series converges:
$$S = \sum_{n=1}^{\infty} (-1)^{n+1} a_n = a_1 - a_2 + a_3 - a_4 + \cdots$$

Furthermore, $S \leqslant a_1$.

The Alternating Series Estimation Theorem(交错级数估计定理)

Let $S = \sum\limits_{n=1}^{\infty} (-1)^{n+1} a_n$, $S_n = \sum\limits_{k+1}^{n} (-1)^{n+1} a_k = a_1 - a_2 + a_3 - a_4 + \cdots + (-1)^{n+1} a_n$, where $\{a_n\}$ is a positive decreasing sequence that converges to 0, then
$$|S - S_n| < a_{n+1}$$

In other words, the error committed when we approximate S by S_n is less than the size of the first omitted term a_n.

Absolute Convergence(绝对收敛)

The series $\sum a_n$ converges absolutely if $\sum |a_n|$ converges.

Absolute Convergence Implies Convergence(绝对收敛则收敛)

If $\sum |a_n|$ converges, then $\sum a_n$ also converges.

Conditional Convergence(条件收敛)

An infinite series $\sum a_n$ converges conditionally if $\sum a_n$ converges but $\sum |a_n|$ diverges.

Computation of Power Series(幂级数的运算)

Theorem 1

Let $f(x) = \sum_{n=0}^{\infty} a_n x^n$ and $g(x) = \sum_{n=0}^{\infty} b_n x^n$, k is any real number and N is any positive integer.

(a) $f(kx) = \sum_{n=0}^{\infty} a_n (kx)^n = \sum_{n=0}^{\infty} a_n k^n x^n$.

(b) $f(x^k) = \sum_{n=0}^{\infty} a_n (x^k)^n = \sum_{n=0}^{\infty} a_n x^{nk}$.

(c) $f(x) \pm g(x) = \sum_{n=0}^{\infty} a_n x^n \pm \sum_{n=0}^{\infty} b_n x^n = \sum_{n=0}^{\infty} (a_n \pm b_n) x^n$.

Theorem 2 Term-by-Term Differentiation and Integration(逐项求导与逐项积分)

Assume that $F(x) = \sum_{n=0}^{\infty} a_n (x-c)^n$ has radius of convergence $R > 0$. Then $F(x)$ is differentiable on $(c-R, c+R)$. Furthermore, we can integrate and differentiate term by term. For $x \in (c-R, c+R)$,

$$F'(x) = \sum_{n=0}^{\infty} [a_n (x-c)^n]' = \sum_{n=0}^{\infty} n a_n (x-c)^{n-1}$$

$$\int_0^x F(t) dt = \int_0^x \sum_{n=0}^{\infty} a_n (t-c)^n dt = \sum_{n=0}^{\infty} \left(\int_0^x a_n (t-c)^n dt \right) = \sum_{n=0}^{\infty} \frac{a_n}{n+1} (x-c)^{n+1}$$

Taylor Series and Maclaurin Series(泰勒级数与麦克劳林级数)

If $f(x)$ has derivatives of all orders at a, then we call the series

$$\sum_{n=0}^{\infty} \frac{f^{(n)}(a)}{n!} (x-a)^n = f(a) + f'(a)(x-a) + \frac{f''(a)}{2!}(x-a)^2 + \cdots + \frac{f^{(n)}(a)}{n!}(x-a)^n + \cdots$$

the **Taylor series for f about** $x = a$. In the special case where $a = 0$, this series becomes

$$\sum_{n=0}^{\infty} \frac{f^{(n)}(0)}{n!} x^n = f(0) + f'(0)x + \frac{f''(0)}{2!} x^2 + \cdots + \frac{f^{(n)}(0)}{n!} x^n + \cdots$$

in which case we call it the Maclaurin series for $f(x)$.

常见的 Maclaurin Series：

$$\frac{1}{1-x} = 1 + x + x^2 + \cdots + x^n + \cdots \quad (-1 < x < 1)$$

$$e^x = 1 + x + \frac{1}{2!}x^2 + \cdots \frac{1}{n!}x^n + \cdots (-\infty < x < +\infty)$$

$$\sin x = x - \frac{x^3}{3!} + \frac{x^5}{5!} - \cdots + (-1)^{n-1}\frac{x^{2n-1}}{(2n-1)!} + \cdots (-\infty < x < +\infty)$$

$$\cos x = 1 - \frac{x^2}{2!} + \frac{x^4}{4!} - \cdots + (-1)^n \frac{x^{2n}}{(2n)!} + \cdots (-\infty < x < +\infty)$$

$$\ln(1+x) = x - \frac{x^2}{2} + \frac{x^3}{3} - \frac{x^4}{4} + \cdots + (-1)^n \frac{x^{n+1}}{n+1} + \cdots (-1 < x \leq 1)$$

Taylor and Maclaurin Polynomials(泰勒多项式与麦克劳林多项式)

If f can be differentiated n times at 0, then we define the nth Taylor polynomial for f about $x = a$ to be

$$P_n(x) = \sum_{k=0}^{n} \frac{f^{(k)}(a)}{k!}(x-a)^k$$
$$= f(a) + f'(a)(x-a) + \frac{f''(a)}{2!}(x-a)^2 + \cdots + \frac{f^{(n)}(a)}{n!}(x-a)^n$$

the **nth Maclaurin polynomial for f about $x = 0$** to be

$$P_n(x) = \sum_{k=0}^{n} \frac{f^{(k)}(0)}{k!}x^k = f(0) + f'(0)x + \frac{f''(0)}{2!}x^2 + \cdots + \frac{f^{(n)}(0)}{n!}x^n$$

The Remainder Estimation Theorem(余项估计定理)

If the function $f(x)$ can be differentiated $n+1$ times on an interval containing the number $x = a$, for all x in the interval, then

$$|R_n(x)| \leq \frac{\max\limits_{c \text{ is between } x \text{ and } a} |f^{(n+1)}(c)|}{(n+1)!} |x-a|^{n+1}$$

for all x in the interval.

参考文献

[1] 美国大学理事会官方网站[EB/OL]. https://apcentral.collegeboard.org/courses/ap-calculus-bc.

[2] Ross L. Finney, Maurice D. Weir, Frank R. Giordana. 托马斯微积分[M]. 北京:高等教育出版社, 2004.

[3] Anton H., Bivens I., Davis S. 微积分[M]. 北京:高等教育出版社, 2004.

[4] 同济大学数学系. 高等数学[M]. 北京:高等教育出版社, 2007.

[5] David Bock, Shirley O. Hackett. Barron' AP Calculus[M]. 12th Edition. New York: Barron's Educational Series, 2013.

[6] Frank D. Demana, Bert K. Waits, Gregory D. Foley, et al. Precalculus Graphical, Numerical, Algebraic[M]. Trenton:Pearson, 2010.

[7] Ross L. Finney, Franklin D. Demana, Bert K. Waits, et al. Calculus Graphical, Numerical, Algebraic[M]. Upper Saddle River:Pentice Hall, 2012.

[8] James Stewart. Single Variable Calculus:Early Trancendentals[M]. Brooks Cole, 2012.

[9] Jon Rogawski. Calculus:Early Trancendentals[M]. New York:W. H. Freeman and Company, 2012.

[10] David S. Kahn. Cracking the AP Calculus AB & BC Exams[M]. TPR Education IP Holding, LLC, 2013.

[11] Flavia Banu, Joan Rosebush. AP Calculus AB & BC Crash Course[M]. Washington, DC:Research & Education Association Inc., 2011.